PLANTS on ISLANDS

PLANTS on ISLANDS

Diversity and Dynamics on a Continental Archipelago

Martin L. Cody

UNIVERSITY OF CALIFORNIA PRESS
Berkeley   Los Angeles   London

University of California Press
Berkeley and Los Angeles, California

University of California Press, Ltd.
London, England

© 2006 by The Regents of the University of California

Library of Congress Cataloging-in-Publication Data

Cody, Martin L., 1941-
    Plants on islands : diversity and dynamics on a continental archipelago / Martin L. Cody.
        p.   cm.
    Includes bibliographical references and index.
    ISBN 0-520-24729-9 (case : alk. paper)
  1. Island plants—British Columbia—Barkley Sound Region.   2. Plant communities—British Columbia—Barkley Sound Region.   3. Vegetation dynamics—British Columbia—Barkley Sound Region.   4. Phytogeography—British Columbia—Barkley Sound Region.   5. Plant ecology—British Columbia—Barkley Sound Region. I. Title.
QK203.B7C63 2006
581.7'5209711—dc22   2005037441

Manufactured in the United States of America

15 14 13 12 11 10 09 08 07 06
10   9   8   7   6   5   4   3   2   1

The paper used in this publication meets the minimum requirements of ANSI/NISO Z39.48-1992 (R 1997) (*Permanence of Paper*).

FOR DARYL ANN, ROSIE, AND TIM,
WITH LOVE AND THANKS FOR ALL OF OUR BAMFIELD SUMMERS

CONTENTS

Preface / ix

1 • INTRODUCTION / 1

2 • ISLANDS IN BARKLEY SOUND, BRITISH COLUMBIA / 6

    Geography and History / 6

    Geology and Topography / 9

    Climate / 10

    More Island Geography and Topography / 13

    Flora and Vegetation / 23

    Methodological Notes / 30

3 • ISLAND BIOGEOGRAPHY: CONCEPTS, THEORY, AND DATA / 36

    The Basic Model / 36

    Some Embellishments / 38

    Incidence Functions / 39

    Colonization and Extinction Data / 41

    Historical Legacies / 50

    Generalities, Specifics, and Modifications of the M/W Model / 55

4 • SPECIES NUMBER, ISLAND AREA, AND ISOLATION / 56

    Equilibrium or Nonequilibrium Species Numbers? / 56

    Variation in Species Counts on Islands / 56

    Cumulative Species Numbers / 62

    Species Richness on Islands / 63

5 • NESTEDNESS AND ASSEMBLY RULES / 74

    Inferences from Species-by-Sites Matrices / 74

    Forest Species / 76

    Shoreline Habitats / 85

    Edge Habitats / 100

6 • SPECIES TURNOVER IN SPACE AND TIME / 116

    Colonization and Extinction, Persistence and Turnover / 116

    Species Turnover in Time, and Island Size / 124

    Species Turnover in Space / 128

7 • DISPERSAL SYNDROMES, INCIDENCE, AND DYNAMICS / 138

    Dispersal in Plants: Options / 138

    Ferns / 141

    Fruiting Shrubs / 150

    Hydrochores: Drifters on the Sea / 161

    Anemochores: Drifters on the Wind / 166

8 • **ECOLOGICAL AND EVOLUTIONARY SHIFTS ON CONTINENTAL ISLANDS** / 180

    Alien Invaders / 181

    Ecological Shifts in Impoverished Biotas / 185

    Evolutionary Shifts in Isolated Populations / 194

9 • **SYNOPSIS: LESSONS FROM A CONTINENTAL ARCHIPELAGO** / 205

    The Barkley Sound Scene / 205

    Is There an Equilibrium out There? / 206

Colonization and Extinction Dynamics / 209

Coexistence and a Potential Role for Competition / 212

Adaptation, Evolution, Conservation / 215

Appendixes / 217

References / 248

Index / 256

PREFACE

This is a book about a field study that just grew and grew; it grew until I was quite sure that, if it were not written up at this "intermediate" stage, it never would be. I was concerned that the patterns would dissolve in the fogs of time, and the lessons I had learned might slip away unless I wrote them down without further delay and interpreted them as best I could while I still had a grasp on them. In one sense this is a broad project on the statics and dynamics of island biogeography and the ecology and evolution of a regional biota; in another it is narrowly constrained to the plants of a single archipelago in British Columbia, Canada. *Plants on Islands* relates various facets of the classical island biogeography literature to geologically recent islands, a few thousands of years old, that house everyday, typical mainland plants (no glamorous endemics here). What does come into focus, I hope, is a scene in which the island plant lists are continually changing with time, generating a turnover of species fueled by rapid colonization rates tempered by the equally rapid extinction of local, small and patchy populations. In a way, this lively, dynamical picture is quite the opposite of what happens on remote (and biogeographically quiescent) oceanic islands; there changes in the biotic roster happen only rarely, and the processes driving ecological and evolutionary shifts are slow, sustained, and perhaps largely already carried to resolution.

In the early years of the work, the questions were fairly specific, somewhat pedestrian, and of the snapshot type—for example, how do species numbers increase with island size?—and the answers were similarly straightforward. As time added dimensionality to the data set in at least one sense, the dynamics soon overshadowed the statics, and the questions became multidimensional and more complex, eventually fragmenting into various subplots, minor themes, and taxonomically qualified sidebars. It was becoming a challenge to keep the islands, the years, the plants, and their comings and goings distinct; clearly it was time for the grand synthesis. While hitherto a few small vignettes have emerged in print, the book is nearly all composed of material with no prior publication history; I have greatly appreciated the luxury of being able to wait to write it until a fairly complete story could be assembled.

The study grew from an initial interest in the ecology of foraging auks (Aves: Alcidae) in the Pacific Northwest. That interest led me first to the islands and marine environments of Barkley Sound, British Columbia, and then to considering the islands' land birds as potential subjects for biogeographic work. Eventually, though, I settled on the land plants. The investigation began with plants on a few islands, then a few dozen, until eventually a few hundred islands were included in the survey. I originally had thought that with data from my first two field seasons (1981, 1982) I might be able to say something useful about island plants and then move on.

Nearly a quarter century later, my thinking on this point is about the same, but with the information accumulated from island visits that averaged two of every three of the intervening years, it is perhaps more justified now. Over time and with the exploration of more and more island bays and banks, coves and crannies, the discovery of some of the rarer species in the system, an appreciation for distributional quirks and morphological variation, and especially the documentation of the initiation, the rise, and the demise of island populations, the picture has become quite compelling. Certainly the research is ever more engaging, and the possibility of calling it quits is even harder to contemplate. After all, we have not yet witnessed a 50-foot tsunami come sweeping up the Trevor Channel, the arrival of West Nile virus into the local berry-spreading crow population, or more than the modest beginnings of the spread of alien plants throughout the Sound. We have seen only one island slug population go extinct in this time, a period over which the achenes of persistent island wall-lettuce populations have evolved measurably larger sizes. In all probability, we have not yet seen the (once in a) 100-year winter storm or summer drought, a really fantastic summer for salmonberries, or a major outbreak of dwarf mistletoe in the conifers. But as we share Rat's sentiment (from Kenneth Grahaeme's immortal tale, *Wind in the Willows*) that "there is nothing—absolutely nothing—half so much worth doing as simply messing about in boats," logical reasons to continue the work are actually superfluous.

It is gratifying to acknowledge that this study has been a family effort at all levels, financial to logistic, intellectual to emotional. This core of support has been supplemented in several different ways from outside sources. The Bamfield Marine Science Center (BMSC) is a superb facility on the southeast side of Barkley Sound that has served, through the auspices of its generous directors and associate directors (Ron Foreman, Alan Burger, Andy Spencer) and supportive staff (especially Cliff Haylock, our bog consultant), as our land base for most of the island work. The National Science Foundation funded the research during two early years, and in one notably busy year, 1987, the Center for Field Studies provided volunteer assistants who were valuably deployed in mapping the islands. Our ultimate authorities for plant identification and taxonomic conundrums were Drs. Adolf and Oluna Ceska, erstwhile of the British Columbia Provincial Museum and the University of Victoria. Both are intrepid bog-walkers with whom we enjoyed many a perilous landing and poppy-seed cake. I acknowledge with thanks Dave Hancock, who first pointed me toward Barkley Sound nearly three decades ago. My thanks and admiration to the late Peter Janitis, who for so many years painstakingly recorded the weather in Grappler Inlet; grateful thanks also to Nina for her hospitality over the years. In three different years I had the benefit of assistance from undergraduate and graduate students in my Island Biogeography classes at BMSC, and in another year from the participants in a MacMillan BioDiversity Workshop at BMSC. Several of my own graduate students assisted in the surveys over the course of the study: Shawn Powell, Dr. Phyllis Nicholson, Dr. Jake Overton, Dr. K. C. Burns. The last mentioned (KC) completed his dissertation on the ecology and distribution of the fruiting shrubs in Barkley Sound, and Jake was a pivotal contributor to our research and subsequent joint paper on short-term evolution of reduced dispersal in island plants. And lastly, my son Tim Cody has provided a great deal of field assistance over the later years: sharp-eyed through fogs and over logs, he devoted summer after summer to piloting the Barkley Sound channels with confidence and finesse, and to finding, identifying, and recording the island plants.

1

Introduction

IN 1846 SIR JOSEPH HOOKER delivered an intriguing paper to the Linnaean Society in London on the flora of the Galapagos Islands, the results of a study he had undertaken at the request of his friend Charles Darwin. Subsequently published in 1849, the paper raised and discussed issues that remain at the heart of most research on island plants. He described the Galapagos Islands as being of rather recent volcanic origin, and the flora as remarkable for its paucity of species, believed at the time to number around 265; over 20% of the species were ferns (Pteridophytes) and sunflower relatives (Asteraceae = Compositae). The paper compared the islands' flora to those of other tropical archipelagoes, which at that time were becoming better known botanically: the Cape Verde and Sandwich (Hawaiian) Islands, St. Helena in the South Atlantic, and the Juan Fernandez Islands and New Zealand in the southern Pacific Ocean.

Hooker singled out for special consideration two components in the Galapagos flora, one of species endemic to the islands and comprising over 50% of the species list, and the other of taxa having wide distributions on other islands and in continental regions over a broad range of tropical latitudes. The means of dispersal by which plants reached an archipelago across nearly 1000 km of ocean was a subject of enlightened speculation; transport by water and wind, by birds, and most recently by humans were seen as the chief options. The implied loss of dispersal ability in weedy Galapagos *Ageratum conyzoides* (Asteraceae), which show reduced awns and scales on the pappus of the achenes and thence presumably a reduced dispersal capacity by wind (anemochory), elicited Hooker's surprise and interest. Indeed, similar morphological changes are found to accompany a loss of dispersal ability in a wide range of island plants. Hooker examined the relationships of endemic species to their nearest known continental relatives and noted the prevalence of endemics in the vegetation at higher island elevations. He reported also the presence of new genera and species of "curious arborescent Compositae which have no near allies in other parts of the globe," foreshadowing the so-called weeds-to-trees theme for oceanic islands that has been widely developed and documented since Hooker's time.

It seemed quite reasonable to Sir Joseph that larger islands, islands more centrally located in the archipelago, and those reaching higher elevations, should support larger numbers of plant species. But the differences in species composition from one island to another, particularly the representation on different islands of endemic genera by different species, were a significant source of wonder, a "mystery which it is my object to portray, but not to explain." An explanation based on the adaptation of local isolates, all derivatives of a single ancestral immigrant to the archipelago, was, however, not long in the offing. In fact, a new theory of species formation via descent with modification from a common ancestor gave Hooker's introductory essay to his *Flora Tasmaniae* (1859) added perspective and insight. This enhanced understanding was deployed just a year after Darwin and Wallace presented the essence of the new theory at the Linnaean Society meetings, and its publication date is the same as that of Darwin's *The Origin of Species*.

Studies of plants on islands, with Hooker's initial contributions providing a foundation, are an ongoing, vigorous, and productive branch of biology, and contribute in major ways to the fields of ecology, evolutionary biology, and biogeography. A key impetus to their synthesis was Sherwin Carlquist's classic *Island Life*, with its wealth of information on origins and adaptations of island plants. This 1965 book remains an invaluable compendium of the novelties produced on remote islands by plants that differ from known or putative ancestors in diverse and often startling ways. Now, of course, most oceanic or partly oceanic archipelagoes, such as the Canary, Galapagos, Hawaiian, and Cape Verde Islands, have their own published floras, and their origins, diversity and endemicity have been examined in great detail. A major attraction of remote islands for both plant and animal ecologists and evolutionary biologists lies in their isolation and their lack of past connections to continents. As a consequence, their floras and faunas are derived from mainland ancestors that successfully reached the island by oversea dispersal and, moreover, succeeded in establishing and persisting there. Further, with increased isolation comes a reduction in the number of successful colonists and thence reduced diversity, followed by increased opportunities for the development of novel morphology and ecology in a new selective regime.

Islands in general have diverse origins and histories. This diversity can be seen even within islands of a single region, such as the Sea of Cortés between Baja California and mainland México, a contact zone between the North American and Pacific Plates (Carreño and Helenes 2002). Variously composed of volcanic or granitic substrates, or of old seafloor or more recent sediments, some were formed by block faulting off the trailing edge of shifting plates, some by seafloor spreading and subsequent uplift, some by fluvial outwash deposits, and some by submarine tectonic and volcanic activity that continued structural deformations inexorably brought to the light and air. In some, past continental connections are ancient, severed by their slower movement relative to that of the parent rocks, while in others continental connections are as recent as a few thousand years. In many instances, the islands still bear the imprint of these diverse origins and histories, which are reflected in present-day differences and discrepancies, as well as similarities, in their floras and faunas (Case et al. 2002).

Most oceanic islands and archipelagos were formed by tectonic activity at the submarine contact zones between major plates and have no history of connection to larger or mainland land masses. In contrast, so-called continental or inshore islands are those that lie on the continental shelves and have experienced past mainland connections, perhaps on many occasions. During glacial maxima, when sea levels were lowered by more than 100 m and the continental ice sheets were at maximum depth and extension, present-day continental islands separated by current channel depths of less than about 130 m were connected to the adjacent mainland by land bridges. Thus the islands were formed, split into smaller islands, and separated sequentially with rising sea levels according to the current submarine depressions between them, until the present island size distribution was reached by about 7000 years BP.

Continental islands lack a number of the features that often distinguish oceanic islands. Their proximity to continents and history of continental connections via land bridges are reflected in their having similar floras of relatively similar sizes living under similar abiotic conditions; in general, these characteristics preclude selection for island novelties, the evolution of endemics, and adaptive radiations. Many oceanic islands, especially larger islands (such as Madagascar, New Caledonia, and New Zealand) and archipelagoes, are conservatories for plant relicts, more primitive species that have been replaced or supplanted on continents by more advanced species of later evolutionary lineages. Among many examples of ancient plants that retain high diversity on islands are primitive gymnosperm families, such as the magnificent podocarp-dominated (Podocarpaceae) forests of New Zealand and the Araucariaceae and Austrotaxaceae of New Caledonia. These same old but large islands are also havens for the more primitive angiosperm families, such as Winteraceae and laurels (Lauraceae), the latter being especially well represented also in the Canary Islands, Madeira, and the Azores. Relicts of more ancient, pregymnosperm, radiations in cycads (Cycadaceae) are perhaps best represented today on islands in the Pacific Ocean (Sachs 1997).

Since the subject of this book is plants on continental rather than oceanic islands, it will say little about endemism, radiations, or relicts, which are classical themes of oceanic islands. But on the other hand, the lower degree of isolation of continental islands and greater proximity to source floras is reflected in much higher colonization rates relative to oceanic islands, and typically the dynamics of colonization and extinction they display are more apparent. As these processes can take place over short-term ecological time, for example, over decades, rather than much longer evolutionary or geological time scales, they can be documented and studied with relative ease. Thus, several of Hooker's concerns, such as species numbers and species turnover between islands, dispersal means, and characteristics and morphological changes on islands, can be studied with much greater resolution on continental islands; the dynamical aspect of island biotas can be revealed only with difficulty and inference on oceanic islands. Further, the proximity of the source floras to large numbers of available and accessible continental islands make for attractive and practicable research programs. Haila (1990) classified islands over space and time scales and pointed out the relevance of different ecological and evolutionary processes over these continua; certain aspects of island biogeography can best be pursued on small, continental islands, and other aspects on oceanic islands, especially large, old, and isolated islands and archipelagoes. On the latter, a range of phenomena occur, from the preservation of relictual taxa often long extinct on mainlands, to adaptive radiations of new taxa; on rather less isolated archipelagoes infrequent but repeated colonization drives taxon cycles that also produce endemic forms (Fig. 1.1). On the former, the dynamics of colonization and extinction rates take precedence, with questions relating to local adaptation, niche shifts, community assembly, and composition lying somewhere between.

This monograph describes a study of plants on the continental islands in Barkley Sound, British Columbia. It can be considered an island biogeography case history, constrained by taxon and region. It covers the statics of the patterns of plant distribution and diversity and especially emphasizes the dynamics of at least some components of the island floras. While the study can illustrate some aspects of local adaptation, niche shifts, and interspecific influences on community assembly, it has little or nothing to say about speciation. The work, in fact, was not begun as a long-term island biogeography study of plants; I first visited the area in 1981 with the notion that it might be suitable for short-term research on the islands' landbirds!

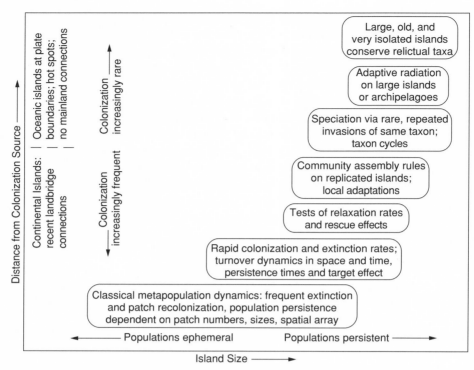

FIGURE 1.1. Phenomenological overview of island biogeographical processes and their realization over island size (abscissa) and isolation (ordinate) scales. Different processes are prevalent and are best studied, on different regions of these two scales.

I soon realized that both analytical scope and statistical power would be much enhanced by switching my focal taxon to plants, since islands of a quarter hectare or so support at most a few pairs of one of two bird species but house dozens of plant species. Even bird-free tiny islets have substantial populations of many different plants. Not only would this change of focus allow me to include in the study the numerous smaller and birdless islands and utilize a species pool more than an order of magnitude larger, but my sample sizes, sampling proficiency, data accumulation rate, and replication potential would all benefit.

Over the years this feathers to fronds, crows to crowberries, and fox sparrows to foxgloves downshift in trophic level has proved increasingly engaging and productive. While the course of the study has incorporated a number of exploratory sidetracks, as will be described later, the basic data set is a matrix of species by islands by calendar years. With an island flora of around 300 species, with over 200 islands surveyed, and with census results from 15 different years, the matrix potentially has nearly one million elements (though actually fewer, as not all islands were visited in each census year). While small portions of the data set have been aired in previous publications, the advantages (and satisfaction) of reporting the study in single monograph seemed obvious from early days, and I hope readers will enjoy a similar satisfaction.*

As a document on a specific case history, this book focuses closely on the Barkley Sound plant data and is in no way intended to be a review, or

---

* "It is the same thing with our publications; they are sown broadcast over the barren acres of Journals and other periodicals which none of us can afford to buy and then weed." J. D. Hooker to T. Huxley, 1856, from *Life and Letters of Sir J. D. Hooker*, L. Huxley (London: John Murray, 1918), 1:369.

overview, of island biogeography, ecology, or evolution, or of island plant biology. A vast primary literature, as well as a burgeoning secondary literature, exists on the ecology, evolution, and biogeography of island taxa. My personal favorites in this latter phylogeny are A. R. Wallace's *The Geographical Distribution of Animals* (1876) and *Island Life* (1881; dedicated to Sir J. D. Hooker), P. J. Darlington's *Zoogeography* (1957), S. Carlquist's *Island Life: A Natural History of the Islands of the World* (1965) and *Island Biology* (1974), R. MacArthur and E. O. Wilson's *The Theory of Island Biogeography* (1967), M. L. Rosenzweig's *Species Diversity in Space and Time* (1995), J. Brown and M. Lomolino's *Biogeography* (2nd ed., 1998), and R. J. Whittaker's *Island Biogeography* (1998). There are of course many additional more specialized texts and compilations, and sparing reference will be made to them, and to other relevant papers, throughout the text.

Following this introductory preamble, I next describe the islands and their history, together with that of surrounding area, the climate of the region, its flora and vegetation, and the approaches this monograph takes to various island biogeographic themes. The results of the research, which spans nearly a quarter century now, are presented in subsequent sections.

2

Islands in Barkley Sound, British Columbia

GEOGRAPHY AND HISTORY

Vancouver Island (Fig. 2.1) lies between 48° and 51° N latitude. The largest island on the west coast of North America, it is part of the western Canadian province of British Columbia and the home of the provincial capital, Victoria. The island was named after George Vancouver (1757–1798), the English navigator who completed its first circumnavigation in 1792. This achievement was not Vancouver's first voyage to the region, though. He had served earlier with Capt. James Cook (1728–1779) on Cook's second (1772–1775) and third (1776–1780) voyages to the Pacific. On this last voyage, Cook sought the fabled Northwest Passage that had been the goal of expeditions for the previous two centuries, but discovered instead Vancouver Island's fabulous sea-otter pelts; the Russians further north were by the 1770s already trading these pelts to China at enormous profit. James Cook and company spent a month in early 1778 (Vancouver was 18 years old at the time) on the outer coast of Vancouver Island in Nootka Sound, likely first seen by the Spaniard Juan Perez in 1774. It was April, too early in the season for the mariners to appreciate the utility of the local plants; "of the Vegetables this place produceth, we benefited by none except the Spruce tree of which we made beer, and the wild Garlic" (James Cook, *Journals*. Pt. 1, pp. 310–311). Cook traded amicably with the natives, exchanging iron and copper for furs, and his reports to the Admiralty on the marvelous quality and value of these pelts triggered a veritable flotilla of English traders to the region. Serious commerce began in 1785 with James Hanna, and subsequent entrepreneurs included James Strange, John Meares, Nathaniel Portlock, and George Dixon. Charles William Barkley arrived in 1787, the Americans John Kendrick and Robert Gray followed in 1788, and the Spaniards Estévan José Martínez in 1789 and Pedro Alberni in 1790. Nootka must have been a very busy harbor in the last decades of the eighteenth century; fascinating accounts of these and earlier coastal explorations are given by Speck (1954) and Arima (1983).

It seems fitting that the Sound, situated about one-third of the way up the outer coast of Vancouver Island, bears Barkley's name, as he was one of the more reputable traders (though not above changing his ship's names and papers

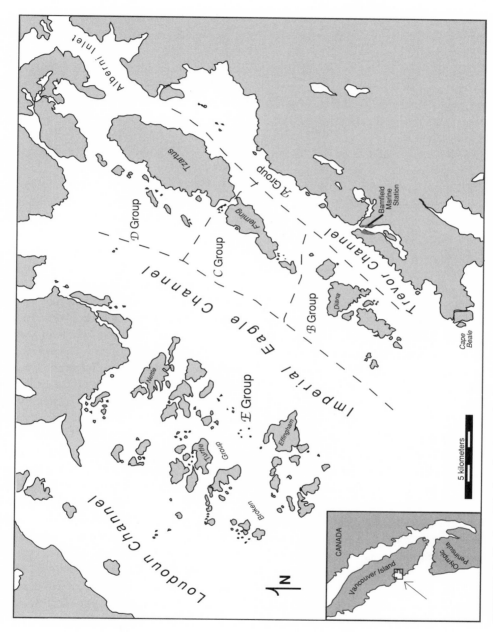

FIGURE 2.1. Barkley Sound islands and environs. Islands are grouped A–E corresponding to isolation from mainland.

for tax evasion purposes). Charles Barkley found and named Barkley Sound in the summer of 1787, and his new teenaged wife, Frances, wrote that "this part of the coast proved a rich harvest of furs." Unfortunately, the voyage was not a financial success, as the Chinese markets were already glutted and the prices depressed by the time Barkley's furs reached them. However, Barkley and his wife (née Frances Hornby Trevor), his ship (*Loudoun*, later *Imperial Eagle*) and officers (Beale, King, Williams) all bequeathed their names to local features such as capes, channels, and islands. A regrettable legacy of the English, Spanish, and American traders was the utter depletion of the sea-otter populations by the beginning of the nineteenth century. In 1817 the French vessel *Le Bordelais* under the command of Camille Roquefeuil found that "the sea otters had long forsaken the area" (Scott 1970). I have yet to see a sea otter in Barkley Sound, though their reintroduction in the 1970s to Checleset Bay on the northwest coast of Vancouver Island was successful, and their range has been slowly extending southward since. By 2003, otters were established as far south as Clayoquot Sound and may be expected to repopulate Barkley Sound within the next decade.

Barkley Sound is centered at 48° 54' N, 125° 17' W and is a roughly rectangular indentation 24 km NW-SE by 18 km NE-SW on the coast of Vancouver Island, open to the Pacific Ocean to the southwest. The Sound encompasses a host of islands (Plate 1; Fig. 2.1), ranging in size from Tzartus (= Copper) at nearly 20 km^2 in area, down to many hundreds of tiny islets of a few or a few dozen square meters. Beyond and below these are rocks that are dry only at low tide, and thence submarine reefs representing drowned islands that were extant when pre-Holocene sea levels were lower. The shorelines of the islands, and the reefs and shallow waters around them, are extremely rich in marine life, from algae and invertebrates to fish, seabirds, and mammals such as seals, sea lions, dolphins, and whales.

Long before Europeans arrived, the region supported large First Nations (or Native American) populations living in complex and sophisticated societies, based in coastal villages with economies largely dependent on the abundant marine resources. The Nuu-chah-nulth (Nootka) tribes occupied the west coast regions of Vancouver Island since postglacial times; the Barkley Sound affiliates belonged to the Huu-ay-aht (Ohiat) band (Arima 1983, Duff 1997). Population densities of indigenous peoples prior to European contact were likely higher in coastal British Columbia than anywhere else in Canada. Captain Meares in 1788 estimated the population of a single village on Effingham Island, a site traditionally occupied for summer fishing activities, at 2000 individuals. William Banfield, the region's first Caucasian settler and the government agent appointed in 1859, estimated the numbers of Huu-ay-aht indigenous peoples at around 550. First Nations populations declined precipitously as a result of the smallpox epidemics of the mid-1800ss and of the more lethal intertribal warfare that followed the acquisition of European weapons. While numbers recovered in the early 1900s (to over 200 Huu-ay-aht by 1963) and may now be close to precontact populations, traditional social, economic, and cultural activities have been largely lost. While the indigenous people originally made wide and ingenious use of many native plants and plant products, they did not maintain cultivated gardens (*fide* Mrs. E. Happynook, Anacla Village resident) and likely did not affect island plant distributions; their current usage of the native vegetation is minimal.

The area's European population grew rapidly in the 1800s by exploiting timber, salmon, and other marine resources. The village of Bamfield (from William Banfield) on the southeast edge of the Sound became the terminus of the Trans-Pacific Cable, a British Empire ("All-Red") project, directed by Sir Sanford Fleming and completed in 1902. The cable provided communications from Bamfield to New Zealand from 1902 to 1959. The Pacific Cable Company's original wooden building was supplemented by a three-story concrete structure in 1929, and this building, in its refurbished state, has housed

Bamfield Marine Science Centre (BMSC) since 1972, under the governance of the Western Canadian Universities Marine Biological Society. The station operates year-round and is particularly busy in the summer when courses in marine sciences, temperate rainforest ecology, and even island biogeography, are offered. The BMSC biologists, along with visitors to Pacific Rim National Park's coastal trails and natural wonders, sport and commercial fishermen, ecotourists, whale-watchers, and sea-kayakers, many of whom use the regular runs of the MV Lady Rose from Port Alberni for transportation, add substantially to Barkley Sound's summer recreational, educational, and commercial traffic.

At present, the islands in the Sound are under a variety of jurisdictions. The islands of the Broken Group lie in the center of the Sound and are part of Pacific Rim National Park Reserve, administered by Parks Canada. Several islands, together with parts of the larger islands, are First Nations reservations with tribal administration, and still others are claimed by the Huu-ay-aht tribe. Many of the remainder are so-called Crown lands (owned by the Province), and a few islands are in private ownership.

GEOLOGY AND TOPOGRAPHY

The western coast of North America, from California to Alaska, has a complex history of plate tectonics, vividly brought to life in John McPhee's (1993) book *Assembling California*. Only the eastern one-third of British Columbia is native North American Plate; the western two-thirds is an amalgam of highly diverse "alien" material, resulting from the formation, migration, and eventual accretion of numerous wandering terranes, essentially minicontinents, with very distant origins (Yorath 1990). Vancouver Island (excluding small southern and western portions) is the largest extant section of the terrane Wrangellia, named after the early polar seas and Alaskan explorer, and later czarist admiral, Ferdinand Petrovich Wrangell (b. 1796). Other parts of this terrane, sheared apart and rifted north, comprise the Queen Charlotte Islands and small areas of southeastern Alaska (Chichagof Island). Wrangellia originated in the ancient South Pacific Ocean, and its odyssey is recounted well by Ludvigsen and Beard (1994). By the early Permian (270 Ma [million years ago]), a substantial terrane had coalesced from volcanic eruptions above an island arc, become overlain with limestone generated by marine invertebrates, and isolated well west of the supercontinent Pangaea. More volcanics were added to the terrane in the Triassic, and granitic rock was intruded through the older volcanics during Jurassic times (200 Ma); accumulating further marine deposits meanwhile, the terrane rode across the paleo-equator into the northern hemisphere. By 130 Ma Wrangellia had completed a 10,000 km journey to collide with western Laurentia at the edge of the North American Plate. Thereafter and throughout the Cretaceous, Wrangellia, deformed by its North American contact, accumulated shales and sandstones in interior basins, and the main structural elements mapped by present-day geologists were assembled.

Most of the islands in Barkley Sound are composed of West Coast Diorites of late Paleozoic or early Mesozoic Era, with younger, mid-to-late Jurassic granodiorite intrusions toward the northeast of the Sound (Muller 1974). While most of the islands look rather uniformly "granitic," light colored with quartz and feldspar, some are bolstered and banded with darker, hornblend-rich volcanic rock into which the diorites have intruded. A few islands at interior locations, such as Nanat on the Trevor Channel's mainland coast, are composed of the characteristically clinkered rock of surface-cooled lava characteristic of the Bonanza Volcanics series. On some islands' beaches, these sharp and abrasive rocks seem almost volcano-fresh.

The Sound is rimmed by mountains reaching 1200–1500 m in elevation, and the vista to the north of the 2000 m peaks in Strathcona Provincial Park is especially impressive. At glacial maximum in the late Pleistocene, southeastern Vancouver Island was covered by a southwest-sloping ice sheet contiguous with the continental ice east across the Georgia Strait and south across the Juan de Fuca Strait (Muller 1974). Ice covered all but the highest mountains, reaching elevations of about 1500 m in the interior and having a thickness of 300 m or so at the outer coast; Barkley Sound was under ice sheets of perhaps 500–600 m thickness. With the passing of the successive glacial maxima, large glaciers descending from the highlands around the Sound incised valleys that are now deep fjords that cut into the interior, the longest of which (the 60 km Alberni Canal) makes Port Alberni a deepwater port and nearly transects Vancouver Island. The glacial sculpting of valleys and basins in Barkley Sound extends topographically in submarine features some 30 km out to sea and covers the inner half of the continental shelf. Here the shelf terminates 50–60 km from the present shoreline, at current ocean depths of around 200 m, and its smoother, outer half lacks the imprint of glacial activity.

Following the retreat of first the continental ice sheets and then the valley glaciers at the end of the last (Wisconsin) glacial episode, beginning around 18,500 years BP (before present), the landscape became revegetated in sequences that at some Pacific Northwest locations have been well studied. By using pollen records from lake-bottom sediments, vegetational stages can be reconstructed and representative plant species identified. First tundra, then parkland, woodland, and finally coastal forest reoccupied previously ice-covered terrain, spreading south to north and lowlands to highlands. The nearest source for the postglacial revegetation of Barkley Sound was the ice-free, lower Olympic Peninsula 150 km to the south, although the higher peaks on Vancouver Island stood as nunataks above the ice sheet and likely served as refugia for some northern plants. For example, the mountains of the Brooks Peninsula on the northwest coast of Vancouver Island gave refuge to several montane species that are disjunct from populations found in wider or more contiguous ranges elsewhere, including both southern disjuncts (with the greater part of the range further north) and northern disjuncts from Olympic Mountains populations to the south (Pojar 1980; Ogilvie and Ceska 1984). Several Brooks Peninsula plant species are otherwise Queen Charlotte Islands (Haida Gwaii) endemics, and the evidence for glacial refugia within the Queen Charlottes is substantial. However, none of these are species that revegetated the islands and coastal areas of Barkley Sound.

CLIMATE

Late Quaternary climates in northwestern Washington, just south of the continental ice sheet, were considerably cooler, by about 4 °C, and 50% reduced in annual precipitation compared to the present regime (Heusser et al. 1980). Given the location of Barkley Sound close to both the seaward and southward edge of the continental ice sheet, glacial retreat must have occurred relatively early, with revegetation following shortly thereafter. A major shift to conditions more like the present occurred around 13,000 years BP, and nearly modern climates have prevailed for most of the last 10,000 years. Recent palynology data were obtained from Whyac Lake on the outer coast of southern Vancouver Island not far south of Barkley Sound. By 11,000 years BP, the major forest tree taxa (*Picea, Tsuga, Abies, Pseudotsuga, Alnus*) already had become abundant pollen contributors, and they have been represented in the pollen record at levels similar to those of modern times throughout this period. An essentially modern forest, by tree species composition, appears to have characterized the Barkley Sound region over at least the last 7000–8000 years (Brown and Hebda 2002; see also Heusser 1989). Note that the reconstitution of the forest flora began shortly after the second major melt-water period of the mid-11,000s BP, and thus it dates from a period when sea levels were about 45 m below present and nearly all of the Barkley Sound islands were then part of the mainland landscape.

The Sound is within Krajina's (1973) "Coastal Western Hemlock" biogeoclimatic zone, where a cool and wet climate supports what is often

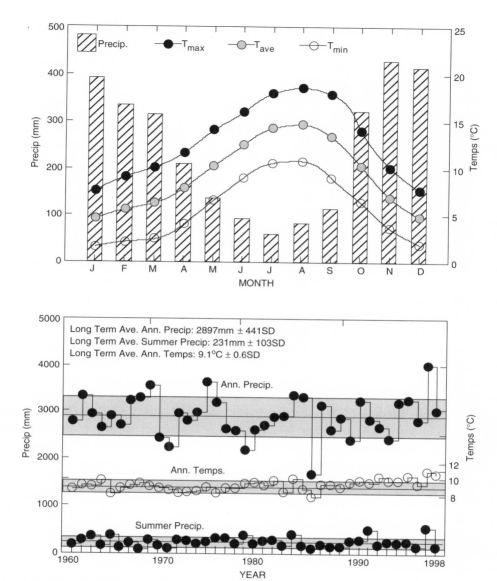

FIGURE 2.2. Climogram (upper) and yearly weather variations (lower) for Bamfield East, Barkley Sound, British Columbia. The vegetation under these climatic conditions is predicted to be temperate rainforest (Holdridge L.R. 1947 Determination of world plant formations from simple climate data. *Science* 105:367–368; Wright et al. 2004).

referred to as temperate rainforest. Indeed, this region's 3 m annual precipitation is similar to that of many tropical rainforests. A climogram summarizing data collected at the Bamfield East weather station, on the southeastern edge of Barkley Sound, is shown in Fig. 2.2 (upper). For six months of the year monthly precipitation averages over 300 mm, but the summer months of June, July, and August are considerably drier, with only 8% of the annual total falling in this period (Fig. 2.2, lower). October through March, inclusively, average 20 days per month with measurable rain, versus 10 days per month in June through August. Winter frosts are common, but snow is unusual and contributes negligible precipitation. Summer months are moderately, even benignly, warm and temperate (see monthly mean minimum, mean, and maximum temperatures in Fig. 2.2, upper). Year-to-year variation in annual precipitation and temperature is shown in Fig. 2.2 (lower), over the 39 years of record. The average annual precipitation is 2897 mm, with about one year in four above or below the standard deviation around the mean (i.e., > 3338, < 2456 mm). The extremes in annual precipitation since 1960 occurred in the drought year of 1985 (1665 mm) and the soggy year of 1997 (4016 mm). Annual temperatures over these years at Bamfield East have averaged 9.1 °C ± 0.6 SD (Fig. 2.2, lower); while the variation around the mean is small, there is a detectable and significant trend of rising annual temperatures by 0.03 °C per year ($R^2 = 0.31$; $p < 0.001$). No such evidence exists for a change or trend in annual precipitation; precipitation variations among years are greater, and as a result it is more difficult to detect trends. Bamfield East is within Environment Canada's Pacific Climatic Region, which extends 1000 km north from southern Vancouver Island and 250 km inland from its outer coast. Over this region as a whole, annual precipitation increased slightly (but not significantly) over the period 1948–1992, but increased more markedly over the longer period of 1911–1992 (Environment Canada 1995). During approximately the first half of the latter period, specifically 1911–1947, 30 of 37 years received less than the 1951–1980 overall average, while during the second half of the period, 1948–1992, 19 of 44 years received less than the indicated average. Thus, over the climatic region as a whole, the first half of the twentieth century was significantly drier than the second half, but an increasing trend cannot be established from measurements in the second half of the century alone, neither region-wide nor at Bamfield East.

Summer rainfall and its variation is particularly relevant for plant growth and reproduction; while wetter summers extend the growing season, for short-lived plants especially, they also delay maturation and diminish dispersal opportunities for plants whose propagules are carried by the wind (anemochores) and require dry and breezy conditions for their effective dispersal. Newly established seedlings of woody plants appear particularly vulnerable to desiccation during dry summers, and their survival must be enhanced greatly by wetter conditions then. As with annual precipitation totals, there is no detectable trend in summer rainfall precipitation over the period shown in Fig. 2.2, although year-to-year variation is high, and its consequences for both plant dispersal and establishment must also be considerable. Note in the figure the particularly wet summers of 1964, 1978, 1983, and 1997, and the especially dry summers of 1965, 1967, 1970, 1985, 1996, and 1998, with the latter two years flanking the wettest summer on record (538 mm) and being over four times drier (130 mm).

Variation in Barkley Sound climates has been revealed by the analysis of microfossils in cores of sediments taken from the seafloor in Effingham Inlet, in the interior, northern part of the Sound. Diatoms recovered from the sediments reveal both longer-term climate shifts, cool-wet to warm-dry, over periods of several thousand years (Chang et al. 2003), and also short-term variations in species composition (Chang et al. 2005). In particular, over the last half-century, diatoms were commoner and larger in sediments dated to years of low sunspot numbers and mirrored the classical 11-year periodicity of the sunspot counts (see NASA's Web site for a graphical display of the 250-year record of sunspot numbers). Sunspot lows are associated with higher values of the North Pacific High pressure index and suppression of the Aleutian Low pressure system; the outcome is greater coastal upwelling off the British Columbia coast, and greater diatom productivity. Years of sunspot lows (1985–1986, 1996–1998) tend to be associated with reduced precipitation and higher T_{MAX} values in the summer months June–August in Barkley Sound; years of sunspot highs (1990–1991, 2001–2002) are generally cooler and wetter, but the limited span of weather data from Bamfield East preclude a more detailed analysis.

Islands differ one from another in terms of the local microclimates they offer plants, with island area and island position within Barkley

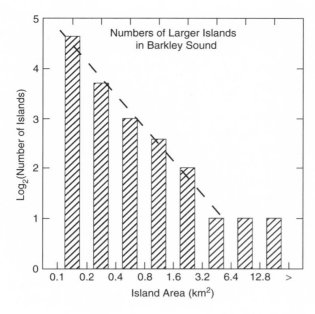

FIGURE 2.3. Numbers of islands decrease with increasing island size. Island numbers are reduced by a factor of 1.68 for each doubling of island size (dashed line: $R^2 = 0.985$).

Sound being the major determinants of this variation. I discuss how plant distributions are affected by island microclimates, particularly relative humidity and temperature variations, at several subsequent points in the book (beginning in the section below on "Flora and Vegetation"); I thought these local climate anomalies would be best brought up in the context of the plant species whose distributions or persistence they affect.

MORE ISLAND GEOGRAPHY AND TOPOGRAPHY

Barkley Sound is divided by three deep channels, Loudoun, Imperial Eagle, and Trevor, running northeast to southwest, more or less orthogonal to the outer coast (Fig. 2.1). Between the northwesterly Loudoun and central Imperial Eagle Channels lies the multitude of islands comprising the Broken Group component of Pacific Rim National Park, and between the Imperial Eagle and the southeasterly Trevor Channel are islands of the Deer Group. The present study divides these islands into five groups or series: Series A islands are those between the center of the Trevor Channel and the mainland (of Vancouver Island) to the southeast; series B, C, and D are islands located, respectively, distally to proximally in the Deer Group, that is, from the outer or exterior to the interior (mainland) areas of the Sound. The more northeasterly islands of the Broken Group in the center of the Sound compose series E. This classification corresponds roughly to the relative isolation of islands from the mainland coasts, from least isolated (series A) to most isolated (series E), with B–D intermediate and ordered most to least isolated.

The islands vary greatly in both area and in isolation, from the mainland coasts and from each other. Islands at the outer edge of the Broken Group are around 12 km equidistant from mainland coasts northwest, northeast, and southeast. The islands most isolated from other islands are Starlight Reef (northwest) and Baeria Rocks (near the head of the Imperial Eagle Channel), each about 2.6 km from other land, and preferred housing for Barkley Sound's largest seabird colonies. As a rough generality, the most isolated of the remaining islands are about a kilometer from their nearest neighbors, but most islands have many, much closer, neighbors. As the size or area of the islands considered decreases, the number of those islands in the Sound increases dramatically. Figure 2.3 shows the frequency distribution over size of the larger islands, those >0.1 km² in area, from the Broken Group southeast across the Sound to the mainland coast. These islands exclude those north of the center of the Loudoun Channel and number 54 in all. Below island areas of 6.4 km², island numbers increase at a near-constant rate as area decreases, such that a one-unit shift on a log (base 10) area scale corresponds to an increase in island number by a factor of five, on a natural log (base e) scale by a factor of two, on a log (base 2) scale by a factor of 1.68. By extrapolation to the many, more numerous islands whose areas are not accurately measured by charts, of which only a sample have been charted as part of this study, we can expect to find more than 100 islands of area 1–2 ha, and more than 500 islands in the size range of 0.1–0.2 ha (1000–2000 m²).

Of the >200 of the islands used in this study, all but the very largest were mapped at the shoreline using a measuring rope (marked at 1 m intervals) and compass. This produced a series of straight-line distance and direction intervals that formed and closed a transit loop around each island and provided accurate maps along which the actual shoreline was drawn in to scale. We used the top of the *Fucus* line, in the upper intertidal zone, to define the shoreline, as this benchmark is constant and readily apparent on all of the islands. Figure 2.4 shows the frequency distribution of island areas, sorted by island location; island areas cover about six orders of magnitude, but islands in the midrange of area were by far the commonest in this study.

Most shorelines are rocky and generally ascend steeply, first to elevations that support the first vascular plants, thence through edge and scrub vegetation to the forested interior (where the island is large enough to support forest). On all but the smallest islands, rocky shorelines are broken occasionally by narrow coves or wider stretches of cobble or shingle beaches. Shelly and sandy beaches are uncommon on the islands, but a few do occur, the former often signifying ancient First Nations encampments, and the latter frequently occupied by modern campers. On sheltered, low-angle beaches, especially on larger islands of 5 ha or more where freshwater seeps reach the coast, the shoreline frequently collects enough muddy, silty soils to support a brackish vegetation of the sort favored by sedges (*Carex* spp.) and rushes (*Juncus*).

The shorelines at the *Fucus* level are convoluted and sinuous. L. F. Richardson in 1961 (elaborated in Mandelbrot 1977) showed that measures of shoreline length are dependent on the scale at which they are measured and vary with the step-length used to "pace off" the shoreline. The apparent length of an island's shoreline increases as step-length decreases; thus shorelines are longer to oystercatchers with 10 cm strides than to humans with 1 m strides. The slopes (b) of the regression lines of coastline length on step-length are roughly constant for Barkley Sound islands (Fig. 2.5), giving an average value of 1.16 ± 0.06 SD for the fractal dimension D ($= 1 - b$) of the shorelines. Note that long step-lengths cannot be employed for small islands (lower left), and larger islands (upper right) are not mapped at scales that allow the use of small step-lengths. Uniformity among the slopes of the different islands (Fig. 2.5) indicates that shoreline morphology does not vary appreciably with island size.

Besides varying in area, Barkley Sound islands vary in elevation, and to a large extent the two variables are clearly correlated. Maximum island elevation (ELEV) for the larger islands, taken from marine charts, is closely related to island area (A): $\log_2 \text{ELEV} = 4.44 + 0.257 \log_2 A$ ($R^2 = 0.92$, $p < 0.001$, $n = 23$); a doubling of island area on average increases its maximum elevation by about 25%. With a much larger sample size ($n = 213$) that encompasses islands in this study, including many much smaller ones, the relation is less strict ($R^2 = 0.54$), though still highly significant statistically ($p < 0.001$).

Elevation contours were drawn at 1, 2, 5, 10, and 20 m intervals on mapped islands; Fig. 2.6 shows how, for islands within a given size category, the area on the island decreases with increasingly higher contours. That part of an island area lying at >1 m elevation increases nearly one to one with increasing island area (coefficient 1.1

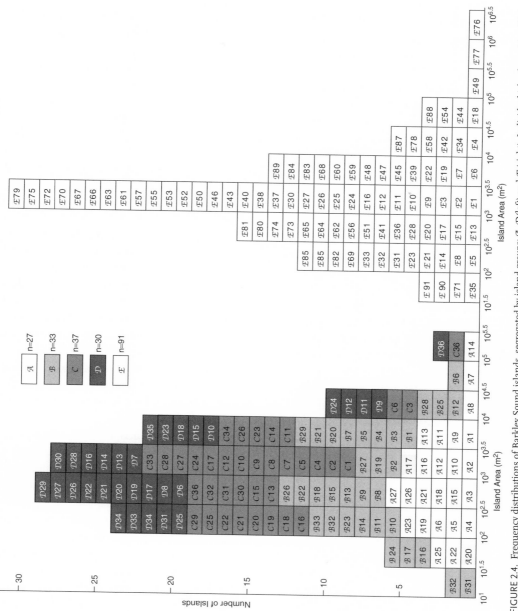

FIGURE 2.4. Frequency distributions of Barkley Sound islands, segregated by island groups: 𝒜–𝒟 (left) and ℰ (right). Individual islands are identified by code letters and numbers (see appendix B).

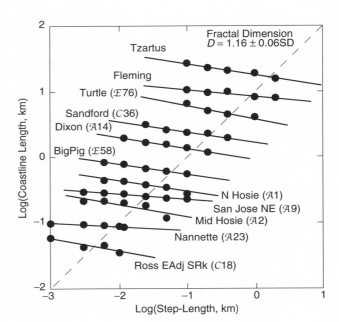

FIGURE 2.5. Apparent coastline length decreases with step-length (unit measure of coastline); regression slopes b give the fractal dimension D of the coastline: $D = 1.16 + 0.06$ SD $= (1 - b)$. Islands are arranged top to bottom by decreasing area, from Tzartus (ca. 20 km^2), through Turtle (1.2 km^2), Dixon (0.012 km^2), Mid Hosie (0.0022 km^2) to $C18$ (0.000131 km^2).

in a log$_2$-log$_2$ regression), but the area of an island lying above higher contour lines increases at increasingly greater rates with island area (coefficients 1.3, 1.5, 1.7, and 2.2, respectively) for areas above elevation 2, 5, 10, and 20 m. Thus the area lying at >20 m increases by a factor of $2^{2.2} = 4.6$ per doubling of island area (measured at the *Fucus* line). These island areas are those that would be exposed if sea levels were 1, 2, 5, 10, or 20 m higher than at present; given the volumes of the remaining ice fields, polar and others, it seems that only the first two increments are within the realms of possibility.

Although elevation is strongly associated with island area, and the areas between successively higher elevation contours increase fairly regularly with island area at the *Fucus* line (Fig. 2.6), there remains a good deal of variation in the shapes of islands of a given footprint. A schematic view of this variation is shown by the "elevations" (or side views) of some islands in the Ross Islets group (Fig. 2.7), where among smaller islands (e.g., $C4$, $C5$) and larger islands (e.g., $C3$, $C6$) similar to each other in map area, the side views differ significantly. Low-elevation islands are more susceptible to the effects of storm and wave surge than taller islands, and low-angle coastlines exacerbate such effects. Tsunamis are examples of rare but surely routine events, with a capacity to inundate coastline vegetation. A relatively recent tsunami occurred on March 27, 1964, the product of an earthquake in Prince William Sound, Alaska, which with magnitude 8.4 was the largest ever measured in North America. Compare this event to the magnitude 9.0 quake in December 2004 that generated the tsunamis in the northeastern Indian Ocean that killed nearly a quarter million people. The Alaska quake's vertical displacement along a submarine fault generated huge waves, which reached Barkley Sound after a four-hour time lag. The second and larger wave, after funneling up the Alberni Canal, was recorded at a height of some 6 m above high-tide levels at Port Alberni. The tsunami drained shallow coastal basins, briefly stranding boats, before the waves crashed ashore and wreaked extensive damage to coastal facilities. In Barkley Sound, the tsunami reached about 3 m above high tide, surely high enough to wash over many low-lying islands, with unknown consequences to the plants and the fauna. Immediately offshore from Barkley Sound is the Cascadia Subduction Zone, where the North American Plate rides over and subducts the Juan de Fuca Plate. A massive earthquake on the Cascadia fault off the Washington coast, with an estimated magnitude 9.0, occurred on the

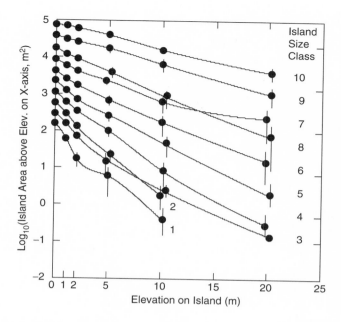

FIGURE 2.6. Areas on islands (ordinate) above elevation contours shown on abscissa. Islands are subdivided by area size classes on a $\log_2$ (successively doubling) scale, from smallest (class 1: area 98–196 m^2) to largest (class 10: area 50 – 100,000 m^2). Samples sizes of island size classes 1–10 are, respectively, 20, 21, 26 35, 33, 20, 15, 11, 4, and 2. Standard error bars around mean values are shown where samples are large enough; data points displaced horizontally to show these bars.

evening of January 26, 1700. The magnitude and origin of the quake are known from historical records of the tsunami it generated, and its arrival time in Japan 8500 km distant. Its destructive consequences are recorded in the oral history of the Huu-ay-aht; Anacla village at the mouth of the Pachena River was destroyed, and doubtless the vegetation on many low islands in the Sound was also obliterated. With the frequency of such Pacific Rim tectonic activities and their proximity to Barkley Sound, tsunamis must have been challenging island plants for millions of years.

Present-day island areas are a function of the current sea level. It is illuminating, and relevant in an island biogeographical context, to examine the histories of the Barkley Sound islands, dating from times at last glacial maximum when sea levels were around 130 m lower than now and the islands were part of the expanded mainland. I term these histories "island phylogenies" and plot them using information on submarine contours from marine charts. As sea levels drop, looking backward in time from the present into the past, present-day islands merge with each other according to the channel depth that separates them. These then join others, and islands become fewer and larger with successively lowered sea levels until, at 18,500 years BP and sea level −130 m, all of the islands are part of the (then southwesterly extended, and ice covered) Vancouver Island mainland. In fact, most of the islands were embedded in the mainland long before this; only a few were extant at sea levels just 50 m lower than present, these exceptions include the couple of isolated and rocky islands that are presently seabird colonies (see above). Of course, some new islands, those that are currently submarine seamounts, are exposed by dropping sea levels, and presumably the overall number and size distribution of islands in the Sound at, for example, −50 m sea levels would not be too dissimilar to the present number and size distribution.

The historical phylogeny of the B-group islands (see Fig. 2.8) is relatively simple, with three islands sharing a common line of ascent (B28, 29, 31: Folger/Leach/Leach N) originating when sea levels climbed above −50 m. Other islands in the group evolved (or, better, were created!) more recently and gained smaller size and increased isolation more slowly. The top row of Fig. 2.8 identifies 35, including four large islands (named on the charts) not included in the study. These islands coalesce to 12 entities with a drop in sea level of 1 m, to 8 after a 2 m drop, and so on (right-hand column) until all merge with the

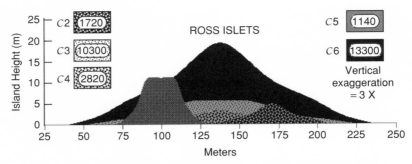

FIGURE 2.7. Despite similarities in area at sea level, island also differ in elevation and shape. This is illustrated diagrammatically in the figure with adjacent islands in the Ross group (C2 – C6). Island areas in m².

mainland. Putting an exact time frame on changes in sea level is somewhat equivocal, as eustatic sea level rise has not been uniform over time. Two widely recognized melt-water pulses (MWP 1A, 1B) occurred during the periods 14.32–14.13 ka BP and 11.5–11.2 ka BP (thousand year units before present) (Liu and Milliman 2003). During these relatively brief intervals of 200 and 300 years, sea levels rose rapidly, 90 and 40 mm/y, respectively, fast enough to drown coral reefs whose maximum vertical growth of 10–15 mm/y was much too slow to keep up with fast-rising sea levels. At the end of the latest (MWP 1B) fast melt period, sea levels were about 45 m below present level, that is, about the level at which the first of the Barkley Sound islands were isolated. I assume that the subsequent rise in sea level from −45 m to the present day level occurred approximately linearly over a 4000-year period post–11,200 years BP. This translates to island ages of around 12,000 years for the Folger-Leach group, of 9200 years for $B1$: Wizard and Bordelais/Bordelais Rocks ($B20$, 25), and 8000 years for Ohiat/Ohiat N ($B3$, 4) and Helby/Blackfish ($B8$); the remainder are likely now as isolated as they have been for the last 7000+ years. In the late Holocene, however, there was activity at the margin of the continental shelf that elevated shorelines and islands, at least those along the outer coast. By examining old wrack lines, where past seas had deposited trees and other flotsam on ancient shorelines, Friele and Hutchinson (1993) determined that the outer central west coast of Vancouver Islands uplifted 3+ m in the period 5000–7000 years BP. Thus, while mean sea levels are now about what they were about 7000 years BP, in the interim relative sea levels were higher, and low-lying islands of elevations less than a few meters would have emerged from inundation only in the last few thousand years. These low islands would have been either effectively submerged or much reduced in area before 2000–3000 years BP; the imprint of higher sea levels on low-lying islands will be identified later, with respect to forest floras.

The phylogeny of the E-group islands of the northern Broken Group is rather more complex (Fig. 2.9). Baeria Rocks ($E84$) were calved from the mainland at −80 m sea levels, and Starlight Reef ($E83$) followed at −40 m, at which time the Broken Group islands were subsumed under two large blocks, north and south. All but two of the 91 study islands in the E group belong to the northern group, a number that would be halved by a 1 m drop in sea level and reduced to 33 by a further 1 m drop.

Looking backward through time, some islands merge rapidly with others as sea level is lowered and soon become minor parts of much larger islands. Others, because of their submarine topography of steep-sided and isolated mountaintops, gain area much more slowly with lowered sea level and grow via mergers correspondingly further back in time. Some of these variations are shown for several B-group islands in Fig. 2.10. It is conceivable that such historical factors are relevant to the present-day plant species composition on the islands. For

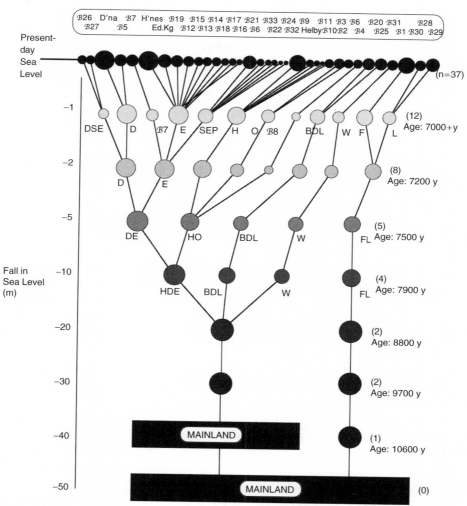

FIGURE 2.8. Historical phylogeny of the B-series islands. Present-day islands (listed at top of figure) are shown successively joining as sea levels are lowered (ordinate) to −50 m (as has occurred over the last 18,000 yBP), reducing the number of islands from 37 (present) to 0 as they become parts of the paleomainland (figures in parentheses at the right). Estimated ages of the island origins are shown at right. Code names for paleo-islands are indicated at the nodes; sizes of present-day islands are indicated by symbol sizes at the top of the figure.

example, islands that have a long and shared history of isolation may share more plant species than islands that were independently colonized or that independently retain species from mainland eras. And further, islands that were relatively recently much larger than their present areas may retain a legacy of past grandeur in terms of larger species numbers, an effect known as "supersaturation."

In addition to area and elevation, island isolation is a factor of considerable importance and is especially relevant to island colonization by new plant taxa. Isolation is somewhat more difficult to measure than island area. This must be why many examples of species-area curves have been published, a few of species-distance curves, but even fewer of species-isolation relationships. Distance to mainland may be the major variable in colonization rates on remote islands, but in archipelagoes where other islands are sources of potential colonization, it is neither the sole nor perhaps the primary one. Distance to mainland,

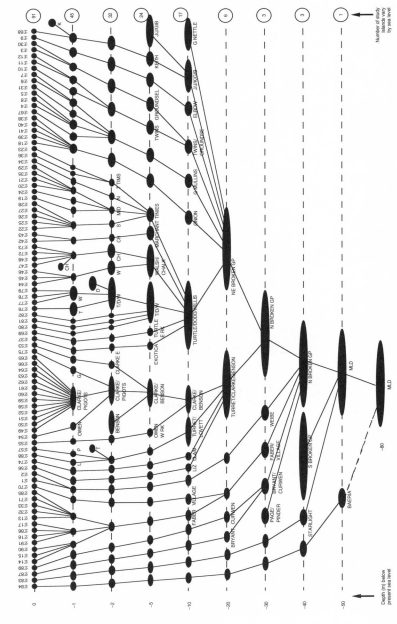

FIGURE 2.9. Historical phylogeny of the Æ-series islands. Code names of 91 censused islands are shown at the top of the figure. The islands coalesce into fewer and larger land masses as sea levels drop. Nodes that terminate within the figure represent extant but uncensused islands.

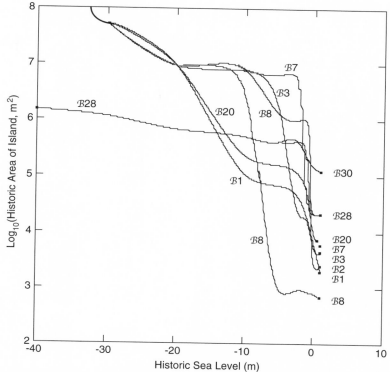

FIGURE 2.10. Historical areas of some islands in the *B*-group as a function of lower sea levels. Some islands (*B*3, *B*8) increase area rapidly as sea level drops, representing either flat hilltops or a rapid merging with neighboring islands. Some islands (*B*28, *B*30) increase in area only slowly as sea level drops, representing steep-sided hills with no near neighbors; others are intermediate. Note that *B*1, Wizard, would be submerged at sea levels of +4 m relative to present.

defined here as "mainland" Vancouver Island surrounding Barkley Sound, varies up to just over 8 km (in *E*-group islands centrally located in the Sound); the frequency distribution of these distances, sorted by island group, is given in Fig. 2.11 (upper).

In archipelagoes of numerous islands with a complex spatial distribution, both the size and the proximity of neighboring islands contribute to island isolation, and I measure this by the angles subtended by neighboring islands on the subject island. These angles are functions of both island size and distance from the potential recipient of colonists from the neighboring sources. This approach seems particularly appropriate for passively dispersing propagules, which appropriately describes most if not all plants. The methodology is illustrated in Fig. 2.12, where angles subtended by neighboring islands are computed for neighbors ≤100 m, ≤250 m, and ≤1 km. Whether or not a neighbor qualifies for inclusion as a potential source is determined by shoreline-to-shoreline separation distance. Given an identified source, the length of its shoreline within a given distance subtends the computed angle. Angles are then summed over the various contributing sources within each of the three distance zones. More distant islands that are "shadowed" partially or completely from the source by closer islands are considered as independent sources; even without a direct line for potential colonists to follow, shadowed islands are thought to contribute potentially via the intermediary stepping stone.

Figure 2.11 (lower) shows the frequency distributions of summed subtended angles for

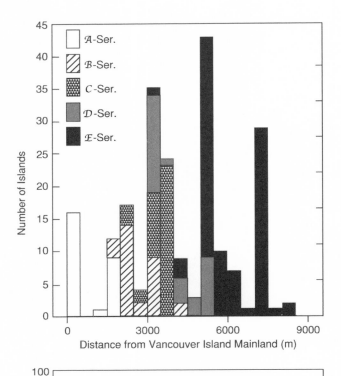

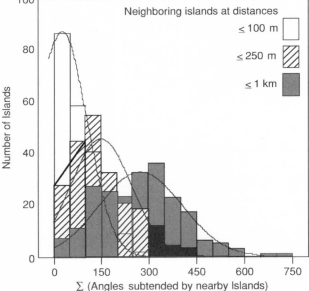

FIGURE 2.11. Measures of isolation in Barkley Sound islands. Upper: frequency distribution of distance from mainland. Lower: frequency distribution of summed angles subtended by neighboring islands, within three distance ranges of 100 m, 250 m, 1000 m.

islands in Barkley Sound. The existence of more and more contributors at increasingly greater distances shifts the distributions from left to right. Note that islands embedded within dense clusters of neighbors may accumulate subtended angles that sum to much more than 360°, to an empirical maximum about twice that value. Median values of the distributions occur at summed angles around 100° at 100 m, 150° at 250 m, and 300° at 1 km, with variance increasing with neighbor distance. These data indicate how far an observer, standing on a given island and turning through a complete circle, can see in various directions before

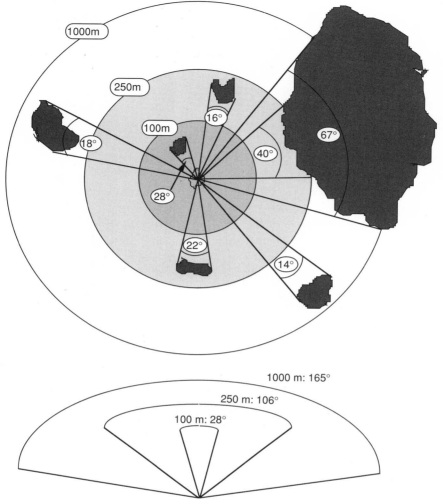

FIGURE 2.12. Upper: Isolation is measured by summing angles subtended by neighboring islands on subject island (in center). Neighboring islands are identified within three ranges of distance: 100 m, 250 m, and 1 km. Summed angles are shown in the lower figure, whose composite represents isolation.

encountering neighboring islands within different distances.

FLORA AND VEGETATION

The Barkley Sound region has mostly highly leached, podzolic soils that are considered relatively infertile, but this by no means limits their ability to support impressive and luxuriant vegetation—the tall, dense Coastal Coniferous Forest for which the Pacific Northwest is justly celebrated. With minor variations, this forest type extends from northern California to southeastern Alaska. It reaches its most impressive development in the central part of this zone, northwest Washington and British Columbia, where coastal rainfall is higher than it is further north and further south. These forests are dominated by western hemlock (*Tsuga heterophylla*), western red cedar (*Thuja plicata*), and Sitka spruce (*Picea sitchensis*). In the protected, old-growth forests of the Quilleute, Queets, and Hoh River Valleys on the Olympic Peninsula and in the Carmanah Valley southeast of Bamfield on Vancouver Island, the trees reach truly spectacular girths and heights, and individuals

>60 m in stature and >3 m in basal trunk diameter are routine. Outside of the Californian redwoods, these are the largest terrestrial organisms on the planet. There is a minor representation of other coniferous trees in the forest, an understory of mainly ericad shrubs (family Ericaceae) especially dense below canopy gaps, abundant ferns, but little herbaceous vegetation, and only a few geophytes such as lilies (Liliaceae) and orchids (Orchidaceae).

The forest composition changes with rainfall and elevation. On the inner coast of Vancouver Island where annual precipitation drops to around 1500 mm, Douglas fir (*Pseudotsuga menziesii*) dominates the forests, and on drier sites (precipitation (<1100 mm) madrone (*Arbutus menziesii*) becomes a common canopy tree. At higher and cooler elevations, beginning around 1000 m, elevational succession replaces western hemlock by mountain hemlock (*Tsuga mertensiana*) and western red cedar by yellow cedar (*Chamaecyparis nootkatensis*).

The Coastal Coniferous Forest zone and its characteristic plant species are well covered by several regional floras, including Hitchcock and Cronquist's 1973 *Flora of the Pacific Northwest*, itself a condensed version of the earlier five-volume *Vascular Plants of the Pacific Northwest* by these authors and their collaborators (Hitchcock et al. 1955–1969). More recently, the eight-volume *Illustrated Flora of British Columbia* (Douglas et al. 1998–2002) provides keys, updated taxonomic treatments and nomenclature, and excellent line drawings. Several field guides further facilitate plant identification with nontechnical descriptions and color illustrations of many of the plants (e.g., Klinka et al. 1989; Pojar and MacKinnon 1994).

For many of the typical, or indicator, plants of the wetter Coastal Coniferous Forest regions, the geographical ranges of the species are essentially that of the vegetation type, spanning latitude from northern California to Alaska, but with an inland expansion to the Rocky Mountains from Idaho and Montana north. This specifies the ranges of the trees such as western red cedar, western hemlock, Pacific yew (*Taxus brevifolius*), shore pine (*Pinus contorta*), red alder (*Alnus rubra*), big-leaf maple (*Acer macrophyllum*), cascara (*Rhamnus purshiana*), Sitka mountain-ash (*Sorbus sitchensis*), and others. However, coastal Sitka spruce is replaced inland by its congener Engelmann spruce (*Picea engelmannii*), and lodgepole pine (*Pinus murrayana*) similarly replaces shore pine in the central and northern Rockies. Many common understory shrubs share the strictly northwestern distribution of the dominant trees (above), including false azalea (*Menziesia ferruginea*), salmonberry (*Rubus spectabilis*), ninebark (*Physocarpus capitatus*), ocean-spray (*Holodiscus discolor*), and western crabapple (*Pyrus* [*Malus*] *fuscus*), while several others occupy just the southern half of this range and extend from southern British Columbia south to California (Manzanita [*Arctostaphylos columbiana*], evergreen huckleberry [*Vaccinium ovatum*], and red huckleberry [*V. parvifolium*]). Salal (*Gaultheria shallon*) and bittercherry (*Prunus emarginatus*) extend somewhat farther north, the former to southeastern Alaska and the latter to the Queen Charlotte Islands.

Some Barkley Sound plants occupy geographic ranges that extend well beyond the northern California–to–Alaska range of the Coastal Coniferous Forest. This extension parallels the degree to which they are not confined to the forest interior but, rather, are variously sun tolerant and dependent on natural gap, edge, shoreline, or bog habitats, or on anthropogenically disturbed or cleared areas such as roadsides and clear-cuts. Islands, as well as mainland shorelines at coast, lake, and river edges, provide a gradient from the forest interior to shrubby edge habitats and beyond, to rocky or other substrates largely lacking soil on which only scattered herbaceous vegetation can survive. In fact, nonforest plants comprise by far the larger part of the flora of the islands and adjacent mainland around Barkley Sound. A plant census of a few hectares of undisturbed forest might provide a list of 25 species or so, but a survey of the same area that encompasses edge and shoreline habitats might easily have double or triple that number. Further, increasing sample area does little to extend the list of forest species, whereas larger sample areas of edge habitats invariably add many new species to the census.

Many of these edge, shoreline, and bog species have geographical ranges extending far beyond that of the northwestern Coastal Coniferous Forest. Many are amphiberingian, meaning that they range north from British Columbia through Alaska and thence west across the Bering Sea into eastern Asia. Some are truly Holarctic, with circumboreal ranges that extend at the higher latitudes throughout North America and Eurasia; many of the plant species that are bog obligates are in this category. A few shoreline species occupy this demanding habitat well beyond the Pacific Northwest (e.g., coast strawberry [*Fragaria chiloensis*] occurs in western North and South America and in Hawaii; red fescue [*Festuca rubra*] in North America and Europe; hairy cinquefoil [*Potentilla villosa*] in western North America plus Asia; seaside arrowgrass [*Triglochin maritimum*], North and South America plus Eurasia). Other common species of rocky shorelines may be found on otherwise similar substrates but in montane habitats inland, including many of the native grasses in such genera as *Agrostis, Festuca, Danthonia, Deschampsia,* and *Poa*. The grass *Hordeum brachyanthemum*, for example, is superabundant on dry coastal rocks in the Sound, but also widespread inland in the mountains and even in dry sagebrush deserts. Other common species with strictly coastal distributions, such pearlwort (*Sagina crassicaulis*), and mistmaiden (*Romanzoffia tracyi*), have close counterparts in the inland mountains (*S. saginoides, R. californica,* and *R. sitchensis*).

It is extremely unusual to find any nonnative plants in the forest habitats of Barkley Sound. But while the aforementioned and broad-ranging edge and shoreline species are native to British Columbia, many other plants commonly encountered in the open, nonforested habitats are nonnatives and have probably lived in the region only since the advent of European traffic encompassed the area. This nonnative ("adventive," "alien," "exotic," "introduced") component of the flora is substantial; it now comprises nearly 25% of the flora of the Olympic Peninsula (Buckingham et al. 1995), whereas a century ago it was far less conspicuous (<3%; Jones 1936). Some alien plants are restricted to chronically disturbed sites around habitation and comprise the weedy commensals routinely encountered outside of airports and hotels nearly worldwide. Others extend from areas of heavy traffic well into the natural vegetation and blend in with the native plants; here disturbance, if there is any, is more subtle and results from natural rather than anthropogenic causes. Yet other alien plants are intermediate, requiring some form of alteration (removal, thinning, degradation) of the native vegetation before they can gain a foothold. Thus there is a progression among the alien plants from those found only in close proximity to garden, lawn, or roadside, to those able to colonize gaps large (clear-cuts) or small (tree falls) in the native vegetation, to those well integrated into the native flora of open sites where natural disturbance such as erosion or wave action precludes the formation of a tall and stable woody vegetation that would otherwise preclude them. In these last circumstances, aliens and native species pursue similar lifestyles, specifically the good dispersal, early maturity, copious reproduction, and short lifespan of the classic ruderal plant. Thus they adapt to and cope with recurrent disturbance and find refuge in often temporary habitats that, without further disturbance, undergo rapid succession to taller and longer-lived plants. In such sites, the aliens are fully naturalized and self-sustaining elements of the vegetation.

Note that nonnative plants may occur in an area as true introductions, in that humans have been instrumental in their overcoming barriers to natural dispersal limits. Alternatively, they might occur only where humans have modified or removed the native vegetation and, thereby, have provided access or opportunity for the establishment of aliens that would otherwise exclude them by the native plant communities.

Often aliens depend on a combination of these factors, with the occurrence of most nonnative plants resulting from often poorly understood combinations of the two sorts of human assistance. One rather well known case history of nonnative plant introduction to the area involves two species of alien sea-rockets (*Cakile*; fam. Brassicaceae), which now occupy sandy shorelines in Barkley Sound. *Cakile edentula* is native to the eastern shores of North America, and *C. maritima* originated in western Europe, but both now enjoy a nearly worldwide distribution following the transport and discharge of their floating siliculas in the seawater ballast of ships. *Cakile edentula* arrived in British Columbia by 1909, having expanded north and south along the Pacific coast of western North America from a San Francisco beachhead in 1882. *Cakile maritima*, first recorded in 1935 near San Francisco and now widespread north and south, arrived much later, reaching British Columbia by 1951. The colonization of Australian beaches by both species has a similar history and comparable time frame (*C. edentula* first appeared in 1863 in Fremantle, Western Australia, and *C. maritima* in 1897 in Victoria, southern Australia). Today the two species share British Columbia beaches, segregated by vertical position on the shoreline, and only occasionally hybridize. But on South Australian beaches, where *C. edentula* was established by around 1900 and *C. maritima* by around 1918, the two taxa show extensive introgression; there plants are perennial rather than annual as elsewhere (Cody and Cody 2003).

Many other plants, alien to both regions, link British Columbian and Australian islands. Abbott (1977) and Abbott and Black (1980) studied the plants on small, rocky limestone islets, at around 32° S, off southwestern Western Australia, with a Mediterranean-type climate. Among the 132 plant species recorded on 121 islands, 50% were plants alien to Australia. About 15% of their island aliens occur also in Barkley Sound, where a few species are considered native (*Atriplex patula*, *Hypochaeris*), but the majority are similarly alien and mostly native to Europe. The shared plants include *Poa annua*, *Vulpia* (= *Festuca*) *myuros*, *Chenopodium album*, *Capsella bursa-pastoris*, *Cakile maritima*, *Stellaria media*, *Trifolium campestre*, and *Sonchus oleraceus*. None except the *Cakile* is especially maritime, but instead the species are generalized weeds of disturbed sites worldwide, and clearly their ruderal habits predispose them very well for island life.

The flora of British Columbia is not at all large by the standards of tropical locations. Nor is it diverse in comparison to such celebrated temperate diversity hot spots as the Cape region of southern Africa. Like the Cape, the world's other four regions with Mediterranean-type climates, at latitudes 30–35° N, S, are generally richer in species than are neighboring higher latitudes; California, with about 5500 plant species in 406,254 km^2, is considered diverse by North American standards (see, e.g., Cody 1986a). British Columbia (area 937,600 km^2) has around 3000 plant species (ca. 23% aliens), about 1600 of which are known from Vancouver Island (area 31,764 km^2). About 1400 plant species are found on the Olympic Peninsula (Buckingham et al. 1995), which, while considerably smaller in area (13,800 km^2) than Vancouver Island, straddled the edge of the last continental ice sheet rather than lying entirely under it. An inland diversity comparison is provided locally by Mount Ranier National Park, with about 700 species in an area of approximately 1000 km^2 (Brockman 1947).

Bell and Harcombe (1973) conducted the most detailed study of plants in Barkley Sound previous to mine, under contract with Parks Canada to provide an inventory of plants in the Broken Group (Æ-group islands), for Phase II of Pacific Rim National Park. This work included relevé counts and detailed descriptions of plant associations and lists 231 vascular plant taxa. Of these, the authors classify 66 species as "extremely rare," 139 as "rare to sparsely distributed," and 26 as "relatively common to abundant." The Bell and Harcombe survey included censuses from all of the larger islands in the Broken Group and also recorded plants from the large shell middens there, relics of the past indigenous occupants, as well as of more recently (nineteenth to twentieth century) abandoned homesteads with various introduced ornamentals. My surveys included few of the largest islands, none of the southern islands in the Broken Group (south of the Coaster Channel), no shell midden sites nor homesteads other than one on Clarke Island, and excluded the marine genera of *Phyllospadix* and *Zostera*. Thus my censuses lack a few species restricted to these sites or habitats, although our lists have around 210 species in common. Some of the taxa listed by Bell and Harcombe specific to the shell middens include native *Elymus hirsutus*, *Urtica dioica*, and *Artemisia suksdorfii*, and alien *Mentha piperita* and *Leonotondon nudicaulis*. Most of the remaining omissions from my lists are garden ornamentals around the homestead sites, but two exceptions are the native nightshade *Solanum nigrum* and the umbel *Sanicula crassicaulis*. My own species roster from censused islands in the northern Broken Group includes 208 taxa, and for the islands in general 270 species; the total is further inflated by species of mainland weeds and mainland bog plants near Bamfield that I recorded on no islands. We can consider that, in general terms, a flora of around 300 plant species occurs on the islands and the adjacent mainland, and this species pool provides the candidates for potential island colonization.

While many mid- to large-sized islands in Barkley Sound appear superficially similar to mainland sites on nearby Vancouver Island, they differ in several important aspects. Of course the islands have shorelines that support specific plants, and the corresponding mainland shores are usually more sheltered from the weather, especially storms, and more closely shaded by larger trees that narrowly limit the edge and shoreline vegetation in comparison to small and more open islands. Further, the islands differ in abiotic conditions interior from their shores, especially in moisture levels. Most islands are in general much drier, and an island of even several hectares is not strictly comparable to a patch of mainland forest of the same size (even apart from the shoreline vegetation). The islands are dry because they lack any run-on from adjacent upslopes, and they lack the ameliorating influence of extensive tall forest. Small islands especially are drier and warmer with their lack of the buffering effects of forest vegetation. In order for the intercepted precipitation to reach the coast as seeps or small creeks, island areas empirically must be at least several hectares; smaller islands lack freshwater seep habitats and the several plant species typically associated with them. An example is skunk-cabbage (*Lysichiton americanum*), abundant throughout the wet mainland forests but restricted to a few of the largest islands in the Sound, those greater in area than one-half square kilometer or so; several other common wetland plants of the mainland, especially rushes and sedges, are similarly restricted to the largest islands.

Figure 2.13 provides an indication of area-related differences among island climates. Temperature and relative humidity data were collected during a two-week period in the middle of July in three separate years, 2000, 2002, and 2003, on islands over a wide size range. While obvious differences in temperature and relative humidity occur among years, the maximum relative humidity and minimum temperatures recorded in the two-week period did not differ among islands of disparate sizes. However, minimum relative humidity and maximum temperature both vary significantly with island size, indicating that smaller islands are drier and warmer.

Small islands, which become warm and dry in the summer, do not support the tall forests of the larger islands. The maximum vegetation height on Barkley Sound islands increases with log(island area)—see Fig. 2.14; an order of magnitude increase in island area corresponds to around a 13 m increase in maximum vegetation height (or somewhat more if the dashed line in the figure is considered an upper bound). Much of the residual variation in this relationship is generated by island position. For example, the island *B*10: "Helby NEE" is a beach satellite

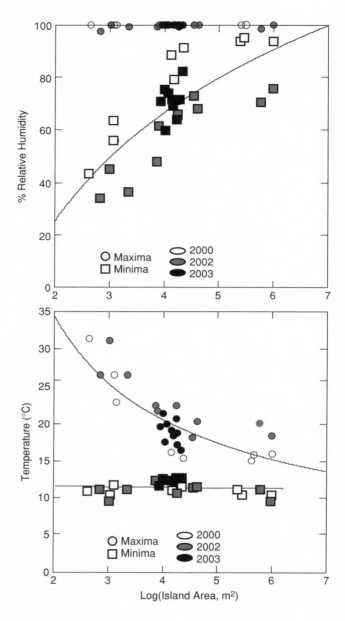

FIGURE 2.13. Relative humidity and temperature in island interiors as a function of island area. Note that minimum relative humidity and maximum temperature deviate on smaller islands.

adjacent to and sheltered by the large island of Helby and has several quite tall trees, whereas the island *B25*: "Bordelais SRks" is an extensive jumble of massive boulders at the extreme outer edge of the Sound, fully exposed to the fury of winter storms and quite treeless. (N.b., I use island names without quotation marks when their names appear on the marine charts; otherwise my working names are enclosed in quotation marks.)

Besides being related to island climate and area, tree heights are presumably related to soil depths on the islands, which vary both among and within islands. Soil depths reach their maximum in the island interiors and are reduced to scattered pockets of shallow accumulations in rock cracks at the island periphery. As an illustration of this variation I use $C4$: "Mid Ross", a half-hectare island measuring 60 m across, shoreline to shoreline. I measured soil depths,

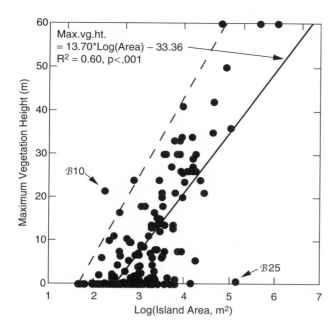

FIGURE 2.14. The maximum height of vegetation increases with island area ($p < 0.001$; $R^2 = 0.60$), such that around 13.4 m of extra vegetation height is gained per order of magnitude increase in island area. Much of the scatter is due to positional effects of islands, such that small but sheltered islands (e.g., $B10$) have disproportionately tall vegetation, whereas very exposed but large islands (e.g., $B25$) are treeless and have only low vegetation.

percent soil moisture, and photosynthetically active radiation (PAR) 1 m above ground level, along a transect traversing the middle of the island. In August of a dry year (1985), moisture content reached 35% in the center of the island, where soil depths were around 0.5 m, vegetation was tall and dense, and PAR at 1 m was reduced to around 5% of that above the canopy. As soil depth dropped to zero toward the island's periphery, its moisture content declined also, to a low of 4%, and vegetation became low and open. This profile is shown in Fig. 2.15.

Plants of drier habitats must be hard pressed in general to find suitable habitat on the outer coast of Vancouver Island. However, while such species are indeed hard to find on the mainland, some are common on the islands, especially on smaller islands and at the south-facing coastal edges of somewhat larger islands. Douglas fir (*Pseudotsuga menziesii*) is an example; it is a common tree on the interior, dry side of Vancouver Island, and it has a wide distribution on the islands off the outer coast, though is very rare on the adjacent mainland. The species is, however, scarce in the Broken Group; this appears to be a consequence of the generally more moderate conditions in the middle of the Sound, where islands are buffered by the broad expanse of ocean waters and remain cooler and more mesic in the summer. In the Broken Group, temperatures are ameliorated and fog banks persist longer owing to reduced land-to-sea ratios and the relatively small land areas to heat and disperse them. Thus islands situated near the edge of the Sound, toward the interior and close to the mainland, are often sunny when the islands in the middle are foggy. A useful indicator plant of these sunny islands is the manzanita *Arctostaphylos columbianum*, which is common on D-group islands, uncommon on A-group islands, but not recorded elsewhere in the Sound.

Local microclimate thus varies with island size and also with island position within Barkley Sound, largely as a function of the proximity of the larger, surrounding land masses. Differences related to island position are illustrated by distributions of many plants that prefer the driest of rocky shores. Shore-pine (*Pinus contorta*), wall-lettuce (*Lactuca muralis*), and the mustard *Arabis hirsuta* all have such preferences, and all show levels of incidence around an order of magnitude lower on the more maritime islands of the B and E groups. The effects of an island's position on its climate are embedded in Fig. 2.13. Of the seven E-group islands for which temperatures and relative humidity data were

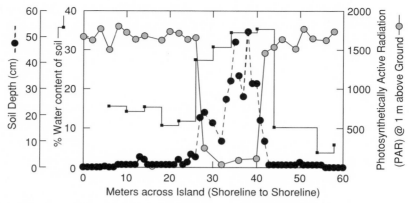

FIGURE 2.15. A transect across a midsized island, C4: Mid Ross, shows how soil depth and its moisture level increases from island edge to center, while the light level at 1 m above ground decreases in the island interior with increasing vegetation height and density.

recorded, six are above the regression line for minimum relative humidity, and six below the regression line for maximum temperature. If we ignore interyear variation and island size effects, this supports the rough impression that islands centrally located in Barkley Sound stay cooler and damper than do peripheral islands.

The final, rather specialized, component of the island flora to be addressed here is that of bogs. Three of the largest islands in Barkley Sound have small sphagnum bogs, namely, Nettle, Turtle (*E*76), and Effingham. This study included surveys of these island bogs, and to provide data for comparison to the island bog surveys, several of the mainland bogs around Bamfield were also censused. Since the bog flora is largely specific to these unique habitats, the island bogs are essentially islands within islands; whereas mainland bogs are islandlike too, many are much larger than the tiny (≤ 2 ha) island bogs and are variously connected by the extensive freshwater drainage systems of the mainland landscape. Many of the more cosmopolitan genera of the region, such as *Carex, Juncus, Vaccinium, Pinus,* have bog-specific obligate species, and many other species are found nowhere else in the vegetation except in the bogs. Some of these are true aquatics and depend on standing, open water; thus they are absent from drier, older bogs that are further along in their long-term succession to forest. Among well-known bog obligates are shrubs of the genera *Kalmia, Ledum,* and *Myrica,* sundews *Drosera,* various lilies and orchids (*Toefieldia, Plantathera*), and herbaceous dicots (e.g., *Apargidium, Gentiana, Hypericum, Lycopus, Sanguisorba,* and *Viola*). With naturally isolated bogs occupied by habitat-specific plants, the possibility arises that the additional isolation of island bogs affects, and perhaps limits, their floras; the phenomenon is explored in detail in a later chapter.

METHODOLOGICAL NOTES

Census efforts. Repeated visits to a very isolated oceanic island are likely to yield little new information on later trips, because such floras (and faunas) change hardly at all over the short term. That is not the case with inshore islands, on which the dynamics of colonization and extinction are expected to show rates far higher than those of more isolated midocean islands. Accordingly, the chief research technique I have used is that of repeated censuses of plants on the same islands over a number of years. Each census year seems unique and produces much new information, and the continued accumulation of new data becomes rather addictive! But eventually a satisfactory level of understanding of the patterns and processes is attained, generating a push to publish before incoming new data cause the extinction of the old. I began this study in 1981 and spent parts of the summers of 1982, 1985, 1986, 1987, 1989, 1990, 1991, 1996, 1998, 1999, 2000, 2002, and 2003 in Barkley Sound. This amounts to an average of about two-thirds of the summers over the 22-year period, and during most survey years I spent around six weeks on the research, usually early July to mid-August.

The research effort was variously allocated to different tasks over the years, but island plant censuses were taken during every summer of residence. During the early years, six weeks was time enough to revisit all of the islands previously censused, but in later years and with >200 islands in the survey, about half were recensused each summer, when islands in either the $\mathcal{A}$–$\mathcal{D}$ series or the $\mathcal{E}$ series were censused. However, a set of between one and two dozen islands was particularly favored and visited during all or nearly all summer census years. Thus a good number of islands have been censused in adjacent years, on multiple occasions, but on others censuses have been several to many years apart. The islands I censused most frequently yield data from 13 different years; most of the 216 islands included in the survey were censused 6–10 times in 22 years. In the case where islands were censused in nonconsecutive years, colonization and extinction events were likely missed, but to some extent these omissions can be calibrated by data from censuses in adjacent years.

Census times. The basic datum of the study is an island census in a particular year. Note that, unlike censuses of island animals, we do not have to try to distinguish between resident individuals and visitors! Censuses were conducted within a particular time frame and following certain protocols. On the smallest islands with only low and open vegetation, virtually every square meter was examined for plants, from coastline to the island center. Most islands larger than a quarter hectare or so have forested interiors; on these the entire coastlines and edge habitats were surveyed thoroughly, and several forays were made into the forest interior at various points. Large islands take much more time to census than small islands, and very large islands of 10 ha or more may require several days to survey thoroughly, with census data accumulated over several trips. Even with the extended census times on large islands, it is difficult to regard these data as having the same reliability as those from small islands. Especially with large islands, like the 1.25 km² $\mathcal{E}$76: Turtle, censused over several years for a cumulative total of >50 h, the census must be regarded as tentative, for even though all of the coastline was covered repeatedly, much the interior has not been surveyed.

All island censuses were timed from start to completion. This was done not in an effort to complete a census within a predetermined time, but to record the time needed for a putatively complete census. In practice, census times for C-series islands fit the regression equation Census Time $(h) = 0.007 * A^{0.532}$, where A = island area in square meters ($p < 0.001$, $R^2 = 0.83$). Empirically this means that, on average, a 1/100 ha island takes <1 minute, a 1/10 ha island 17 minutes, a 1 ha island about an hour, a 10 ha island 3.2 h, and the largest of the C-series islands, C36: Sandford (29 ha), nearly 6 hours. Some variation in census time is generated by the variability of the coastlines and the difficulty or otherwise of traversing them. These times are roughly comparable to those required by Roden (1998) to census lake islands in western Ireland, and somewhat shorter than the census times employed by Nilsson and Nilsson (1982) to census plants on Swedish lake islands.

Another representation of census effort and its standardization is depicted in what I call "plant discovery curves." Besides timing the censuses, in most censuses I used specific symbols on the census sheets to represent data from consecutive time intervals, using 5-minute intervals on small islands and 10- or 15-minute intervals on large islands. These data are plotted for representative C-series islands in Fig. 2.16a, the cumulative number of plant species encountered as a function of time spent on the census. Logarithmic curves fitted to these data

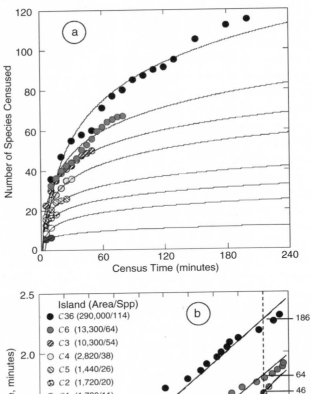

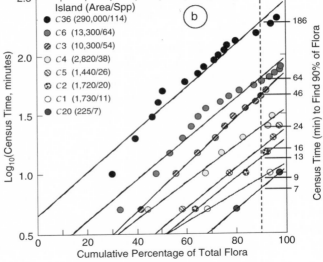

FIGURE 2.16. "Discovery functions" for island plant censuses. a. The cumulative number of species increases, at a decreasing rate, with census time, with the rate declining sooner and faster on smaller and less species-rich islands. b. The census time required to "discover" a given proportion of the species on an island, log-scaled, increases at similar rates over islands and is more dependent of flora size that island area, viz. areas of $C_1 > C_2 > C_5$ with decreasing species numbers. Data are averages of timed censuses in two separate years for island C_{36}, for 4–6 years for other islands.

rapidly flatten out with increasing time, especially on smaller islands. Their continued rise beyond the data points indicates that new species continued to be discovered throughout the census (until its completion after full coverage of the island) and should not be taken as an indication of a potential for adding further new species with a hypothetically increased census time. With cumulative species numbers expressed as a percentage of the total island flora (on islands where that total is well known), the time required to discover a given proportion of the flora increases linearly, on a logarithmic ordinate scale, at a similar rate on islands of different areas and floras (Fig. 2.16b). Islands with steep and indented coasts require longer census times and have larger floras; this accounts for the differences among the three islands C_1, C_2, and C_5 in the figure, whose floras' sizes are in reverse order to the island areas. There is a tight linear relationship between the time required to discover 90% of the flora (Fig. 2.16b, right ordinate) and species number SPP: $\log_{10}[\text{Time}_{90\%}] = 0.842 + 0.1013 * \text{SPP}$ ($p < 0.001$, $R^2 = 0.977$). This corresponds to 10-minute censuses for 12-species islands, 14 minutes for 24-species islands, 29 minutes for 48-species islands, and 123 minutes for 96-species islands, in order to discover 90% of their floras.

Sources of error. In general, census times were accorded to islands in a standardized fashion in hopes that no resident species were missed. However, there is a natural asymmetry between ascribing presence and absence to species in any survey, as recorded species are certainly present and known with certainty, whereas species absence is only assumed, following failure to find the species in a census. Thus absence is necessarily less certain than presence.

Species actually present on an island, but missed by the census taker, contribute to an underestimate of the island's total species count and, perhaps more importantly, contribute to errors in the measurement of species turnover, namely, the difference in species composition between successive censuses. A missed species, one actually present but mistakenly recorded as absent, is tallied as a false extinction if missed in the second census but recorded in the first, and as a false colonization if missed in the first but recorded in the second. Thus missed species inflate the apparent turnover rate and contribute a "pseudoturnover" component to it (a term coined by Lynch and Johnson [1974]). By comparing replicated censuses on islands in Lake Möckeln, Sweden, by different observers within the same years, Nilsson and Nilsson (1982, 1985) showed that pseudoturnover significantly inflated true turnover in their study.

While missing species that are in fact present inflate turnover rates, increased intervals between censuses, intervals during which species might colonize and go extinct without being noticed, can deflate them. Techniques exist to hone estimates of turnover rates when censuses are conducted at irregular intervals, and to calibrate them as functions of census interval (Clark and Rosenzweig 1994; Diamond and May 1977; Rosenzweig and Clark 1994), and I refer to their application below.

Species may immigrate to islands more frequently than the colonization records indicate, since some immigrants must reach islands that are already occupied by that species. Such events generally go unnoticed and therefore contribute to underestimation of colonization rates. These redundant colonization events perhaps have significance at the genetic level, for instance by swamping island gene pools and precluding local adaptation there. They will not much affect patterns of incidence, at least not on islands where the species is present in large numbers and persistent over the longer time, but on smaller islands more precarious populations might be "rescued" by recurrent colonization that thereby prevents or reduces extinction (see following chapter). Extinction rates are underestimated whenever a new colonist arrives synchronously with the extinction of the preexisting population.

Being quite aware of the importance of census completeness, I employed several techniques designed to maximize census accuracy; I believe that missed species were very few on islands up to a couple of hectares in size, despite some island populations being represented by few or even a single individual plant. Firstly, beyond the challenges of negotiating often steep, rocky coasts, the islands are not really difficult to census, as the low vegetation is almost invariably quite open. That is, one is seldom faced with the challenges of discerning different plant species in continuous, intertwined mats of greenery, but rather plant individuals are spaced out in soil-holding cracks and crevices. In interior forests, growing sites are not nearly as patchy, but there plants are large, widely spaced, and easily assessed. The most careful scrutiny is demanded by the edge vegetation, the zone between the open shore and the forest, where plants of intermediate stature are often crowded closely together and individuals can be quite cryptic.

Secondly, census accuracy was enhanced by the fact that the data I report here were all collected by me, alone or in conjunction with a field-hardened assistant (although others, students especially, practiced data collection on the islands). Over the years, I gained a certain facility in knowing where to look and what I was looking at. Early in the study, when I was less sure of my competence, I conducted censuses on the same islands in conjunction with (but independently of) experienced, professional field botanists, namely my colleagues Drs. A. and O. Ceska, both of whom are intimately acquainted with the local flora. One or another of us sometimes missed a species that others found, usually represented by a single individual; these were, for me, reassuring exercises. I then began the practice of noting, either on the census sheets or the island map, the locations of rare or solitary plant individuals. If, during the course of a standard census, plants present in a previous census were not found, actual past locations were searched for the plant; in the absence of such notes on specific locations, potential sites were searched, and if encountered, the plant was noted as the product of a "directed search." With practice and over the years, I was able to recall practically all of an island's locations where specific rare plants or those with restricted island distributions were to be found, and I paid particular attention to such places. This practice perhaps served to reduce pseudoextinctions but would be less effective in discovering obscure new colonists.

Thirdly, the flora is relatively simple and includes very few "cryptic species" in the usual, taxonomic sense. Thus plant identification on the islands is fairly straightforward, such that students in my field classes usually became remarkably able census takers within a couple of weeks. My usual practice was to take any material that seemed questionable back to the laboratory for a more formal examination. This was required for some *Carex, Agrostis,* and *Poa* species, but little else. Occasionally I discovered a new species (i.e., a new island record) in the form of a plant that had clearly arrived earlier than the year it was discovered, but it had been missed in previous censuses. Such an occurrence, for example, might be the discovery in an island's interior of a single individual thimbleberry (*Rubus parviflorus*), already a half-meter tall, and clearly already a few years old. In early years and before my eye was as keen as it eventually became, I missed a few Pacific yews, which even on rather large islands are often represented by single individuals. Such individual yews are quite cryptic in the sea of conifery provided by the cedars, spruces, and hemlocks, and many are clearly ancient (indeed, they may look all but dead). Some biennial or perennial herbaceous species were missed as prereproductive juveniles, and first recorded as adults (second year). Thus, in these few instances, I felt justified in back-dating censuses to reflect the presence of species missed earlier as juveniles and actually first recorded later.

On islands where successive censuses were several years apart, I undoubtedly missed some colonists that went extinct without my seeing them. But because the vast majority of the plants are perennials, and even the great majority of the herbaceous species are at least biennial, census intervals of two years actually missed very few species as a result of a plant's short island tenancy. As mentioned above, apparent colonization rates on similar islands with different intercensus intervals can be compared and, if different, used to correct for variable census intervals.

Population sizes. In addition to census data on species' presence or absence, I noted the population sizes of certain species in various circumstances when they seemed of interest. I counted numbers of individuals in some species over several years subsequent to their initial colonization of an island; of species that were rare or restricted; of species that appeared to be increasing or decreasing in population size or distribution; or of species whose numbers seemed to be changing at the expense of or as a consequence of changes in other species' population.

Taxonomic notes. There will be a few more methodological notes of a general nature I shall mention when I discuss measuring anemochore dispersal or fern frond morphology, but the remaining point in this section concerns taxonomy. I mentioned above that I encountered very few problems with field identification of the plants, but in some instances the correct name to apply was unclear. In other words, a taxon was readily and repeatably identifiable as taxon "X," but it was not obvious which name in the flora should be applied to taxon "X." I did not spend an inordinate amount of time sorting out the few taxonomic anomalies; rather, I usually relied on A. Ceska to supply the best name to dubious material. Sources of confusion included (a) hybridization (e.g., among plantains, *Plantago* species both native and alien; grasses such as *Hordeum* and *Elymus*; *Polypodium* ferns, etc.); (b) species that on the islands shift in morphology or lifestyle, for example, from annual to perennial, obscuring their specific identities (e.g., *Hypochaeris, Arabis* forbs); (c) species with an inherently difficult and poorly understood taxonomy, like that of some *Agrostis* and *Poa* grass species; (d) species that are variable in morphology among island populations but still fall within a single species epithet (*Rosa, Ribes* shrubs). While I did not let nomenclatural niceties detract from the biogeography focus of the study, I do acknowledge their existence; a plant appendix (appendix A) is provided, and some points of taxonomic interest are elaborated there. Beyond the fact that a given taxonomic entity was recognized and recorded as a constant on the islands, I relegate these questions to others.

3

Island Biogeography

CONCEPTS, THEORY, AND DATA

IN THIS CHAPTER I present a brief overview of the theory, concepts, and analytical tools of island biogeography, and illustrate them where possible with my own examples as I go. It can be brief because the topic is well covered, and in much detail, in several recent treatments such as Rosenzweig (1995), Brown and Lomolino (1998), and Whittaker (1998). Mine is also a narrow overview, in that it is concerned chiefly with aspects of colonization and extinction dynamics appropriate to continental islands in a setting such as Barkley Sound. This necessarily will leave a lot of interesting biogeography unmentioned. Thus, I shall need to say little about the relevance of historical legacies and ancient land-bridge connections, as I would if discussing, for example, the long evolutionary history of New Zealand's present biota. Nor will I require input from molecular biology, as I would if the mainland ancestors of island endemics, such as Hawaiian tarweeds (Baldwin et al. 1991) or honeycreepers (Johnson et al. 1989) were in question. Islands with somewhat more isolation in space and time that those in Barkley Sound, such as the Sea of Cortés islands (Case et al. 2002), necessitate such an expanded view, with consideration of these additional components as well as others (Case and Cody 1987). The Barkley Sound islands represent, in a way, island biogeography at its simplest.

THE BASIC MODEL

The modern era of island biogeography studies was initiated by a paper (1963) and later a monograph (1967) by R. H. MacArthur and E. O. Wilson, modeling species number on islands as a dynamic equilibrium between balanced immigration and extinction rates. Since its inception, the "equilibrium theory of island biogeography" has promoted a great deal of field research and generated further theoretical development of the basic model, as well as considerable modification of and embellishment to it. It also, especially in the early years when new and revolutionary, attracted a good deal of sharp, critical comment (e.g., Sauer 1969 and much thereafter). The "M/W" theory at its most basic is represented by colonization rates as declining and extinction rates as increasing functions of the number of species on an island

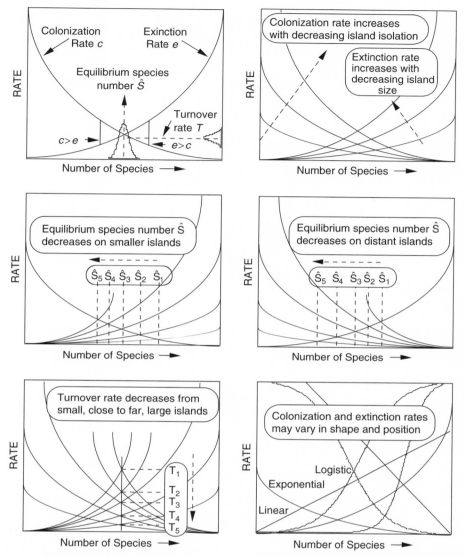

FIGURE 3.1. Thumbnail graphical sketch of the main elements of the MacArthur-Wilson (M/W) equilibrium theory of island biogeography. Throughout, colonization rates are decreasing and extinction rates increasing functions (ordinate) of the number of species on the island (abscissa), and these rates vary with island isolation and size, respectively. The intersection of these curves determines an equilibrial species number $\hat{S}$, and the height of the intersection above the abscissa determines the turnover rate of species on the island. See text for discussion.

(Fig. 3.1, upper left). The curves intersect to define a dynamic equilibrium, $\hat{S}$, around which there will likely be some variation, as indicated by the Gaussian curve below, because of temporal and spatial variations in colonization and extinction rates, or "turnover noise." To the right of this intersection, extinction rate e exceeds colonization rate c, and species number declines; to the left, the reverse is true, and thus the equilibrium is a stable one. M/W considered that colonization rates are primarily a function of island isolation and decrease with distance, whereas extinction rates are primarily functions of island size (a surrogate for population size) and decrease with increasing island area (Fig. 3.1, upper right). Thus equilibrial species

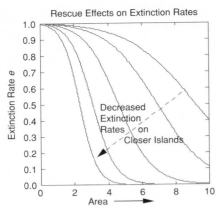 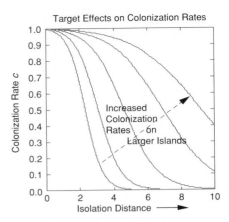

FIGURE 3.2. Extinction rates fall off as area increases (left), but closer islands may receive more colonists, which, by enhancing population size, may reduce extinction rate; thus rescue effects alleviate extinction on closer islands. Colonization rates (right) decrease with distance of island from the course, but larger islands provide easier "targets" for passively dispersing propagules, and thus colonization is enhanced on larger islands.

numbers decline with island size (generating a species-area curve of positive slope; Fig. 3.1, left center) and decrease with island isolation, for a species-distance curve of negative slope. A critical feature of the model is a predicted species turnover through time, the magnitude of which is determined by the height above the abscissa of the intersection of the colonization and extinction curves. Turnover is expected to be high in small, close islands and low in far, large islands (Fig. 3.1, lower left).

The colonization and extinction curves might be variously affected in shape and position by attributes of the islands and of the taxa under consideration; three alternatives are given in Fig. 3.1 (lower right). Nonlinear rates are thought, a priori, to reflect species interactions of various sorts, such as extinction rates increasing and colonization rates decreasing more steeply, as space and resources become scarce on islands with increasing numbers of species. Clearly resource preemption by earlier arrivals could severely impair the success of later potential colonists. The extent to which earlier arrivals can expand their niches and increase their population densities, and the order of arrival of species with varying propensities to benefit from species-poor habitats in this way, will affect the shapes of colonization and extinction curves.

SOME EMBELLISHMENTS

Rather than regard colonization rate as simply a function of island isolation or distance D, and extinction rate simply a function of island area A, it seems logical that both rates will be affected to some extent by both of these island properties. As discussed above, "supplementary immigration" (so termed by Whittaker [1998]) to islands that already possess the taxon in question has the potential to reduce its extinction rate there, called a "rescue effect" by Brown and Kodric-Brown (1977). Rescue effects might be important in some circumstances, for example, on small islands with limited ability to sustain persistent populations, but within the dispersal range of source populations producing a plentiful supply of propagules. This is illustrated in Fig. 3.2 (left), where extinction rates $e \Leftarrow (\mathbf{A}^{-1}, D)$—boldface indicating the supposed primary effect of area, along with a secondary effect of distance. A similar modification of colonization rate would include a supplementary effect of island area, namely, a "target effect" (Fig. 3.2, right; $c \Leftarrow (\mathbf{D}^{-1}, A)$). Target effects have been shown relevant for aerially drifting spiders (Toft and Schoener 1983) and waterborne plant propagules (Buckley and Knedlhans 1986). They are potentially important components of colonization for a wide variety of passively dispersed organisms, including windborne (anemochorous) as well as waterborne (hydrochorous) plants. Target effects may also characterize zoochorous (animal-transported) plants, whose seeds are voided by fruit-eating crows and thrushes that are preferentially attracted to larger islands.

INCIDENCE FUNCTIONS

A convenient way to arrange island census data is in matrix form, with islands as columns and species as rows; such species × sites matrices (SSMs) become of course three dimensional with repeated censuses, time being the third axis. The discussion thus far in the chapter has been "island centric," looking at SSM column totals: how many species does a given island support, and why? The other view looks at row totals and asks: on how many islands, and on what sorts, does a given species occur? This information is best presented as an "incidence function," a term coined in a landmark paper by Jared Diamond (1975) and used there to describe and analyze the distribution patterns of birds in the Bismarck Archipelago. Incidence J is the proportion of islands in a given category that a taxon occupies, and incidence functions of different shapes are shown in Fig. 3.3 (left). The more usual form is that of decreasing incidence from levels at or near one (all islands occupied) to zero, over islands of decreasing species richness. Decreases in incidence may appear variously logistic or exponential in form, or may abruptly decline to zero when a particular threshold or critical minimum value is reached on the abscissa. Besides these, Diamond (1975) showed a quite different form of incidence exemplified by his so-called "supertramp" species, which were absent from the largest, species-rich islands but reached their highest incidence on small islands (the quasi-Gaussian curve in the figure). These are species that presumably lack the ability to withstand strong competitive pressures but possess the ability to reach and make use of the small, isolated, and likely species-poor islands.

At this point, other shifts in perspective are useful. Up until now, colonization and extinction rates were considered island specific but can be regarded with somewhat more resolution (and biological reality) to be also taxon specific. Further, rather than regard these rates as functions of area and distance indirectly, through the medium of species number (the abscissa in the previous figures), they can be considered directly as functions of island area and isolation. Thus in Fig. 3.3 (center) incidence is shown as a declining function of island area (abscissa), with curves shifted right or left to reflect variation in island isolation. Incidence declines on islands of smaller area, and also on islands of increasing isolation. Incidence calculated in this way is a mean value, the average proportion of islands occupied, over some subset of islands (sorted by size or isolation). Over time, some of the islands within a size or isolation class will lose the taxon through extinction, and others gain it through colonization. Thus the incidence (of islands per species) is another dynamic equilibrium (like the previous dynamic equilibrium in the number of species on an island, where colonization balances extinction). Incidence computed at one point in time for a particular island class will generate a value that will vary somewhat with the vagaries of colonization and extinction, just as species numbers on islands do. And through time there will be a turnover in the islands that possess the taxon in the same way that there is turnover of species (names) on a given island among repeated censuses. If a proportion J of islands in a particular class are occupied at one point in time, over an extended time period a proportion $J * e$ lose the taxon through ongoing extinction and a proportion $(1 - J) * c$ gain the taxon through colonization (see Fig. 3.3, right). Incidence J is the dynamic balance of these two processes, $J * e = (1-J)c$; $J = c/(e + c)$; the overall number of species on an island, time-averaged, is then the sum of the incidences, ΣJ_i, over all candidate species i. See Gilpin and Diamond (1976, 1981) for further discussion.

The simple expectations from M/W theory are that (a) at equilibrium, species numbers on an island will be determined by a balance between equal extinction and colonization rates,

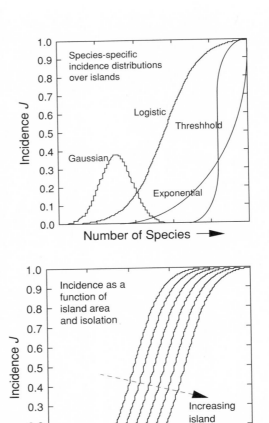
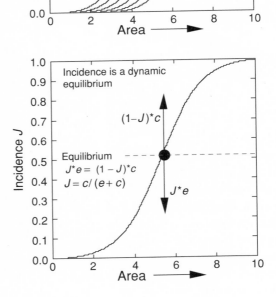

FIGURE 3.3. Incidence J is the proportion of islands in a certain class on which a certain taxon occurs. Incidence generally declines with species richness (upper) or island area (center), but some incidence functions show modal values at intermediate island size or richness (Gaussian form, upper). As species numbers generally depend on both island area and isolation, incidence is expected to decline with both island area and distance from colonization sources (center). Note that incidence is a dynamic equilibrium, with some island within a class (e.g., size class, lower) gaining the taxon via colonization and others losing the taxon through extinction over time. At equilibrium, $J = c/(e + c)$, i.e., J increasing with colonization rate c and decreasing with extinction rate e.

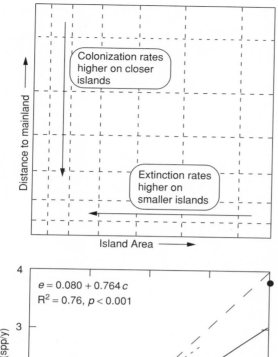

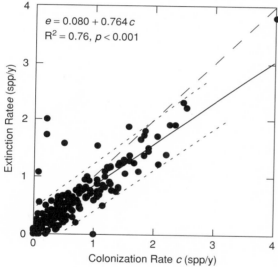

FIGURE 3.4. Above: simple expectations from the M/W model suggest that extinction and colonization rates will be equal at equilibrium, with the former highest on small islands, the latter highest on close islands. Below: extinction and colonization rates on Barkley Sound islands vary, on different islands over two decades, from zero to nearly four species per year. Each data point represents a particular island ($n = 188$ islands in all). Note that extinction and colonization rates are roughly equivalent, indicating a relative constancy of species numbers per island despite ongoing extinction and colonization; however, the regression equation reveals a colonization rate significantly higher than the extinction rate ($p < 0.01$ that slope $b = 0.764$ might be 1).

with (b) the former increasing on smaller islands, and the latter increasing on closer islands (Fig. 3.4, above). With a shift in this balance, with higher or lower extinction and colonization rates with changing area and distance, equilibrium species numbers will increase with island area and decrease with island isolation. This will give rise to a species-area curve of positive slope (more species on larger islands), and a parallel species-distance curve with negative slope (more species on islands closer to the colonization source). There is abundant literature, data, and discussion on species-area curves, but relatively little has been written about species-distance relationships.

COLONIZATION AND EXTINCTION DATA

I leave further discussion of species-area-distance relations until the next chapter and elaborate further on colonization, extinction, and incidence in this section. Empirically, the numbers of colonization and extinction events (C, E) recorded on individual Barkley Sound islands are measured in units of species per year (spp/y). I will refer here to these numbers as "rates" of colonization and extinction, although more refined estimates (symbolized by lower case c, e) are made below. The numbers vary over islands within a range of zero to four and are roughly equal (Fig. 3.4, below). However, over the 20-plus years of this study and averaged over islands, colonization rates have exceeded extinction rates by a factor of 4/3. Note that the observed difference between extinction and colonization rates seems to be real, as potential biases associated with missed species would record higher extinction and lower colonization rates, the opposite of what is observed. Despite the significant difference between the two, extinction and colonization rates do approximately match and are simultaneously high or low on different islands.

I relate observed colonization and extinction rates to island size and isolation in Fig. 3.5. Both rates are much more strongly correlated to log(area) than to area per se, and to distance rather than log(distance); thus units along the axes in Fig. 3.5 are distance in kilometers and $\log_{10}$(area-meters^2). Colonization rate varies as we might expect, as higher rates are recorded on closer islands, and also on larger islands. In fact, the effect of island size dominates that of island distance, emphasizing the importance of the target effect. As island size and distance are essentially uncorrelated ($r = 0.086$, $p \gg .05$), both effects are significant and independent contributors to colonization rate: distant islands receive significantly fewer colonists than close islands, while larger islands receive significantly more than small islands.

Extinction rate is related to island size and isolation in Fig. 3.5 (lower), but with quite different results. In fact, the upper and lower parts of the figure look quite similar, and indeed this is what we expect since colonization and extinction rates are about equal over islands (Fig. 3.4, lower). However, the M/W expectation was that extinction rates would increase on smaller islands, not decrease as the figure shows, and an added rescue effect would moderate extinction rates on close islands, whereas extinction rate actually decreases with increasing distance (see

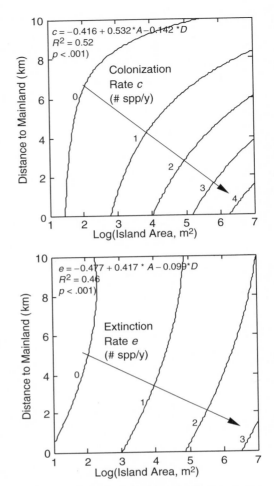

FIGURE 3.5. Observed colonization and extinction rates on Barkley Sound islands are related to both island area (abscissa) and distance to mainland (ordinate) although the former is the dominant variable. Quadratic functions are shown in the figures, while the multiple linear regression equations are also given. See text for discussion.

figure). Note that the contribution of distance to this regression is even less than it was to colonization rate. It appears that the dominant aspect of the dynamics of species on these islands is a very strong target effect producing high colonization rates on larger islands, and a corresponding large and matching extinction rate on the same islands.

The anomaly of higher extinction rates on larger and closer islands is resolved in part by strong target effects, and by the additional fact that only a portion of the island floras is dynamic.

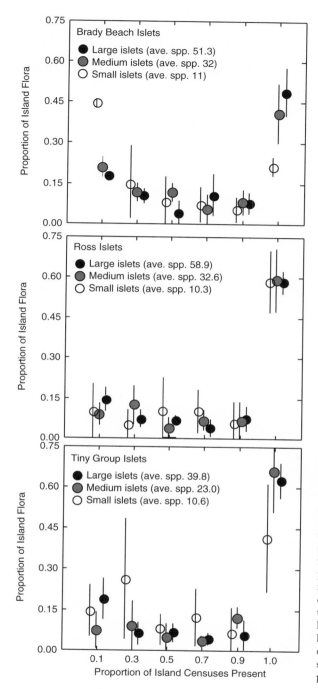

FIGURE 3.6. For three groups of islands, Brady Beach Islets (series 𝒜, close to mainland), Ross Islets (series 𝒞, intermediate isolation) and Tiny Group Islets (series 𝔈, most isolated), the floras of large, medium, and small-sized islands are partitioned into components (by proportion of the total flora, ordinate) reflecting the persistence of species on the islands over time (J_r; abscissa). Note the lower persistence on small islands, where a high proportion of the floras are present for short time periods and a low proportion for long time periods.

Most plant species on larger islands are in fact persistent over time, and these species have remained as permanent island occupants for the duration of the study. The persistence of plant species over time is measured as another sort of incidence function, that which describes what proportion of the cumulative flora has been recorded on every census occasion ($J = 1$), or recorded on increasingly smaller proportions of the total number of censuses. Time incidence is shown in Fig. 3.6 for three groups of islands that are very close to the mainland (Brady Beach

Islets, series $\mathcal{A}$), moderately isolated (Ross Islets, series C) or very isolated (Tiny Islands, series $\mathcal{E}$) from the mainland. In each island set, islands are segregated into three nearly equal groups by size and mean species number; mean and standard error bars are shown. The minimally isolated Brady Beach Islets (Fig. 3.6, upper) average just 36 m from the adjacent mainland; there are 13 censuses spread over 22 years. Here the larger islands ($n = 3$) support an average 49% of their cumulative floras as permanent residents, medium-sized island 41%, and small islands 21%. On the same islands, 18% of the floras of the largest islands were recorded in $\leq$20% of the censuses, with 21% of the floras of medium-sized islands and 45% of small-island floras similarly recorded in a fifth or fewer of the censuses. Given the sizes of the respective floras, these numbers translate to 9.2, 6.8, and 4.9 itinerant species (e.g., 0.18 * 51.3, etc.), and thus there are more species coming and going on the larger islands, even though these itinerants are a larger proportion of the floras on smaller islands.

The Ross Islets (averaging 11 censuses over 22 years, $n = 7$ islands in each of three size groups) show a rather different picture, as there are similar proportions of permanent residents versus those of shorter residence times among the three island size groups. Perhaps the increased isolation of these islets, located 3.61 km from the mainland and presumably with correspondingly reduced colonization rates, reduces the potential for differences in island size to generate differences in colonization rates. The Tiny Islands are still further isolated from the mainland (average 5.42 km distant); with $n = 4$, 5, 5 islands of large, medium, and small sizes, respectively, they were censused six times over 21 years. Here again there is a tendency (though not statistically significant) for smaller islands to have a larger proportion of their floras as nonpermanent residents (Fig. 3.6, lower). In all three island groups, the numbers of species that are short-term vagrants are higher on the larger islands (Brady Beach: 9.2, 6.8, 4.9; Ross: 8.5, 3.0, 1.0; Tinies: 7.8, 1.7, 1.6), the absolute numbers decreasing with increasing isolation distance.

Plant incidence on islands varies over space and time, and incidence patterns are also to some extent habitat and taxon specific. Some of this variation is displayed in Figs. 3.7–3.10. Incidence varies widely in four shoreline species, plants that are found within or close to the splash zone on the upper beaches. Pearlwort *Sagina crassicaulis* (or *S. maxima* ssp. *crassicalis*; Caryophyllaceae) is a tiny mosslike plant of damp rock cracks and has a nearly linear incidence (Fig. 3.7 upper left) declining with island size, with no incidence differences related to island isolation (this is true for all four shoreline plant species figured; chi-squared tests: $p \gg 0.05$). It is highly ephemeral and persists poorly except on the largest islands. I recorded 103 colonization events (C) and 104 extinction events (E) for the taxon, indicative of rapid turnover dynamics. Another caryophyll with slower turnover ($C + E = 31$) is *Honkenya peploides*, typical of sandy or gravelly beaches and closely restricted to larger islands where these habitats are more prevalent. Incidence fall steeply on smaller islands in the near-ubiquitous rocky shore plantain *Plantago maritima*, which by contrast to the foregoing species is a good persister with slow dynamics ($C = 11$, $E = 9$; note the shift in the ordinal scale in lower part of Fig. 3.7 to the right of *Sagina*). Lastly, the hydrophyll *Romanzoffia tracyi* shows a uniformly low incidence across a wide range of island sizes, and it also has a rather dynamic colonization-extinction pattern ($C + E = 36$). It appears that, while these shoreline species are all perennial, potentially at least, the vagaries of life near the surging surf mitigates against persistent populations in most plants that occupy the lower shoreline, and favors a good recolonization ability.

The similar variety of incidence curves is seen in forest plants (Fig. 3.8), and incidence is contrasted between closer, inshore islands (series $\mathcal{A}$–$\mathcal{D}$) and the more isolated $\mathcal{E}$-series islands. Like the shoreline plants just discussed, some species show little or no effect of island

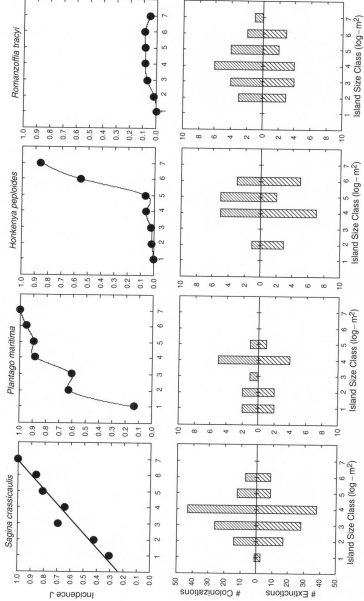

FIGURE 3.7. Incidence functions of four shoreline plant species (above), with numbers of colonizations and extinctions recorded for the taxa (below). Note the differences in shape among the incidence functions, the overall balance between colonization and extinction, and the variation in turnover dynamics (total colonizations and extinctions) among the four species (viz. five times higher in *Sagina* versus *Plantago*).

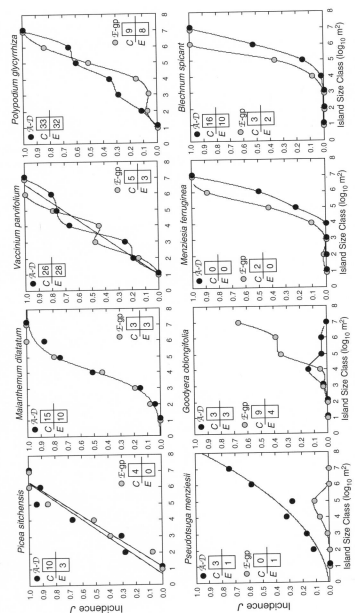

FIGURE 3.8. Incidence functions for eight species of forest plants. From the left are two tree species (Sitka spruce, above, Douglas fir, below), two monocot forbs (a lily, above, and rattlesnake orchid, below), two ericad shrubs, and two ferns (licorice fern, above, and hard fern, below). In the upper row, species have similar distributions over closer islands (series A–D) and distant islands (series E), but in the lower row species' incidences differ, sometimes dramatically, with island incidence. The boxed numbers are observed colonization (C) and extinction (E) events, and these also vary widely among taxa.

isolation (Fig. 3.8, upper row), whereas others show such effects strongly (Fig. 3.8, lower row). The figure contrasts related taxa (left to right) of coniferous trees, forest floor monocot forbs, understory ericad shrubs, and ferns. Of these species pairs, the lower, but not the upper, species of the taxon displays the effects of isolation. Sitka spruce (*Picea sitchensis*) (upper far left) has roughly linear incidence with no distance effects; Douglas fir (*Pseudotsuga menziesii*) (lower far left) incidence falls at a decreasing rate on smaller series A–D islands but is represented on only a few series E islands of intermediate size. Note the excess of colonizations over extinctions in general, and higher rates of colonization on closer islands of series A–D. The false lily-of-the valley *Maianthemum dilatatum* (upper midleft) shows a smooth logistic incidence declining with island size, whereas the orchid *Goodyera oblongifolia* declines steeply on E-series islands, with lower incidence overall and highest on islands of intermediate size on A–D islands (the opposite of Douglas fir). The common high-bush huckleberry *Vaccinium parvifolium* has a similarly wide distribution on near and far islands, whereas incidence in the confamilial false azalea *Menziesia ferruginea* falls more rapidly on smaller islands and is shifted to lower values on A–D islands relative to E islands. Lastly the two ferns, licorice fern *Polypodium glycyrrhiza* and hard fern *Blechnum spicant*, also differ in incidence, with the latter "falling off" sooner than *P. glycyrrhiza* with decreasing island size. While *Blechnum* persists better on E-series islands than A–D islands, *P. glycyrrhiza* maintains a somewhat higher incidence on A–D islands, at least over an intermediate size range. More extended discussions of the forest flora, relative to nestedness, relictual distributions, and dispersal characteristics, are given later.

The third functional group of Barkley Sound plants is "edge species," composed of generally weedier plants, less habitat specialized and geographically more widely distributed, that inhabit the zone between the islands' shorelines and their interior forests. Indeed, this is the largest component of the flora, and one from which we might expect a diversity of responses to island size and isolation. To illustrate this diversity, I choose pairs of related perennial species in Fig. 3.9 with distributions in series A–D islands given in the upper row and on series E islands in the lower row. In these figures, incidence is represented in histogram form and subdivided in four components, representing islands on which (a) the taxon persisted throughout this study, (b) colonization occurred, (c) colonization and extinction were both recorded, and (d) extinction alone was observed. The fifth component in these bar graphs, the uppermost, unshaded portion, represents the proportion of islands per size class permanently unoccupied by the taxon.

Two forbs in the family Asteraceae, yarrow *Achillea millefolium* and pearly everlasting *Anaphalis margaritacea*, are shown in Fig. 3.9a (left). *Achillea* is extremely widely distributed on near and far, large and small islands; its populations are conspicuously long-term persistent, and very few colonization events were recorded ($C + E = 7$ on A–D islands, $= 9$ on E islands). In stark contrast, *Anaphalis* is much more strongly confined to large islands, is much commoner on A–D islands than E islands, and is persistent only on islands of the largest size class. It occurs on smaller islands for short time periods, as vagrants arrive frequently and go extinct equally rapidly. Further, there is far more action in *Anaphalis* incidence on the closer A–D islands, where $C + E = 53$, a total that includes several repeated colonizations and extinctions on the same island; on E islands, $C + E = 4$ and the species has a much lower incidence profile.

Two edge shrubs in the family Caprifoliaceae, the berry-producing and zoochorous twinberry *Lonicera involucrata* and red elderberry *Sambucus racemosus*, are shown to the right in Fig. 3.9a. These species show differences comparable to those of the composites (above), with the former very widely distributed and with modest turnover dynamics ($C + E = 25$: A–D; $= 18$: E). The elderberry has a much reduced persistence with a very high

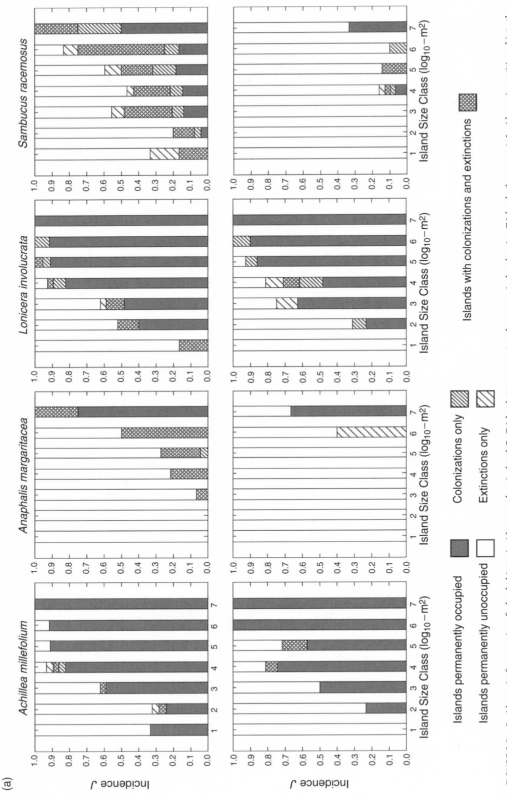

FIGURE 3.9a. Incidence in four species of edge habitats; incidence on less isolated *A–D* islands (upper row) and on more isolated series *E* islands (lower row). Incidence is partitioned into the four components reflecting persistent populations and transient populations as indicated, with upper (clear) part of histogram depicting empty islands in the size group. The two left-hand species are herbaceous Asteraceae, the two right-hand species are zoochorous shrubs.

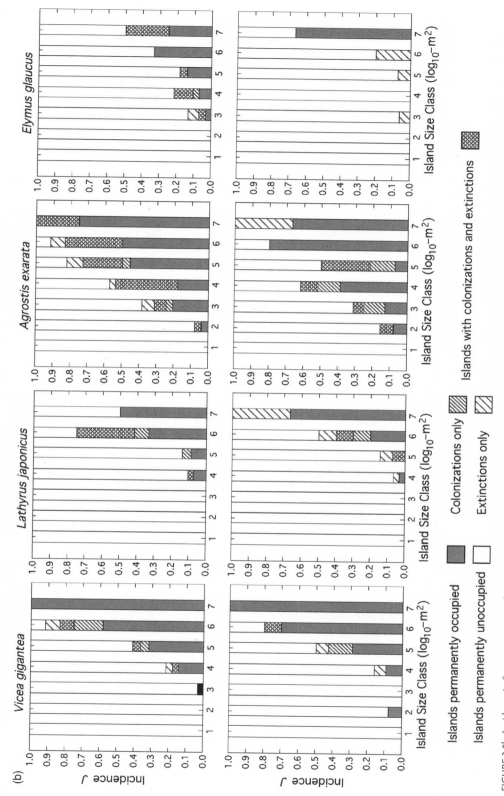

FIGURE 3.9b. Incidence in four more species of edge habitats; incidence on less isolated A–D islands (upper row) and on more isolated series F islands (lower row). Incidence is partitioned into the four components reflecting persistent populations and transient populations as indicated, with upper (clear) part of histogram depicting empty islands in the size group. The two left-hand species are herbaceous Fabaceae, the two right-hand species are grasses Poaceae.

turnover ($C + E = 110$: $A–D$), and is much more narrowly restricted on E islands (where $C + E = 10$).

Two pea species (Fabaceae) are relatively common edge plants, with the vetch *Vicea gigantea* more so than the beach pea *Lathyrus japonicus*, and the latter in rather fewer sites and further from the shaded forest edge. *Vicea gigantea* has similar incidence and similar turnover on $A–D$ versus E islands (see Fig. 3.9b; $C + E = 11$, 8, respectively), and occupies similar proportions of the islands in the two groups (which total 126 and 91 islands, respectively). *Lathyrus japonicus* is somewhat less common and with proportionately fewer persistent populations, but again there appears no significant disparity between $A–D$ and E distributions. This is ascertained by a chi-squared test: the proportions of P (permanently occupied), C/E (itinerant populations), and Q (permanently unoccupied) islands in the A group are computed as 0.079, 0.056, and 0.865. These proportions, call them p_i, predict numbers of E islands in the three categories (i.e., $p_i * 91$) of 7.2, 5.1, and 78.7, which compare favorably to the observed numbers of 5, 7, and 79.

Lastly, two edge-habitat grasses are compared in Fig. 3.9b (right), the bentgrass *Agrostis exarata* and blue wildrye *Elymus glaucus*. This particular *Agrostis* is a common grass in the system and shows a fairly orderly decline in incidence with decreasing island size. It has been recorded on 52% of the $A–D$ islands, and as permanent resident on 21% of the islands in those groups. On E islands, the equivalent figures are 48% and 29%; it persists, therefore, on 40–55% of the islands in the two groups. *Elymus*, in contrast, maintained persistent populations on 11 $A–D$ islands, versus just 2 E islands, and colonization plus extinction events were four times (16 vs. 4) more frequent on $A–D$ islands; *Elymus glaucus* definitely has a smaller presence on the more distant islands.

Before returning to the major theoretical themes, the roles of island area versus isolation in determining incidence will be examined a little more closely. Many of the species illustrated and discussed above show island incidence responding to island size in a regular, orderly fashion, which begs the question: what additional input to incidence is expected from isolation? The reader will recall that island size and isolation are uncorrelated. For the sample 20 species discussed above (Figs. 3.7–3.9), incidence is plotted as a function of isolation distance in Fig. 3.10. Species depicted in the top row of the figure, which includes the four shoreline species, the composites, a fern, and the orchid, show no systematic effect of distance. In the second row of the figure, species again show no systematic effect of distance, yet their incidences show similar or parellel variation over distance. Taking into account the numbers of islands in the different ($A–E$) groups contributing to the island isolation classes, it seems likely that there are spatial or positional effects on incidence. The distribution of the different island groups over isolation distance is included in the lower right panel of Fig. 3.10. Thus, islands 2–4 km from the mainland come largely from the B, C, and D island groups, and it is apparently a relative species impoverishment in these groups that reduces incidence on islands of those two distance classes. Their imprint seems to affect most of the species in the lower two-thirds of the figure. In the bottom row (Fig. 3.10B) are four species that show significant distance effects, but incidence declines monotonically with distance only in the grass *Elymus glaucus*, with the other three species showing an incidence dip largely attributable to scarcity on the $B–D$ islands.

HISTORICAL LEGACIES

The basic M/W model depicts the number of species on an island as a balanced and stable equilibrium, but an extensive literature describes circumstances in which such an equilibrium is not yet attained or is perhaps unattainable (Whittaker 1998). In one sort of historical legacy, past land-bridge connections are invoked to explain the presence of a particular species on a particular island; here we are more

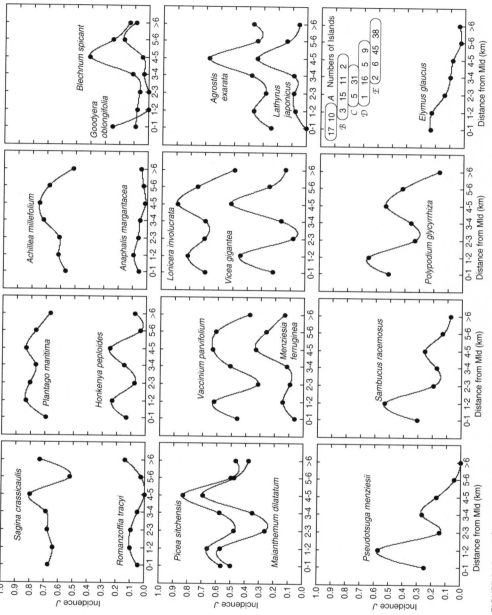

FIGURE 3.10. Incidence (ordinate) as a function of distance from the mainland (abscissa). Top row: species showing no influence of island isolation; center row: species with similar effects of island position on incidence; bottom row: species with declining incidence with island isolation. In lower right panel, the numbers of islands in different isolation groups contributing to the sample size for isolation distance.

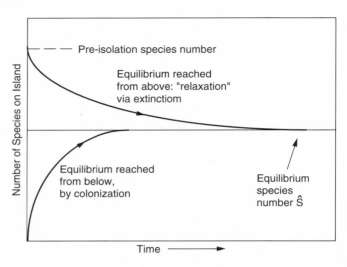

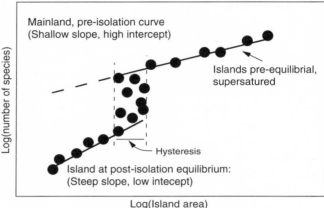

FIGURE 3.11. Species numbers on islands may not be in equilibrium, islands may be in approach to an equilibrium (upper) from below, if newly formed or previously sterilized, or from above, if relaxing from supersaturated levels of prior land-bridge connections. Islands with past land-bridge connections may have equilibrated to present isolation status if small, but may still retain pre-isolation species richness if large (lower).

concerned with historical legacies that affect species numbers on the island. There are in general two sorts of circumstances in which we might not expect species numbers to be at equilibrium between colonization and extinction rates. Firstly, on islands either newly created by tectonic activity or older but sterilized by such activity, colonization may still exceed extinction. Possibly an equilibrium exists but it is not yet reached; it is being approached from below (with $C > E$), and in time it may be reached ($C = E$). Similarly, but less dramatically, islands newly created by rising sea levels may still be gaining new species appropriate to their newly acquired edge and shoreline habitats; these components of the flora again may be on approach to, but still below, an equilibrium.

Secondly, land-bridge islands, which when connected to the mainland had a full mainland species complement, may still be in the process of losing species ($C < E$) by sequential extinctions. In this case, an equilibrium is approached from above, and reaching it may require more time than has so far elapsed since rising sea levels isolated the islands in question (in our case, more than 10 or so millennia). Figure 3.11 (upper) illustrates these notions; species-area curves might show the larger islands supersaturated at pre-isolation species numbers (upper line, Fig. 3.11, lower), small islands on a lower line representing post-isolation equilibrium species numbers, and intermediate islands scattered between these two bounds.

52 ISLAND BIOGEOGRAPHY

The classic example of recolonization of islands in the Krakatau group by plants and animals is an unfolding story avidly followed by island biogeographers, ever since defaunation and defloration of the islands after the cataclysmic explosion and eruption in 1883. These islands are tropical and lie in the Sunda Strait some 40 km from the large potential source biotas of Sumatra and Java, to the north and south, respectively. The recovery speed of the flora and fauna, and the rapid buildup in species richness, constitute a dramatic exhibition of colonization potential, involving many and widely different taxa (Thornton 1996). Diamond (1972) showed that exploded (defaunated) volcanoes around New Guinea attained substantial vegetation cover and impressive bird species richness in a matter of couple of centuries or less, realizing of course that the birds have to wait for the recovery of the vegetation. Recolonization is a function both of an island's isolation and a taxon's dispersal ability; on tiny, near-shore mangrove islets that were experimentally sterilized, reconstitution of the insect fauna by colonization proceeded very rapidly. Predefaunation species richness was attained in little more than a year, although adjustments in species composition went on for some time thereafter (Simberloff and Wilson 1969, 1970; Wilson and Simberloff 1969).

The story is different, in the time dimension especially, with islands equilibrating from above, as land-bridge islands may retain the imprint of past mainland connections, at least in certain taxa, for surprisingly long time periods, even millennia. On islands in the Sea of Cortés, channels separating the land-bridge islands from the mainland have become inundated by rising sea levels at different times during the Holocene. Species loss from these islands is a function of both time and island area but varies also among different species groups (Case et al. 2002). The variables that determine equilibration rates, and potentially maintain or erase historical legacies, are colonization and persistence rates, and these are very much taxon specific. Figure 3.12 summarizes intertaxon differences,

empirically determined, in colonization and persistence for the Sea of Cortés islands. Colonization abilities are highest in plants and birds, intermediate in reptiles and insects, and very low in mammals; persistence rates are lowest in mammals and birds, high in reptiles, and highest in plants and possibly beetles. Species that colonize well but persist poorly will generate the fewest island endemics (e.g., birds), and vice versa, such that poor colonists but good persisters readily form endemics (e.g., reptiles, beetles). The islands will be most impoverished in species that both colonize and persist poorly, and these taxa show the greatest "island effect" (mammals). Islands will be most mainlandlike in species that both colonize and persist well (plants), and good persisters will show the effects of past land-bridge connections for the longest time periods after isolation.

The time required for an equilibration of species numbers on "new" islands is quite disparate in terms both of taxonomic characteristics and the direction from which equilibrium is approached. Empirically, equilibration times (e.g., to 90% recovery or relaxation) are relatively short in mobile taxa with high colonization rates equilibrating from below, mostly on the order of a few centuries, but much longer, several millennia at least, when equilibrating from above via excess extinction over colonization (Diamond 1972). This means that the times scales for ecological processes on supersaturated islands are of the same order of magnitude, more or less, as the geological processes that drive the rise and fall of sea levels. In this context land-bridge islands are particularly interesting, since they may be losing species in the habitats they shared with the previously contiguous mainland, forest species in our case, while simultaneously gaining species in new habitats, for example, open edge and shorelines, not previously represented there. Further, the extent to which small islands have lost most or all of their original forest habitat is proportional to the space available to colonists drawn from the weedy, open-edge plant species excluded from forested areas. These components

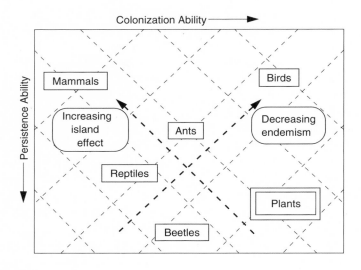

Birds
    Endemics: Weak subspecies, most on large, oceanic islands, in 10 species, ca. 2% of island populations
    Isl. Spp. #s: Mainlandlike except for smallest islands
    LB Relicts: Several relicts on largest LB island of Tiburón

Plants
    Endemics: Species and subspecies, ca. 4% of island flora
    Isl. Spp. #s: Mainlandlike except on smallest islands
    LB Relicts: Few, but not easily distinguished from size, distance effects; relicts also on non-LB islands.

Mammals
    Endemics: Endemic mouse spp., several subspecies
    Isl. Spp. #s: Much reduced diversity, esp. on oceanic islands
    LB Relicts: Many large mammals occur on LB islands only

Reptiles
    Endemics: 1 endemic genus, several endemic species
    Isl. Spp. #s: 10 km^2 island with 1/3 (LB) to 1/4 (oceanic) of mainland species numbers
    LB Relicts: LB islands ca. 35% richer, esp. notable in snake spp.

Ants
    Endemics: 4 spp. endemic, 6% of island, 4% of peninsular spp.
    Isl. Spp. #s: 2/3 of central peninsular spp. occur on islands
    LB Relicts: Army ants plus wingless ponerines on LB islands only

Tenebrionid Beetles
    Endemics: 1 genus and 25 spp. (24% of island fauna)
    Isl. Spp. #s: Ca. 27% of peninsular fauna on islands
    LB Relicts: No effect on island species numbers

FIGURE 3.12. Endemism, species richness, and the occurrence of land-bridge (LB) relicts (box, below) on islands in the Sea of Cortés, interpreted from taxon-specific differences in colonization and persistence abilities (above). Adapted from Case and Cody (1986) and Case et al. (2002).

of the flora will thus interact, indirectly, in their potential for island representation.

GENERALITIES, SPECIFICS, AND MODIFICATIONS OF THE M/W MODEL

The M/W model has served, and continues to serve, to focus critical attention of various aspects of the diversity and dynamics of island species, as reviewed by Rosenzweig (1995). It is most notably deficient in accounting for patterns on what Whittaker (1998) refers to as "disturbed islands," meaning essentially those with strong historical legacies that contribute to supernumery or subnumery species lists. Further, the M/W theory can contribute little to the biogeography, ecology and evolution of the biotas on very remote oceanic islands and archipelagoes, where colonization events are supremely rare and in the intervening long intervals opportunities exist for a great deal of adaptation and evolution. On large and isolated islands, opportunities exist for autochthonous (or in situ) speciation, which may be a far more important contribution to local species richness than immigration from elsewhere. Thus the main source of new anoles on larger Caribbean islands ($>3,000$ km^2) is not immigration but within-island, autochthonous speciation (Losos and Schluter 2000). The main questions that stem from the model's original formulation, and upon which the Barkley Sound study can be brought to bear, are summarized as follows:

a. Are plant species numbers on Barkley Sound islands at equilibrium? The M/W model predicts that species numbers will remain constant, a product of balanced colonization and extinction rates. Although we found (above) that colonization and extinction rates are roughly equal, does this necessarily imply that a stable equilibrium obtains?

b. Is colonization rate simply determined by island isolation and extinction rate by island area? We discovered already that this is not the case and that target effects, the contributions of island size, are predominant in plant colonization events, although distance remains a significant though minor contributor. Further, the expectation that area alone determines extinction rate was not supported, and rescue effects alone do not reconcile observations with expectations. These relationships will require a more detailed exposition.

c. The basic M/W model does not differentiate among species, but are all species really equal in terms of persistence and colonization potential? On the contrary, it might be expected that, even though we are considering but a single taxon, plants, that there might be many and diverse variations, among species, in these characteristics. Certainly we will document species turnover on islands through time, but do all species participate in these dynamics?

d. Questions of historical legacies certainly apply to the Barkley Sound land-bridge islands. They are the more intriguing because we expect one component of the flora, the forest species, to be in the process of relaxation to reduced species numbers concomitant with their relatively new isolation status of the islands, while other components, edge and shoreline floras, are in the process of building up to occupy newly available habitat. Are these legacies still apparent on the islands, and can they be disentangled?

e. David Lack (1976) drew one major conclusion from his studies of the birds of Jamaica, that adaptations evolve among successful colonists that serve to reduce extinction rates and slow species turnover through time. Such adaptations might be regarded in the context of mutual coadaptations amongst community members that enhance persistence, reduce the invasibility potential of the island, and generate assembly rules. Or they might operate entirely within the context of a single species, as its phenotypic characteristics are tested in the new abiotic environments of islands. Thus we might look for the products of such selection in plant populations on Barkley Sound islands and relate them to population longevity. Reduction in dispersal potential and variation in fern frond morphology are examples that are elaborated below.

4

Species Number, Island Area, and Isolation

EQUILIBRIUM OR NONEQUILIBRIUM SPECIES NUMBERS?

In this chapter I consider the numbers of plant species on the Barkley Sound islands and evaluate the extent to which species numbers vary over time. I assess whether the data support an equilibrium value for island species counts, and how these counts are related to the more obvious variables of island area and isolation, and less obvious variables such as island topography and position within Barkley Sound. However, as discussed above, the number of species on at least larger continental or land-bridge islands may not be at equilibrium in terms of the longer view, owing to the historical legacy of past contiguity with mainland species pools and the long time periods required to lose supernumerary species. Further, newly created islands will possess habitats not present pre-isolation, namely, edge and shoreline habitats, and these new habitats also may not have equilibrated because of a time lag in the colonization process of appropriate new species. Thus the island plants may still be in the process of tracking changes driven by both their new isolation status (species loss) and new habitat opportunities (species gain).

Neither of these potentially equilibrating trends seems likely to be uncovered by a study of a mere 20 years duration. Thus the question becomes, given the high rates of both colonization and extinction on the islands, whether there is evidence that, in the shorter term of the study, species numbers within islands remain relatively constant. Thence we can investigate whether variations in counts among islands can be understood in terms of island characteristics that influence plant colonization, persistence, and local extinction.

VARIATION IN SPECIES COUNTS ON ISLANDS

There are, in general, two sorts of variations in island species numbers over time that might bring into question the notion of equilibrium. Species numbers might vary widely and wildly around the mean value, such that their central tendency is unconvincing, or temporal changes in numbers may indicate a trend, to either larger or smaller species numbers over time.

Census data show clearly that colonization, extinction, and species turnover are ongoing processes on these islands (chapter 3). Repeated censuses over time generate an average number of resident species for the island (±SD, SE, or CV), a time series of species counts over the course of the censuses that could provide evidence for a trend, and also a cumulative species number that represents the collective total of all species seen on the island. The cumulative species number will climb steadily over time but, conceivably, at a rate that varies; for instance, it may slow, or less likely increase, over time.

To illustrate the nature of these data, cumulative and average species numbers are shown in Fig. 4.1 for six Brady Beach islets ($A16–A21$) censused repeatedly (but not in every year) from 1981 or 1982 through 2003. The numbers of recorded colonization and extinction events between successive censuses are also shown. Note that there are years in the late 1980s and then again post-1998 when colonizations generally exceeded extinctions on most of these islands, whereas in the intervening period the reverse is true. I will examine this sort of synchronicity, and the time lags between colonization and extinction spurts, in a later chapter on species turnover. For now, I note that, over the two decades, colonizations are a reasonable numerical match to extinctions for these islets collectively (149 vs. 135 overall) and also balance well on individual islands (see Fig. 4.1).

Because of temporal variations in colonization and extinction events, year to year and decade to decade, and the expectation of their approximate correspondence only in the longer view and not necessarily within short time intervals, species numbers on islands vary with time. Thus, the stochastic nature of colonization and extinction contribute to "turnover noise," expected to be more pronounced where their rates, and corresponding intersections of the curves (above the abscissa in Fig. 3.1), are higher.

While standard errors around mean species numbers, included in Fig. 4.1, are narrowly confined, variations over time in species counts are substantial. These data are shown in Fig. 4.2. for all eight Brady Beach islets ($A15–A22$). Testing for trends with simple linear regression, there is but one that is statistically significant, on the largest island, $A17$, which has gained an average of 0.283 species per year over the more than two decades of the study period. The remaining data are without demonstrable trends and support the general notion of a quasi-constant equilibrium species number that differs, of course, among islands.

The Brady Beach islets are located just off a mainland beach and are minimally isolated. Numbers versus time trends in progressively more isolated islands are illustrated in Figs. 4.3 and 4.4. Across the Trevor Channel in the B group, 4 of 10 islands at the southeast entrance to the Dodger Channel ($B12–19$, $26–27$) also show significant increases in species numbers over time (Fig. 4.3, upper left), including the two larger islands $B12$ and $B19$. The former ("Haines SE") is particularly variable but has gained on average 0.76 species/y ($R^2 = 0.44$, $p = 0.05$). In the D group, 4 of 16 Stud Islands ($D18–30$, $32–34$) show significant increases over time, including 2 of the 10 islands in Fig. 4.3 (lower left). Among the islands with increasing species counts, gains are from 0.27 to 0.42 species/y.

Fourteen islands in the Tiny Group (series E), located in the center of the Sound, have been censused on five or six occasions over 22 years (Fig. 4.3, right). Of these islands ($E19–29$, $34–37$), none shows significantly increasing species numbers, 6/14 regression coefficients are negative, and two of these reflect significant decreases of -0.30 ($E22$; $R^2 = 0.91$, $p = 0.003$) and -0.22 ($E35$; $R^2 = 0.89$, $p = 0.016$) species/y.

Two other islands groups from the E series are shown in Fig. 4.4. Those in the Pigot group are heavily forested and adjacent to the much larger $E49$: Clarke; most have been censused just three times. Smaller islands in the group show no time trends in species numbers, but the larger three islands, $E58–60$, show declining species numbers between censuses (paired

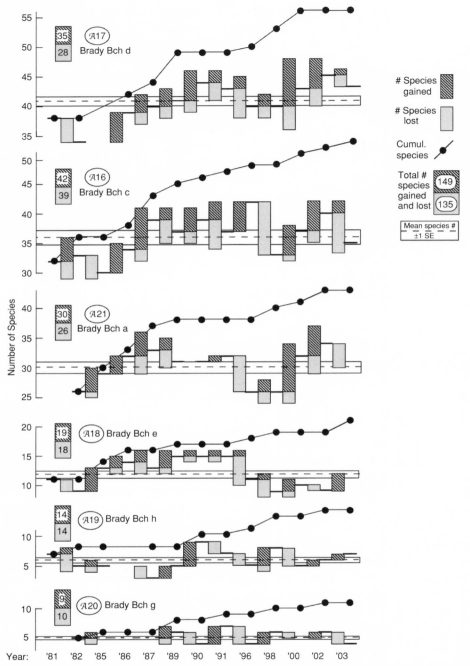

FIGURE 4.1. Plant censuses over successive years reveal both colonizations and extinctions. The number of species present varies above and below the mean values as indicated. The cumulative number of species for the islands continues to rise over time, such that it far exceeds the average number present. Overall, the number of colonization and extinction events are evenly balanced, within islands and over all islands combined. Islands are ranked from highest average species number (above) to lowest (below).

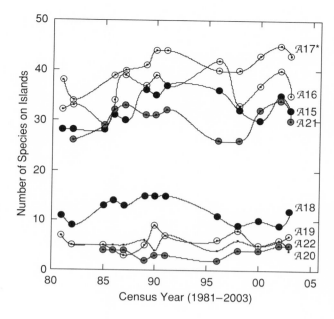

FIGURE 4.2. Species numbers vary over time on these Brady Beach islets, but on only one island is there a significant trend: on $A17$ (see asterisk) species numbers tend to increase over time (regression analysis, $b = 0.283$, $p = 0.013$).

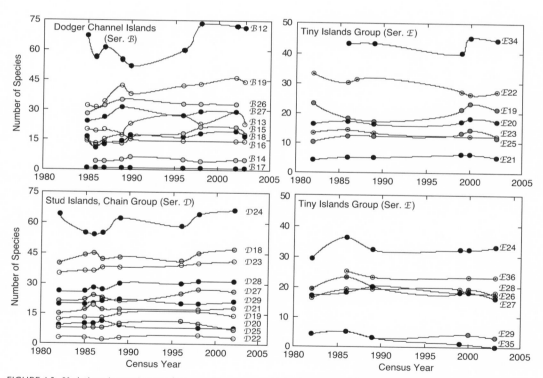

FIGURE 4.3. Variations in species numbers over time in several island groups. Of the islands depicted, $B12, 15, 18, 19$, and $D23, 27$ show significantly increasing species numbers over time, while $E22, 23, 35$ show significant decreases over time. The remaining 26 islands show no species trends over time.

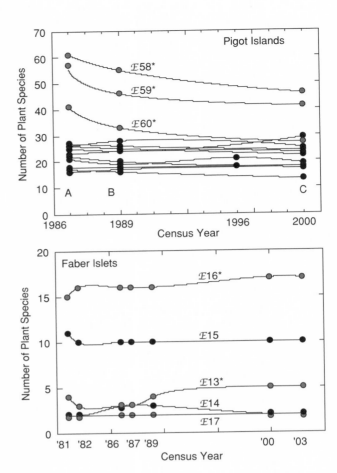

FIGURE 4.4. Trends in species numbers over time on two archipelagoes in the 𝐸 group, the Pigot and the Faber Islands. The three largest islands in the Pigot group (above), 𝐸58, 𝐸59, 𝐸60, all individually show significantly declining species numbers over time (paired t-tests between years A-B: $p = 0.029$; A–C: $p = 0.028$; B–C: $p = 036$). The smaller Pigot Group islands, 𝐸51–53 and 𝐸61–66 (unlabeled lines) show no trends. Among the tiny Faber Islets (below), two (𝐸13, 𝐸16; note asterisks) show significantly increasing species over time.

t-tests: $p < 0.05$). The Fabers are tiny, rocky, and sparsely vegetated islets in Village Bay in front of Effingham Island, and five of them (𝐸13–17) have been censused seven times (Fig. 4.4, lower). Two (𝐸13, 16) show increasing species number over time, gaining 0.073 (𝐸16; $R^2 = 0.83$, $p = 0.004$) and 0.086 (𝐸13; $R^2 = 0.67$, $p = 0.025$) species/y, respectively.

Lastly I illustrate time trends in species counts in group C's Ross Islands ($C7$–16, 19–22; Fig. 4.5, top), most of which were censused 11 times over 22 years. Six of these islands show significantly increased species counts over the census period ($C22$ is not illustrated), by 0.08–0.42 species/y. One, $C16$, decreased in species numbers over time ($b = -0.13$ species/y; $R^2 = 0.83$, $p < 0.001$). From this figure, it is apparent that species counts do not vary independently among the islands. On the contrary, there is rather obvious covariation among islands, such that increases and decreases occur at similar points in time across the archipelago. I show this by computing standardized species counts for the islands and display the means, across islands, of these standardized counts (Fig. 4.5, center). The common trend shows initially low counts (1981–1982) rising to a 1987 peak, followed by steep decreases to 1989 lows, then a gradual recovery to high counts again in 2002–2003. The Ross Islands are closest, among the 𝐴–𝐸 islands of Figs. 4.3–4.5, to the Brady Beach islets, and standardized counts for this latter group also are given in Fig. 4.5 (lower). In fact there is a strong correspondence in their year-to-year synchronicity and variation, with an overall correla-

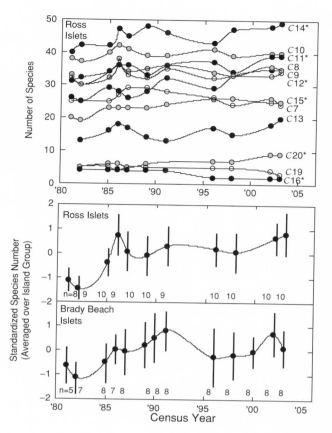

FIGURE 4.5. Trends in species numbers over time, Ross Islets. Top: of 12 islands in group, 5 (C11, 12, 14, 15, 20) show significant increases in species numbers over time, while C16 shows a significantly decreasing trend, and the other 6 no trend. Variations in species numbers tend to be synchronized among islands, as shown (center) by standardized species counts averaged over islands (vertical bars: SD). Most islands had low species numbers in 1982–1983, peaked in species number in 1986, then fell (1989) and subsequently climbed again to high levels 2002–2003. Similar data on standardized species counts from the Brady Beach islets (Fig. 4.2) are also shown (bottom); there is a close correspondence in trend synchrony between these two island groups.

tion of $r = 0.79$ ($n = 11$, $p < 0.01$) among year-specific standardized means. Note that there is no such correspondence between yearly variations in other, more distant, island groups. Besides a degree of synchronicity in time, there are clearly regional aspects to variation in species counts, as colonization and extinction rates tend to be similar across space, that is, on islands in the same region of the Sound, during specific time intervals.

Variations in species numbers on these islands are primarily driven by the vagaries of colonization and persistence of short-lived plants of open habitats, the "edge" species, rather than by the much slower dynamics of the forest-associated species. In Fig. 4.6 I plot the coefficient of variation (CV) of species numbers on islands, for all islands with >5 censuses, as a function of the proportion of open habitat on the island. There is a significantly higher CV on islands with more open habitat, with CVs more than twice as high on islands of >90% open habitat compared to islands of 0–30% open habitat. Overall, nearly one-quarter of the islands with >5 censuses ($n = 40$) show significant trends in the numbers of species over time. On $\mathcal{A}$-series islands closest to the mainland, 27% show increasing and 4% show decreasing trends in species numbers, with the remaining 69% showing no significant change. With increasing island isolation, these percentages change systematically, to fewer islands with increasing counts (18%) and more islands with decreasing counts (9%) in the $\mathcal{E}$-series islands. Note that only one-third (33) of the $\mathcal{E}$-series islands meet the criterion of ≥5 censuses; there are others with fewer censuses that would certainly inflate the proportion of distant islands losing species over time (e.g., those of Fig. 4.4, upper). $\mathcal{B}$-, $\mathcal{C}$-,

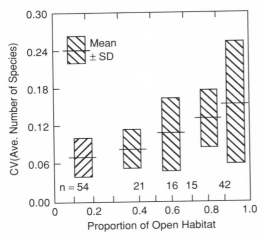

FIGURE 4.6. Variation in species numbers over time, measured as the coefficient of variation (CV; ordinate) increases with the proportion of open habitat on islands (abscissa; $R^2 = 0.27$, $p < 0.001$). Residuals from the relationship are positively related to the number and proximity of nearby islands, and negatively related to distance from mainland.

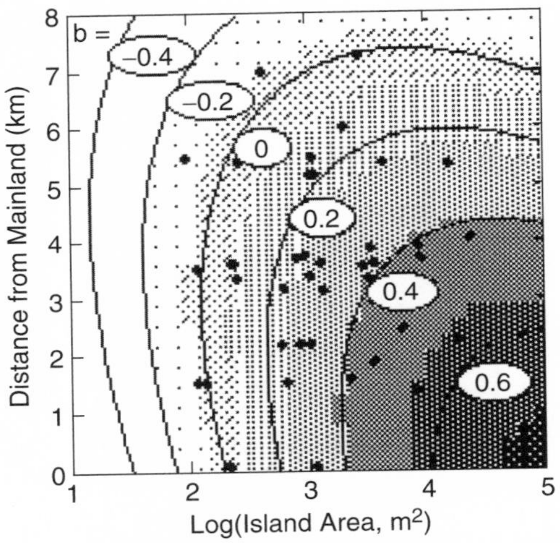

FIGURE 4.7. On 40 islands, 26% of those censused in five or more years, there is a significant trend in species numbers over time. The slope b of these regression analyses is highest on large, close islands (lower right), and negative on small and/or distant islands (left; top).

and $\mathcal{D}$-series islands are generally intermediate (with an average of 22% increasing in species over time).

The trend to increasing species numbers over time follows what was apparent earlier (chapter 3) and corresponds to the chief determinants of colonization rate. Colonization rates are highest on near islands and on large islands; these are exactly the islands on which species numbers have increased over time. The slope b of the regression equations measures numbers of species gained per year; the gain rates peak on large, close islands and fall to negative values on both small and distant islands (Fig. 4.7). Multiple regression analysis of these factors show both isolation (D_{MLD}) and island area (Log $AREA_{FL}$) are significant contributors to regression slopes, as also is the proportion of open habitat (P_{OPEN}): $b = 0.998 + 0.369$ (Log $AREA_{FL}$) $- 0.053(D_{MLD}) + 0.036(P_{OPEN})$, $R^2 = 0.50$, $p < 0.001$. Another measure of local isolation, ANG_{250} (summed angles of neighboring islands within 250 m), also enhanced b values, but just outside the bounds of statistical significance; islands with many close neighbors seem to have increasing species counts also.

In summary, year-to-year variation in species numbers is especially pronounced on islands with a high proportion of open habitat readily open to colonization by herbaceous edge species. Large, open islands close to mainland colonist sources are those that are more likely to show trends of increasing species richness over time. Small and/or distant islands, with low colonization potential, are more prone to lose species over time, but in addition, distant, forested islands in the middle of Barkley Sound have also lost a significant component of their species quotas over the last two decades. It appears that equilibration from above and below are both detectable and ongoing processes within the Sound, as isolated islands lose forest-associated species and accessible islands that provide easy targets for dispersing propagules gain new species in open habitats suitable to herbaceous, adventitious plants.

CUMULATIVE SPECIES NUMBERS

While the number of species on an island at any one time might remain more or less constant, the species list for the island continues to

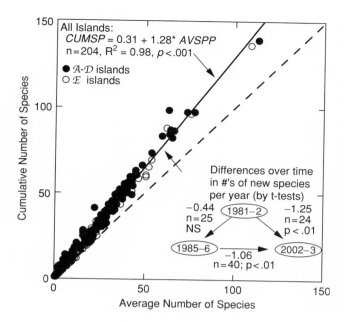

FIGURE 4.8. Cumulative species numbers on islands ($CUMSP$) are proportional to mean species numbers ($AVSP$), and average 28% above the latter over two decades. The regression equation is not significantly different in islands close ($\mathcal{A}$ series) or distant ($\mathcal{E}$ series) from the mainland ($p > 0.05$). Inset lower right: comparison in numbers of new species between paired early (1981–1982), later (1985–1986) and recent (2002–2003) censuses in adjacent years shows significantly larger gains between early censuses (1981–1982, 1985–1986) than between later censuses (2002–2003); i.e., relatively fewer new species were recorded between adjacent years in recent censuses, relatively more in the earlier years.

lengthen over time. New colonists, species hitherto unrecorded on the island, are discovered in nearly every new census, and they accumulate steadily (Fig. 4.1). Cumulative species counts are directly proportional to mean species counts over Barkley Sound islands regardless of size or isolation (Fig. 4.8); over the period of the study, cumulative species numbers exceed mean numbers by an average of 28%, or about 1.5% per year. The question of whether cumulative species numbers increase at steady, increasing, or decreasing rates is obscured by the fact that censuses were not conducted in every year and were conducted in partially nonsynchronous years on different islands. However, there is a substantial sample of islands in which censuses were conducted in several pairs of adjacent years, namely, 1981–82, 1985–86, and 2002–03. For paired t-tests on numbers of new species in the second year of these adjacent censuses, new species declined significantly in 2002–03 relative to both earlier periods (Fig. 4.8, inset). Thus there appears to be a decline in the rate at which new species are accumulated on islands over time, a trend that presumably reflects an upper limit to the total number of species in the pool that comprises the candidates for island colonization.

SPECIES RICHNESS ON ISLANDS

AREA EFFECTS

Several of the largest islands in Barkley Sound support well over 100 plant species, and on larger but uncensused islands the number of species is expected to be somewhat higher. At the other extreme, a number of likely looking rocks were censused (and also mapped; five in all) because they appeared capable of supporting plants, although in fact none were found there. Species richness on the islands is potentially governed by a number of factors including area, isolation, elevation, and other topographical variables, positional factors that moderate exposure to storms or to colonists, and also historical concerns such as age since isolation. Of these, I first discuss area effects, noting that Olaf Arrhenius, a Swedish plant ecologist, was first to point out, in 1921, the regularity with which species numbers increased with sample area. Over a wide range of vegetation types and with quadrat sizes from 0.1 to 10 m², Arrhenius found that ratios of sample area, y_1/y_2, were related to ratios of species numbers according to $(x_1/x_2)^n$, where n is a constant specific to the vegetation type being sampled. In most vegetation types around Stockholm, n lay between 2.0 and 3.3 (geometric mean of all sites = 3.04). The data produce a species-area curve of the form $S = bA^z$, where the exponent of area z, averaging 0.33 for Arrhenius's data, measures the slope of the relationship on logarithmic axes.

Henry Gleason (1922) took strong exception to Arrhenius's suggested generality for log-log species-area curves, calling it "wholly erroneous" and a "fallacy." He produced his own data on the plants of aspen woodlands in Michigan, which with either scattered or contiguous quadrats (240 × 1 m²) showed a linear species-area relationship on semi-log axes (a logarithmic-scaled axis for area only, but not for species number). In turn Arrhenius (1923) found Gleason's critique, which emphasized the inaccuracy of predictions of species numbers on large areas extrapolated from a few, small areas, of "very little value," and maintained his previous stance. Arrhenius further pointed out that different slopes are obtained from scattered versus contiguous quadrats, as well as from different vegetation types. Thus the large literature on species-area relations began rather contentiously, and it has continued to be a topic of energetic debate among ecologists.

Detailed reviews of the species-area relationship have been published (e.g., Connor and McCoy 1979; Schoener 1976; Kelly et al. 1989), and a thorough evaluation of the Arrhenius-Gleason schism and of botanical data in general was made by Dony (1970; reviewed in detail in Williamson 1981). While a wide array of data conforms to a linear species-area relationship when the axes are logarithmically scaled, many studies have found improved linearity on semi-log axes. More recent overviews are available in the synoptic literature (e.g., Rosenzweig 1995; Whittaker 1998). In this section my concern is less with the comparative generality of the Barkley Sound species-area data, and more with examining how such plots can inform us of the nature and diversity of island vegetation and the area- and isolation-specific colonization and extinction processes that maintain it.

There are in general three sorts of processes that contribute to increasing species numbers with increasing area: larger areas provide larger samples, enhanced access, and a wider range of habitat types. It is best to distinguish at the outset between data collected from nested versus nonnested quadrats or sampling areas. If quadrats are censused in a nested fashion, that is, areal samples are expanded from smaller quadrats nested within increasingly larger quadrats, species number cannot decrease, only remain constant or increase. It will increase if some species are recorded only in the larger samples, either because they are rare (low density), or if not rare then patchily distributed, such that smaller samples miss them. But with increasing area, different microsites, with different edaphic or microclimatic or other abiotic characteristics, are sampled, and eventually substantially different habitat or vegetation types are included. Thus are species added with increasing area; note here we are dealing with cumulative species numbers over space, not over time as above. Further, as areal samples expand regionally beyond major topographic and other distributional barriers, whether current or historical, the addition of ecological counterparts, or vicariant species, can continue to bolster the species count. These increments to species numbers with increasing area were quantified and interpreted as diversity components by Cody (1973, 1983a). Within-habitat species counts, of species that coexist locally, measure α-diversity; thence species that occupy different habitats, often congeneric replacements, add a β-diversity component, and ultimately the geographical counterparts within habitats, the vicariant species, contribute a γ-diversity component.

Species counts in patchy habitats separated by very different terrain, or on real islands, generate species-area curves from nonnested data. Patches or islands have reduced access to potential residents, and their own site-specific colonization, persistence, and extinction rates. Thus patches of comparable areas but differing in isolation, position, or some other variable that affects those rates, may be somewhat or quite different in the species they support at

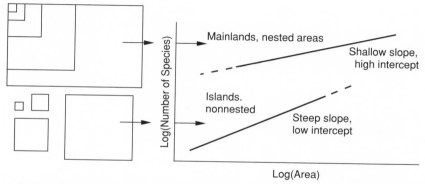

FIGURE 4.9. Hypothetical (but typical) species-area curves for nested areal samples on the mainland versus nonnested counts on islands.

any one time. Species counts from such discrete habitat islands often accumulate faster than those in nested samples in contiguous habitat, and generate steeper species-area curves. On true islands, species counts apply necessarily to discrete "patches," and only if larger islands contained all of the species present on smaller islands would the species count be strictly cumulative with area. I discuss this quality of "nestedness" further in the next chapter.

Island data points generate species-area curves with steeper slopes than in cumulative "mainland-type" species-area curves, owing to both noncontiguous areal samples and to so-called small-island effects contributing to steeply falling species numbers on islands of smaller areas (Fig. 4.9). Whitehead and Jones (1969) first called attention to the qualitative differences, beyond reduced areas, of small islands; included in the notion of small-island effects potentially relevant in Barkley Sound are harsher climates (e.g., hotter, drier, conditions; see above), unbuffered by tall vegetation that is precluded by shallower soils and lowered productivity, and perhaps a higher susceptibility to disturbance by storms.

A broad view of the species-area relationship for Barkley Sound islands in shown in Fig. 4.10, where for comparison similar data are plotted for Sonoran Desert islands in the Sea of Cortés. Only the larger, more diverse islands (with >40 species) show a linear relationship in these log-log plots (Barkley Sound: $\log S = 2.01 + 0.15 * \log A$; $n = 35$, $R^2 = 0.58$, $p < 0.001$; Sea of Cortés: $\log S = 1.86 + 0.18 * \log A$; $n = 19$, $R^2 = 0.67$, $p < 0.001$). These lines, perhaps fortuitously, are not significantly disparate in slopes ($p > 0.05$ that they really differ). In both cases, the large-island regressions substantially underestimate regional plant species totals. However, when adjacent Barkley Sound islands are grouped into local archipelagoes, with higher (combined) area and higher species counts, a new regression line extrapolates quite accurately to the regional totals for Vancouver Island and for British Columbia (see Fig. 4.10, upper). In both island sets, there is a steep fall-off in species numbers on increasingly small islands; on the hot, dry desert islands in the Sea of Cortés this collapse occurs on islands below a square kilometer or so in area, but in Barkley Sound the cascade is shifted to the left, beginning on islands of around 0.01 km^2 or smaller. On the desert islands, major components of the vegetation are lost when arroyos, bajadas, and alluvial fans are absent, as is the case on small islands that do not intercept enough of the infrequent precipitation to form these features (Cody et al. 2002). In Barkley Sound, where precipitation is over an order of magnitude higher and temperatures much more moderate, the small-island effect is ameliorated on midsized islands, but even here small islands become warm and dry in the summers, especially relative

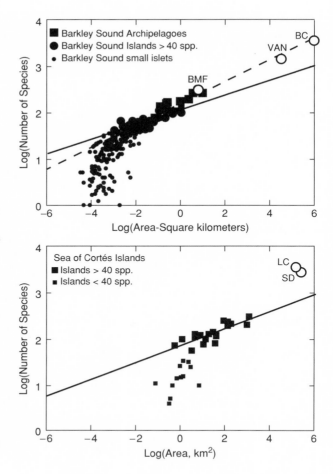

FIGURE 4.10. Species-area curves for Barkely Sound islands (upper) and islands in the Sea of Cortés (lower; after Cody et al. 2002). Solid lines are regressions based on islands with >40 species. In each case these lines significantly undershoot regional species-area data points (BMF: Bamfield mainland; VAN: Vancouver Island, BC: British Columbia; LC: Lower California, SD: Sonoran Desert). If Barkley Sound islands are grouped into archipelagos with cumulative species counts and areas, the new regression line (dashed, above) more accurately predicts the regional counts. Note how small islands fall increasingly below the large-area regressions in both sites, but this cascade occurs sooner, with decreasing area, on the desert Sea of Cortés islands.

to the temperature and humidity tolerances of temperate rainforest plants.

Species-area curves for the Barkley Sound islands are more orderly in semi-log "Gleasonian" plots, which handle the small-island effects better. Series $\mathcal{A}$, C, and $\mathcal{E}$ data are plotted in Fig. 4.11 (left-hand side), where the abscissa is the logarithm of the total area of the islands covered by vascular plants (AREA$_{VP}$). Using this measure of island area rather than the total area of the island above the *Fucus* line (AREA$_{FL}$) improves R^2 values by up to 5%; AREA$_{VP}$ excludes areas of low and exposed rocks, quite extensive on some islands, that contribute to total island area but not to plant habitat. The two area measures are of course closely related (log AREA$_{VP}$ = −1.02 + 1.19 * log AREA$_{FL}$; R^2 = 0.91, p < 0.001). The use of AREA$_{VP}$ partially offsets the small-island effect, as on small islands only a minor proportion of total area (AREA$_{FL}$) supports vascular plants, $<1/4$ for a 100 m^2 island, whereas on larger islands the proportion is very much larger ($>4/5$ for a 100,000 m^2 island). The regression coefficients in Fig. 4.11 do not differ significantly between $\mathcal{A}$- and C-series islands, but on $\mathcal{E}$-series islands there is a modestly (but significantly) lower slope (p < 0.05 that the slopes might be the same as $\mathcal{A}$-series or C-series slopes). The regression lines on the left are replicated in the right half of the figure as dashed lines. On the right-hand graphs, species are summed and areas (AREA$_{VP}$) are accumulated, as islands are added up (as if contiguous) from smallest to largest islands. The discrepancies among the island groups are more easily seen in these cumulative data, with larger

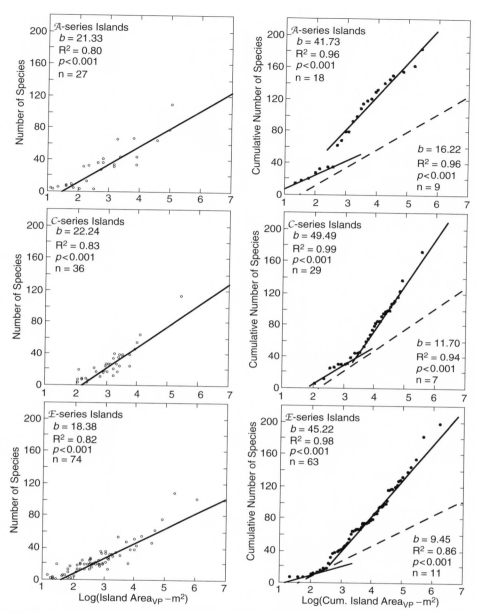

FIGURE 4.11. Regression lines for species-area curves are plotted on semi-log axes for 𝐴-, 𝐶-, and 𝐸-series islands (right). These lines are replicated (dashed) on the right, and compared with cumulative species-area curves, fitted by two-stage regression lines. See text.

species totals per accumulated area in the 𝐴-series islands, intermediate in the 𝐶 series, and lowest in the 𝐸 series. These differences are assessed with regression analyses, and all pairwise comparisons 𝐴–𝐶, 𝐴–𝐸, 𝐶–𝐸 are significantly different. The striking feature of the cumulative curves is the break or kink observed as initially species are accumulated linearly at low rates, but past a certain threshold the accumulation rate, still linear, becomes much higher. Thus on small islands species are initially accumulated at a rate of 9.45 (series 𝐸) to 16.22 (series 𝐴) species per 10-fold increase in total area, but beyond the threshold the rates are

3–4 times greater. This threshold is a point beyond which islands have a substantially new and different floral component included in the cumulative tally. It is the point, on all three island sets, at which the first trees and shrubs such as red cedar *Thuja plicata* and the ericads *Gaultheria shallon*, *Vaccinium ovatum*, and *V. parvifolium* join the species lists. Here also other common trees (*Tsuga heterophylla* and *Picea sitchensis*), shrubs (*Amelanchier alnifolia* and *Sambucus racemosus*), and the fern *Polypodium scouleri* are added in, on around two-thirds of the islands. Clearly the vegetation changes in character at the threshold, with the addition of the first woody species and with many other plants added shortly thereafter. The change comes at values of $AREA_{VP}$ between 114 and 148 m^2 on $\mathcal{A}$-series islands (with the addition of 28 new species), between 69 and 156 m^2 on C-series islands (+16 species), and earlier still, between 70 and 72 m^2, on $\mathcal{E}$-series islands (with 14 new species). That is, the shrubs and trees take root sooner, with respect to increasing island size, in the middle of the Sound, where the conditions on small islands may be somewhat ameliorated by lower temperatures and higher humidities.

Note that, while the cumulative species curves in Fig. 4.11 (right) lie well above the noncumulative species area curve for $\mathcal{A}$-series islands, there is not such great disparity on $\mathcal{E}$-series islands, and the C-series islands again are intermediate. Clearly small islands with sparse, open vegetation harbor more of the weedy species typical of them on $\mathcal{A}$-series islands, and these are species with colonization rates impaired by isolation and therefore comparatively underrepresented on the $\mathcal{E}$-series islands. Indeed, the dashed species-area curve for the $\mathcal{E}$-series islands intersects the cumulative curves around $\log(AREA_{VP}) = 2.5$, such that a single island of 400 m^2 $AREA_{VP}$ will have about the same number of plant species as a dozen small islands that sum to the same size. As there is no such intersection point on the $\mathcal{A}$- and C-series islands, no single island amongst them will have as high a species count as will several islands combined. These observations recall the "SLOSS" debate (Diamond and May 1981)—whether to conserve species with a "Single Large Or Several Small" islands; clearly at an investment size of 400 m^2, several small islands would conserve more species in the $\mathcal{A}$ group, but the reverse is true in the $\mathcal{E}$ group. The answer to SLOSS is more obvious for larger areal investments, >>400 m^2, for which many small islands accumulate considerably more species than single islands of the same size. Of course (and this is an important caveat), the single large island may harbor species absent from the smaller islands and, in particular, may harbor forest species of particular conservation concern and of far more intrinsic interest than the weedy, edge species prevalent on small islands.

ISOLATION, TOPOGRAPHY, HISTORY, AND OTHER EFFECTS

Not surprisingly, area is the predominant factor in determining numbers of plant species on the islands, but there are others that also contribute in statistically significant ways. An all-island multiple regression includes island elevation (ELEV), distance from mainland (D_{MLD}), and density of neighboring islands within 250 m (ANG_{250}) as significant contributors: $SPP = -19.28 + 15.19 * \log AREA_{VP} + 0.40 * ELEV - 0.68 * D_{MLD} + 0.012 * ANG_{250}$ ($n = 196$, $R^2 = 0.83$, $p < 0.001$). Thus species numbers increase with island elevation and with the density and proximity of neighbor islands, and decrease with isolation. However, the net increase in explanatory power is minor, with an increment (over island area alone) of just a few extra percentage points in the accounting of variation in SPP.

The plant species typical of the three distinct island habitats, forest, edge, and shoreline, might be expected to respond to determinants of species richness in different ways, and the extent to which this occurs is shown in Fig. 4.12. On log-log plots of species numbers versus $\log(AREA_{VP})$, it is the forest plants in particular that show the cascading small-island effect discussed above (Fig. 4.12, upper left). The edge and shoreline plants (Fig. 4.12, center and lower

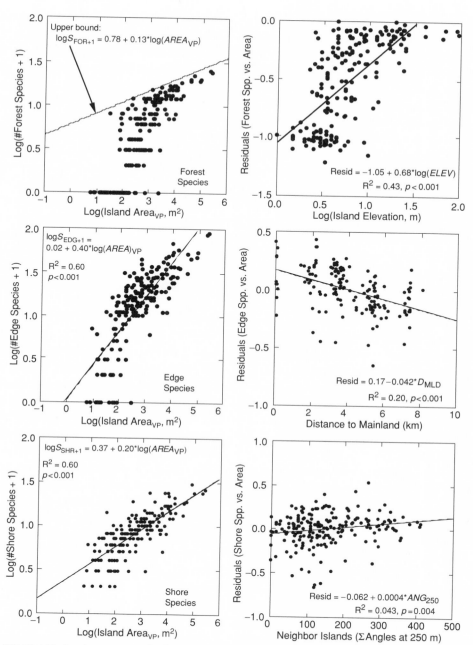

FIGURE 4.12. Species-area curves for forest (upper), edge (center), and shoreline species (lower). For forest species, islands show an upper bound (fitted by eye), and residuals below this bound are significantly greater on lower than taller islands (upper left). Residuals from regressions for edge and shoreline species are significantly related to distance from mainland and to density of neighboring islands, respectively.

left) display reasonably linear relationships, with a much less obvious small-island fall-off in species numbers. Clearly it is chiefly the forest plants that find conditions increasingly inhospitable on smaller islands, while edge and shoreline plants are trimmed out proportionately and more linearly with decreasing island area.

I draw an upper bound to the numbers of forest plants (Fig. 4.12, upper left), fitted by eye, which approximates the upper limit of the data set over about four orders of magnitude of island areas. After computing the deviations of points below this bound, these residuals are cast as dependent variables relative to other abiotic measures. The residuals are unrelated to various measures of isolation ($p > 0.05$), but there is a strong influence of island elevation (Fig. 4.12, upper right; $R^2 = 0.43$, $p < 0.001$). Numbers of forest species increase, toward the upper bound, with $ELEV^{0.68}$, such that islands 35 m high would lie on the upper bound, two species would be lost on 12.5 m islands, and a further two on islands 4.5 m high. For edge species, there is no effect of island elevation, but a marked species loss with increasing island isolation (Fig. 4.12, center right; $R^2 = 0.20$, $p < 0.001$). This amounts to a two-fold reduction in the numbers of edge species between the least and most isolated islands of comparable size.

Shoreline species are those found on island coasts within a few meters of the *Fucus* line, plants characteristic of open, rocky habitat, sandy or cobble beaches. Not surprisingly, and unlike forest and edge species, they display little or no small-island effect on log-log axes (Fig. 12, lower left). Their numbers also are unaffected by island elevation or isolation ($p < 0.05$). There is a significant, but rather minor, influence of nearby islands, as measured by ANG_{250} (summed angles subtended by neighboring islands <250 m distant). This measure of local isolation averages 148 ($\pm$ 96 SD), and ranges to over 400 (Fig. 4.12, lower right). In effect, one or two species are added to the shoreline count on islands positioned in areas with high densities of neighbors, relative to neighborless islands. The dependence of shoreline plants on floating propagules via the complex ocean currents within the Sound identifies island target size as the most important determinant of their number, and enhanced numbers on islands with close neighbors perhaps reflects a potential for such island clusters to "trap" the propagules in currents eddying amongst them.

Certain islands stand out from the others as being particularly species rich or species poor, relative to what is expected from their areas, elevations, and isolation. Residuals from the four-factor regression equation (above) can be used to identify these outliers; at least some of the anomalies can be resolved, and for others, the circumstances related to their deviations from expected species counts can be the subject of useful speculation. There are 16 islands in each category of residuals >10 and <10 species, and they are listed in Table 4.1. Of species-rich outliers, two large islands (*D*24, *E*49) have wide, sandy beaches and are popular camping sites; they are especially rich in alien weeds. Three others are also weed rich (*B*12, *D*11, *E*68); the first was a historic homesite, occupied until the 1940s, and it is still largely cleared of shrubs and trees. *D*11: Friend, is privately owned and gained 10 plant species after it was colonized by humans in the late 1990s. *E*68: "Keith W" is a much older habitation site, being part of an ancient First Nations midden site. Some very small rock stacks (*D*32, 34) are beach satellites closely adjacent to much larger islands (*D*24: "Big Stud" in this case) and apparently manage to retain species in excess of predictions based on area, elevation, and neighborhood by dint of close proximity to very large neighbors. Several other species-rich islands lie in particularly sheltered positions, that is, they are relatively small islands with much larger islands nearby and often on either side. Islands *A*2 and *A*3 lie between *A*1 to the north and *A*7 to the south; islands *C*3, *C*4, and *C*6 lie in the narrow channel between the very large *C*36: Sandford to the south and Fleming to the north, a channel crowded with many other islands.

Note that islands in positions like these, dwarfed by large and close neighbors and

TABLE 4.1
Species-Rich and Species-Poor Islands

	AVERAGE NO. OF SPECIES		
	OBSERVED	PREDICTED	REASON

Species-rich Islands (Residuals >10)

A1: N Hosie	66.3	45.9	?
A2: Mid Hosie	65.2	34.8	Sheltered
A3: Hosie Rks a	35.2	25.2	Sheltered
B12: Haines SE	63.0	42.2	Old homestead
B19: Haines SERks g	39.0	25.5	?
C3: Ross SSE	54.1	38.0	Sheltered
C4: Mid Ross	37.8	26.7	Sheltered
C6: Ross/Sandford Adj	63.9	48.5	Wide beaches
C36: Sandford	113.2	98.5	Wide beaches
D11: Friend	73.5	52.5	New homestead
D24: Big Stud	59.8	46.0	Camping beach
D32: Big Stud NE Rk	21.7	11.3	Beach satellite
D34: Big Stud E Rk	22.9	12.2	Beach satellite
E49: Clarke	109	93.8	Camping beach
E68: Keith W	52.5	32.1	Indian midden

Species-poor Islands (Residuals <−10)

A7: S Hosie	64.0	76.5	Steep coast
A11: Nanat	33.7	49.2	Steep, shaded coast
A18: Br.Bch.e4th	12.0	23.6	?
A26: Dixon W Rk	2.4	14.3	Low, flat rock
A27: Dixon SW Rk	2.0	21.3	Low, flat rock
B20: Bordelais	29.0	46.3	Exposed, isolated
B25: Bordelais S Rks	21.2	36.1	Exposed, isolated
C17: Ross E Adj	23.1	38.8	Steep coast
C23: Fleming E	23.7	46.1	Steep coast
D16: Meade NN Rk	3.3	13.9	Exposed
E18: Onion	45.6	56.3	?
E19: Tiny Mid E	20.5	40.4	Steep coast
E67: Otter	13.0	35.3	?
E70: Village Rf*	0	21.2	Low, exposed, isolated
E83: Starlight Rf*	4.0	38.9	Seabird colony
E84: Baeria Rks*	5.0	37.7	Seabird colony

Note: Islands with high positive residuals (upper) and high negative residuals (lower) in four-factor species versus area/elevation/isolation regressions, identified by comparison of average numbers observed versus numbers predicted from regression. Possible reasons for status are noted.

simplistically labeled "sheltered," may be species rich for several reasons that in general are difficult to dissociate. They may enjoy shelter from storms, which may preserve small populations from major disturbance effects, and they may also enjoy ameliorated microclimates because of the proximity of nearby, large islands. Further, they were in most cases a part of much larger islands, when sea levels were lower and they were actually connected to these neighbors; and lastly, the large neighbors support large species pools that generate a close and steady source of colonists to the "sheltered" islands in question. One further point concerns three islands $A1$, $A2$, and $A7$ that are close neighbors in the Hosie group. Note from Table 4.1 (upper and lower) that these three islands have very close to the same average species numbers, 66.3, 65.2, and 64.0, respectively. This is all the more surprising because their areas (spanning 1.3 orders of magnitude) and elevations (from 8.5 to 67 m) are all quite different, as will be obvious from the differences amongst them in predicted species numbers (45.9, 34.8, and 76.5, respectively). A fourth island in the group, the smaller $A3$, (Table 4.1 upper), is also very rich in species for its size and elevation. It is tempting to hypothesize some synergistic effect amongst such neighbors, such that their joint proximity results in a shared species pool that largely overrides differences in area; indeed, they seem to act like a "meta-island" with respect to their plants.

With species-poor islands (Table 4.1 lower), there are some cases in which low numbers are attributable to some rather obvious factor, but in others there is no readily apparent reason for impoverishment. $E70$: "Village Reef," for example, was censused on several occasions, but no vascular plant species were recorded; given that its $Fucus$ area is 2120 m^2, it is 2.2 m high, and 155 m^2 of the island are above 2 m in elevation, it seems reasonable, a priori, that a few shoreline species should have found root there. By comparison, $E37$: "Exotica," is 2190 m^2, reaches 2.8 m elevation, and supports an average of 21.4 species. Size and topography are similar in the two islands, but there are differences chiefly in their relative positions with respect to neighbors. "Village Reef" is extremely isolated (ANG$_{250}$ = 0, ANG$_{KM}$ = 22°) and exposed to storms rolling in from the adjacent Imperial Eagle Channel. Within a kilometer of "Exotica," on the other hand, are many large islands (ANG$_{KM}$ = 288°), including $E76$: Turtle just to the west and >1 km^2 in area. On two other islands, the mutual near neighbors $B20$: Bordelais and $B25$: "Bordelais Rks," low species counts seems quite obviously related to their position on the outer (oceanward) edge of the Sound, and their extreme exposure to storms and wave bashing. Both islands are high rocks, and a navigational beacon is conveniently situated atop the 24 m high Bordelais. The outer sides of these islands have no vascular plants below around 8 m up from the tide line, testimony to the challenging conditions that plants face on rocks fully exposed to the Pacific Ocean breakers; only above 15 m on Bordelais do plants reach normal densities.

Two other islands, $E83$ and $E84$, also have spectacularly low plant species counts (see Table 4.1), but for different or at least additional reasons from "Village Reef" and Bordelais. These two islands also are very isolated, and there is no other terrain within a kilometer of either. Perhaps owing to this feature of extreme isolation, both are busy seabird colonies and support high densities of breeding Glaucous-winged Gulls (*Larus glaucescens*). Not only do the gulls lay down a generous layer of guano over the islands, but they also pull up any plant material they are able to, for use as nesting material. Indeed, these islands are adequately censused by evaluating the composition of a few gull's nests! $E84$ is the only island where *Lasthenia minor* is recorded, a species that apparently has a great tolerance for guano; this taxon was previously known as *Baeria maritima*, hence the island's common name of Baeria Rocks.

Several of the remaining species-poor islands share a common feature of steep coastlines, such that they are extremely difficult or even impossible to circumperambulate. Steep cliffs are relatively inhospitable to the edge species, which comprise the largest component of the island floras and are restricted to the open, coastal areas. It seems likely that very steep coastlines support fewer species than do more moderate slopes; this is a topographical feature that could be quantified, but as yet I have not done so. $E67$: "Otter," for example, has rather steep cliffs; the island was named for the otters (*Lutra americana*) that we routinely saw there. The otters access the top, treed part of the island via many well-scrubbed paths and often leave by way of the numerous slides that cut down the steep banks. These activities might possibly contribute to a reduced species count on the island.

Attempts to isolate a purely historical component to present-day species richness are fraught with ambiguities similar to those described above, where small, satellite islands lie close to large neighbors and can benefit from this proximity in several quite different ways. Islands in the Sound with the longest history of isolation are also those that are most isolated in space. Thus they may have reduced species richness because of their long history as islands, and a longer time for relaxation and a closer approach to equilibrium. Alternatively, because of the correlated spatial isolation, they may be species poor because of reduced colonization rates and higher levels of exposure to storms. Likewise, the islands with the shortest histories, those that are separated from neighbors or parent larger islands by minimal channel depths, may be species rich because insufficient time has elapsed since they were calved by rising sea levels for them to lose supernumerary species. Alternatively, such islands are invariably very close to the larger parent, which can provide shelter from disturbance, a moderated climate, and a ready source of colonists. Even with my large data set, these elements are difficult to disentangle. For example, $B28$ and $B29$ have a long history as islands (Figs. 2.8, 2.10) and are moderately species poor with -9.3, -7.6 species deficits, respectively (from the four-factor regression, above). However, these negative residuals are composed of species deficits in forest, edge, and shoreline species. Although all three species groups have negative residuals on these islands, history may qualify as the major factor contributing to their relative impoverishment, as species deficits in forest species outstrip those of the other two species groups.

Some newer islands such as $B26, 27$, minimally separated from the large (and uncensused) island of Diana by channel depths of <1 m, are relatively species rich, by $+7.2$ and $+3.7$ species, respectively, over regression predictions. But they are species rich not because they have retained forest species since their isolation (negative forest residuals for both islands), nor because they have accumulated shoreline species (essentially zero residuals). These islands are species rich because they have more edge species than predicted, and this in turn may be attributable either to their proximity to Diana or, more likely, to their lying just 80 m across the Dodger Channel from the weed-rich islands of $B12$: "Haines SE" and its satellites. While historical contributions to variation in species richness seem to be at best minor, they may still help to explain distributional anomalies and patterns of shared species relicts.

5

Nestedness and Assembly Rules

INFERENCES FROM SPECIES-BY-SITES MATRICES

A common representation of distributional data, species occurrences over a range of sites, is in the form of a species-by-sites matrix (SSM). Conventionally, species are listed in the rows of the matrix and the sites in columns; generally SSM entries are presence-absence data, namely, 1's or 0's. SSM databases are standard format for many sorts of ecological and biogeographical studies and are amenable to a wide range of analytical and statistical techniques. These include, among many others, the ordination or classification of species and sites to reveal distributional patterns among species and species richness or diversity patterns among sites (e.g., Gauch 1982). Thus, species may be ranked from broad to narrow in distribution, sites from rich to poor in the species they support. Typically rows and columns in the SSM are rearranged with the species ranked top to bottom in decreasing order of the number of sites occupied, and sites ranked left to right in decreasing order of the number of species present.

If the sites are islands spanning a range of size, topography, and position relative to colonization sources, SSMs with ranked columns will order islands by those properties that are associated with high species richness, namely, by decreasing size and elevation, and increasing isolation. Often this process produces a matrix with most entries concentrated in the upper-left triangle, a property known as nestedness. Lizard and bird distributions are nested on islands in the Sea of Cortés (Case and Cody 1983, 1987; Case et al. 2002), the mammals less convincingly so (Lawlor et al. 2002). Area is a dominant factor in these island rankings, for several likely reasons, and most conspicuously there is an increasingly wide range of habitats on larger islands, with bajadas, arroyos, and outwash plains being added successively to the narrow habitat range of small islands, which are essentially just rock piles. Further, there is a higher colonization probability and an enhanced persistence of populations on larger islands. Among many other illustrations of the nestedness phenomenon, the strongly nested fish species in islandlike spring-fed pools in the central Australian desert are particularly striking (Kodric-Brown and Brown 1993).

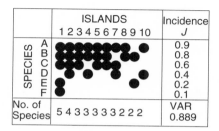

FIGURE 5.1. Ten islands (columns) are ranked from most (left) to fewest (right) species present, and species (rows) are ranked from most (top) to fewest (bottom) islands occupied. With a fixed incidence vector J and a random distribution of species over islands (i.e., within rows; null model), the expected number of species on islands is $\Sigma_i J_i$ with variance $\Sigma_i J_i (1 - J_i)$. Despite the ordering of islands and species by ranks, there are three possible outcomes: a random matrix (upper left), a nested matrix (lower left), and a matrix with disjunct distribution of species (lower right).

While larger island areas routinely correlate with higher species counts, nested SSMs indicate specifically that the biotas of smaller islands are subsets of those on larger islands (assuming for the purposes of discussion that island size is the predominant factor in species richness). Thus, large island biotas are built from those of small islands by a simple accretion of additional species, and in general species are added or subtracted in a predictable sequence as island size increases or decreases. If area dominates the ranking of islands in nested SSMs, then species loss, in continental islands relaxing from above, must depend strictly on area and have but a minor stochastic component. Species gained from colonization, likewise, must be area dependent, in that target effects dominate isolation effects, and again the stochastic component must be minimal.

But note that nestedness is not a necessary consequence of rank ordering species and sites within the matrix. Figure 5.1 shows ranked matrices with 10 islands and six species with fixed incidences J. Using a test involving the expected variance of species/island (bottom row of the matrix) relative to that following the null expectation (D. Ylvisaker, pers. comm.; see also Gilpin and Diamond 1981), one matrix shows a random distribution of species over islands, a second has a significant degree of nestedness, and a third departs from random in a different way, as it shows species' distributions with either large-island or small-island skew.

A considerable literature has accumulated on the measurement and meaning of nestedness in SSMs (e.g., Patterson and Atmar 1986; Simberloff and Martin 1991; Lomolino and Davis 1997; Cook and Quinn 1998; Wright et al. 1998). Sanderson et al. (1998) provide definitive guidelines for the construction of null matrices to which observations may be compared. Nestedness per se could be a useful metric for conservation purposes and inform the SLOSS debate (see above) for reserve design. If there are no species unique to small islands, then a conservation strategy that protects the largest island, with a full complement of species, might be superior. Even so, this ignores other considerations of potential importance, such as population vulnerability to disease or disturbance, which might be ameliorated with more, smaller reserves that collectively house the entire species quota. On this topic, too, a variety of opinion has been expressed, some of it rather forcefully (e.g., Boecklen and Gotelli 1984), and it seems unlikely that optimal conservation strategies will be illuminated by consideration of single factors in isolation, such as local species richness (Diamond 1975b; Saunders et al. 1991).

I regard the property of nestedness per se as having a rather limited potential to provide insights into island biogeographical processes, and in this regard it seems to have been overemphasized. SSM data on species' distributions are often more transparently analyzed by incidence functions and their patterns of congruence or disjunction. Variation in species richness over sites, the SSM column totals, can be assessed with direct reference to site characteristics such as area, elevation, isolation etc. (chapter 4). However, the more useful attention to SSM analysis has addressed questions of whether species are distributed independently of each other, or whether the presence of one on certain islands influences the occurrences of another. These questions fall under the general rubric of "assembly rules" (Diamond 1975a; Keddy and Wieher 2001). A simple example is that of "checkerboard" distributions, in which one species' presence is significantly associated with another's absence, and vice versa (see e.g., Diamond and Gilpin 1982; Gilpin and Diamond 1982). In a more general way, SSM data can provide evidence for more complex coincidences and disjunctions of suites of species and lead to questions such as which islands support one or the other species' suite and why.

With multiple censuses of islands over time, the Barkley Sound plants provide more than presence-absence information since, while some island species have been permanent residents for the duration of the study, others have been recorded for various proportions of the time period. Thus matrix entries can be written as incidence over time J_T, which take on values $1 \geq J_T \geq 0$. Further, it will prove interesting to construct SSMs for different components of flora, such as forest, shoreline, and edge species, and for various subsets of the islands, such as small, isolated rocks, large-island satellites, or all islands within a limited size range. These and other aspects of the island flora that relate to species co-occurrence are examined in this chapter.

FOREST SPECIES

The most conspicuous component of the island flora is certainly the forest species, which comprises the codominant trees, the understory shrubs, and the ferns and herbaceous plants of the forest floor. Assigning species to this category is relatively straightforward, although there are some judgment calls. Douglas fir, for example, is occasionally found in interior forests with spruce, cedar, and hemlock, but on most islands in Barkley Sound it occurs more often in open sites near the island's forest fringe. Shore pine, in contrast, grows strictly at the island edge, when it is present. Making a few such judgment calls, I have assembled a list of 39 forest species in Table 5.1, which collectively occur on 147 (of 217) islands. The list includes the four tall canopy trees (*Picea, Thuja, Tsuga, Pseudotsuga*) and four short, subcanopy trees (*Acer* [two spp.], *Taxus, Rhamnus*). In addition, there are nine fern species, the two ericad shrubs commoner in the interior than on the edge (*Menziesia ferruginea, Vaccinium ovalifolium*), two geophytic orchids (*Goodyera, Listera*), and numerous forest floor herbs (*Boykinia, Maianthemum, Galium, Streptopus, Tellima, Tiarella, Tolmiea*). The list also includes the ericad saprophyte *Hemitomes congestum* and the parasitic *Boschniakia hookeri* (among others not mentioned here).

The species are ranked in the table from most to fewest islands occupied (at least part-time, reviewing all island records cumulatively: $J_T > 0$; col. 2). The three forest trees, spruce, cedar, and

TABLE 5.1
Forest Species and Their Incidence on Islands in Barkley Sound

FOREST SPECIES[a]	NUMBER OF ISLANDS[b]						INCIDENCE[c]	
	CUM (RANK)		>50% (RANK)		100% (RANK)		$J_I =$ CUMULATIVE	$J_T =$ OVER TIME
Picea sitchensis	126	(1)	121	(1)	114	(1)	0.581	0.95
Thuja plicata	108	(2)	106	(2)	100	(2)	0.498	0.97
Tsuga heterophylla	100	(3)	91	(4)	85	(4)	0.461	0.897
Maianthemum dilatatum	99	(4)	93	(3)	87	(3)	0.456	0.929
Polypodium glycyrrhiza	92	(5)	76	(5)	54	(5)	0.424	0.753
Polystichum munitum	85	(6)	66	(6)	54	(6)	0.392	0.765
Boschniakia hookerii	64	(7)	43	(8)	8	(16)	0.295	0.53
Taxus brevifolius	58	(8)	50	(7)	47	(7)	0.267	0.814
Blechnum spicant	50	(9)	34	(11)	28	(10)	0.23	0.594
Pseudotsuga menziesii	43	(10)	34	(10)	33	(9)	0.198	0.839
Menziesia ferruginea	39	(11)	37	(9)	36	(8)	0.18	0.967
Athyrium felix-femina	38	(12)	21	(13)	9	(14)	0.175	0.407
Rhamnus purshiana	29	(13)	24	(12)	20	(11)	0.134	0.713
Goodyeara oblongifolia	26	(14)	20	(14)	11	(12)	0.12	0.727
Dryopteris austriaca	16	(15)	9	(19)	7	(18)	0.074	0.465
Vaccinium ovalifolium	15	(16)	11	(15)	10	(13)	0.069	0.597
Adiantum pedantum	15	(17)	10	(17)	8	(17)	0.069	0.515
Galium trifidum	11	(18)	10	(16)	5	(22)	0.051	0.658
Linnaea borealis	10	(19)	9	(18)	9	(15)	0.046	0.902
Galium triflorum	9	(20)	6	(23)	5	(23)	0.041	0.4
Streptopus amplexifolius	9	(21)	7	(22)	4	(24)	0.041	0.72
Tolmiea menziesii	9	(22)	8	(20)	6	(20)	0.041	0.786
Aruncus sylvester	9	(23)	8	(21)	7	(19)	0.041	0.757
Acer glabrum	7	(24)	5	(24)	5	(21)	0.032	0.64
Listera caurina	6	(25)	3	(25)	2	(27)	0.028	0.357
Tiarella trifoliata	4	(26)	3	(27)	3	(26)	0.018	0.714
Tellima grandiflora	3	(27)	2	(28)	1	(32)	0.014	0.625
Hemitomes congestum	3	(28)	1	(35)	1	(35)	0.014	0.222
Boykinia elata	3	(29)	0		0		0.014	0.235
Lysichitum americanum	3	(30)	3	(26)	3	(25)	0.014	1
Oenothera sarmentosa	2	(31)	2	(30)	2	(28)	0.009	1
Asplenium trichomanes	2	(32)	2	(29)	1	(31)	0.009	0.667
Puccinellia pauciflora	2	(33)	1	(34)	0		0.009	0.4
Selaginella wallacei	2	(34)	1	(33)	1	(34)	0.009	0.286
Cinna latifolia	1	(35)	0		0		0.005	0.111
Cystopteris fragilis	1	(36)	1	(31)	1	(30)	0.005	1
Dryopteris felix-mas	1	(37)	0		0		0.005	0.286
Acer macrophyllum	1	(38)	1	(32)	1	(29)	0.005	1
Lycopodium selago	1	(39)	1		1	(33)	0.005	1
Number of species	39		36		35			

[a] Species are listed (39 in all) in order of incidence (proportion of total 217 islands) over all islands.
[b] The number of islands is given for which a species is recorded ever (CUM), or in >50% or 100% of the censuses. Species rank, in number of occupied islands, is also given, for each criterion.
[c] Incidence over time is the proportion of years a species was observed on the islands on which it occurs.

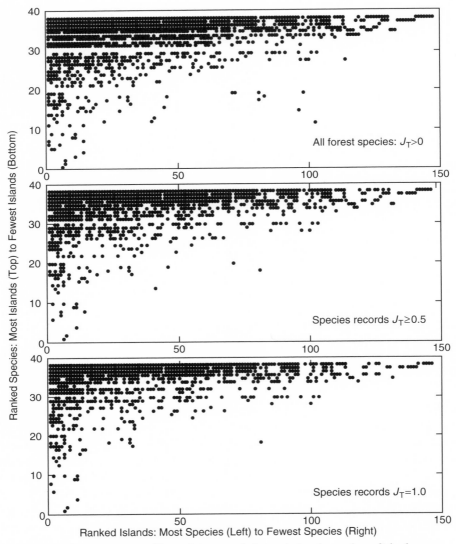

FIGURE 5.2. Thirty-nine species of forest plants are ranked top to bottom by numbers of islands occupied, 147 islands are ranked left to right by numbers of forest species present. Upper matrix: cumulative records; center: incidence over time $J_T \geq 0.5$; bottom: permanent residents ($J_T = 1$). See text.

hemlock, rank at the top of the table, with occurrences on 128, 108, and 100 islands, respectively, and five species at the bottom of the table are single-island occurrences. Numbers of occupied islands generally decline when species presence for at least half the time is the criterion ($J_T \geq 0.5$; col. 3), and decline further on islands where species were permanent residents throughout the study ($J_T = 1$; col. 4). Note that there are minor changes in species rank as different criteria for island presence are used. Incidence over all islands, J_I, is shown in column 5. The average proportion of censuses in which the species is recorded, for islands on which a species does occur or has occurred, in shown in column 6; this is a measure of the permanence of the species' island populations. Note that several species ranking low in the table, and thus present on very few islands, are nevertheless permanently resident on those few islands ($J_T = 1$).

Forest species are arranged in SSM form in Fig. 5.2, where rank (from Table 5.1) is kept

TABLE 5.2
Observed and Expected Occurrence of Spruce, Cedar, and Hemlock on Small Forest Islands in Barkley Sound

	EXPECTED	OBSERVED	CHI-SQUARE
Islands with <4 Forest Species ($n = 52$)[a]			
No. islands with			
3 spp. (pqr)	2.74	7	6.62
2 spp. (pqr′, pq′r, p′qr)	15.93	7	5.01
1 sp. (pq′r′, p′qr′, p′q′r)	23.93	29	1.07
0 spp. (p′q′r′)	9.4	9	0.02
Totals	52	52	12.72
Islands with <3 Forest Species ($n = 39$)[b]			
No. islands with			
3 spp. (pqr)	0.38	1	1.01
2 spp. (pqr′, pq′r, p′qr)	8.17	4	2.13
1 sp. (pq′r′, p′qr′, p′q′r)	19.49	26	2.17
0 spp. (p′q′r′)	10.96	8	0.8
Totals	39	52	6.1

[a] Crit. chi square for df = 3: 7.82; $p < 0.05$.
[b] Crit. chi square for df = 3: 7.82; $p > 0.05$.
Note: Species incidences for islands with fewer than four species are spruce (p) = 0.615; cedar (q) = 0.404; hemlock (r) = 0.212. Species incidences for islands with fewer than three species are spruce (p) = 0.538; cedar (q) = 0.35; hemlock (r) = 0.051.

constant but residency criteria are varied from top ($J_T \geq 0$) to middle ($J_T \geq 0.5$) to bottom ($J_T = 1$). Island rank is also preserved from the upper to lower matrix, although some changes in actual island rank do occur with the exclusion of temporary residents. As expected, islands' ranks are most strongly associated with island area and island elevation (upper matrix: $r = -0.76, -0.62$, respectively, Spearman rank correlation coefficients, $p < 0.05$). With Wilcoxon rank sum tests (Siegel 1956), all three matrices are significantly nested (z values 7.4–9.3, $p < 0.05$), with departures from nestedness most notable in individual species that occupy single islands. Successive exclusion of species records by residency time, upper to lower, results in tidier matrices and tighter nestedness, with fewer entries located away from the upper-left concentration of the data.

The three codominant forest trees, spruce, cedar, and hemlock, all show retreats from species-poor islands (right-hand columns) when shifting from inclusion of temporary island residents to inclusion of permanent residents only (matrix A to matrix C). For these trees, z-values increase by 68%, 25%, and 23%, respectively, A to C. As all of these species coexist routinely on the largest islands, I examined the possibility of interspecific influences on species presence on the smallest islands, those with the most limited potential to support trees. I computed the incidences of the three trees on the subsets of islands with ≤4 and ≤3 forest species, with $n = 52$ and $n = 39$ islands, respectively (see Table 5.2). Under the assumption that the species are distributed independently on these islands, the expected numbers of islands with all three, with two, one, or no species are easily derived from incidence values and then compared to observed numbers using chi-squared tests. The results (Table 5.2) show that the trees are distributed independently on the smallest islands (≤3 spp.), but not on the subset of somewhat larger islands (≤4 spp.). However, the deviation from expected in this latter group is toward coincident rather than disjunct occurrences of the trees, with more islands than expected having all species. Clearly, some small islands are particularly predisposed to support trees, and on such islands all of the trees tend to co-occur. These islands are not distinguished by area or elevation, but by position. All three-tree islands in the subset with ≤ 4 forest species are relatively small and low, but closely adjacent to much larger islands ($B11, C18, C24, D30, D35, E61, E89$). They are large-island satellites that may benefit from either biotic or abiotic shadows of their larger neighbors; they might support trees by dint of their close proximity to colonization sources, or to ameliorated and tree-conducive temperature/humidity environments in their larger neighbor's ambience.

Given the dominant effect of island area in the ranking of islands by species richness in Fig. 5.2, further insight might be gained by removing much of its influence and considering only islands of similar size. Following an earlier analysis of this restricted island set (Cody 1999), I look at forest species with $J_T \geq 0.5$ on midsized islands, a subset of $SIZE = 4$ islands with $1360 \leq AREA_{FL} \leq 2920$ m^2 and varying in size by a factor of about two or less. This produces a subset of 35 islands with vascular plant areas $1.44 \leq \log AREA_{VP} \leq 3.33$, or 28–2138 m^2. These 35 midsized islands are ranked by species richness of forest plants, 16 species in all, in Fig. 5.3 (upper matrix), where species number per island ranges from 11 ($E46$) to zero (9 islands); here I have included shore pine in the forest species list, for reasons that will become obvious below. For the $n = 13$ islands with >3 forest species, area is no longer a predictor of species richness ($R^2 = 0.05$, $p = 0.40$) although it is if the more depauperate islands are included.

In an attempt to discover why some midsized islands are relatively rich in forest species and others not, I group species by growth form in Fig. 5.3 (center matrix). No patterns emerge, and there is no apparent nestedness within growth form groups. However, islands do fall into categories associated with the co-occurrences of certain species, as shown in the lower matrix of Fig. 5.3. Ten of the 35 islands have one or more of the three trees, Pacific yew (*Taxus brevifolius*), shore pine (*Pinus contorta*), and Douglas fir (*Pseudotsuga menziesii*). Of these, the islands with yew and those with shore pine are nearly disjunct sets: just one island ($B2$) has both yew and shore pine (although the two species coexist on 13 larger islands). Further, all but one (4/5) of the islands with shore pine also have Douglas fir. Of the expected combinations by assumption of independent distribution, islands (collectively) with just yew or with both pine and fir, are overrepresented when expected and observed numbers are compared (EXP: 3; OBS: 7), and islands with other combinations of yew, pine, and fir are underrepresented (EXP: 7; OBS: 3), a significant divergence from random assortment ($\chi^2 = 7.62$, df = 1, $p < 0.01$). I use this distinction to classify islands into four classes (Fig. 5.4 upper): yew islands ($n = 5.5$), pine-fir islands ($n = 4.5$), other species-rich islands with >1 species ($n = 12$), and species-poor islands with one or no forest species ($n = 13$).

In Fig. 5.4, rows two to four, I examine the factors that determine which midsized islands will fall into the classes 1 to 4. The class 4, species-depauperate islands are certainly smaller than those of the other classes and are significantly overrepresented in the lower half of the size range, underrepresented in the upper half. They tend also to be predominantly low islands, as are pine-fir islands, whereas yew islands are inclined to be tall (statistically not significant). The main distinction between yew and pine-fir islands, termed "sheltered" or "exposed" in Fig. 5.4 (lowest row), is a function of island position within Barkley Sound on a gradient from the inner, mainlandward region of the Sound to the outer, oceanward region more exposed to the wind and weather. B- and C-group islands, distally in the Deer Group, and E-group islands, oceanward of the Thiepval Channel that splits the central Broken Group islands, comprise the "exposed" islands, on which yew is extremely scarce, but pine and fir are

Top matrix: ranked species and islands; center matrix: species grouped by growth forms; bottom matrix: islands grouped by those with Pacific yew, those with shore pine/Douglas fir lacking yew, and others.

	E46	B2	E24	C4	A2	C10	E67	E63	C8	E27	E50	E72	C9	E75	D7	E11	E61	E37	C27	C28	E12	E57	E2	C2	D14	E53	C1	E1	E10	E16	E25	E26	E43	E66	E70	#Isl
Picea sitchensis	•	•	•	•	•	•	•	•	•	•	•	•		•		•	•		•	•		•		•	•	•										
Thuja plicata	•	•	•	•		•	•	•	•	•	•	•	•		•		•			•	•															21
Maianthemum dilatatum	•	•	•	•	•	•	•	•			•	•	•	•	•	•					•	•														16
Tsuga heterophylla	•	•	•	•		•	•	•	•	•	•		•				•																			16
Polypodium vulgare	•	•	•		•							•			•	•		•	•				•													12
Taxus brevifolius	•	•	•		•					•		•																								10
Boschniakia hookerii	•		•							•	•			•																						6
Pinus contorta		•		•		•			•				•																							5
Pseudotsuga menziesii		•		•	•	•			•																											5
Polystichum munitum		•		•	•																															5
Menziesia ferruginea	•	•						•																												3
Goodyeara oblongifolia			•				•																													3
Athyrium felix-femina																		•																		2
Rhamnus purshiana	•																																			1
Blechnum spicant	•																																			1
Vaccinium ovalifolium	•																																			1
Number of Species:	11	10	8	7	6	6	5	5	5	5	5	5	4	3	3	3	3	2	2	2	2	2	1	1	1	1	0	0	0	0	0	0	0	0	0	1

Matrix nested by ranked species and ranked islands

	E46	B2	E24	C4	A2	C10	E67	E63	C8	E27	E50	E72	C9	E75	D7	E11	E61	E37	C27	C28	E12	E57	E2	C2	D14	E53	C1	E1	E10	E16	E25	E26	E70	E43	E66	#Isl
Picea sitchensis	•	•	•	•	•	•	•	•	•	•	•	•		•		•	•		•	•		•		•	•	•										
Thuja plicata	•	•	•	•		•	•	•	•	•	•	•	•		•		•			•	•															21
Tsuga heterophylla	•	•	•	•		•	•	•	•	•	•		•				•																			16
Taxus brevifolius	•	•	•		•					•		•																								12
Pseudotsuga menziesii		•		•	•	•			•																											6
Pinus contorta		•		•		•			•				•																							5
Menziesia ferruginea	•	•						•																												5
Rhamnus purshiana	•																																			3
Vaccinium ovalifolium	•																																			1
Polypodium vulgare	•	•	•		•							•			•	•		•	•				•													1
Polystichum munitum		•		•	•																															10
Blechnum spicant	•																																			3
Athyrium felix-femina																		•																		1
Maianthemum dilatatum	•	•	•	•	•	•	•	•			•	•	•	•	•	•					•	•														1
Boschniakia hookerii	•		•							•	•			•																						16
Goodyeara oblongifolia			•				•																													5
Number of Species:	11	10	8	7	6	6	5	5	5	5	5	5	4	3	3	3	3	2	2	2	2	2	1	1	1	1	0	0	0	0	0	0	0	0	0	2

Species grouped by growth form: Trees

Shrubs

Ferns

Forbs

	E46	E24	A2	E27	E72	B2	C4	C10	C8	C9	E67	E63	E50	E75	D7	E11	E61	E37	C27	C28	E12	E57	E2	C2	D14	E53	C1	E1	E10	E16	E25	E26	E43	E70	E66	#Isl
Picea sitchensis	•	•	•	•	•	•	•	•	•		•	•	•	•		•	•		•	•		•		•	•	•										
Thuja plicata	•	•		•	•	•	•	•	•	•	•	•	•		•		•			•	•															21
Tsuga heterophylla	•	•		•		•	•	•	•	•	•	•	•				•																			16
Taxus brevifolius	•	•	•	•	•	•																														12
Rhamnus purshiana	•																																			6
Menziesia ferruginea	•					•						•																								1
Vaccinium ovalifolium	•																																			3
Pseudotsuga menziesii			•			•	•	•	•																											1
Pinus contorta						•	•	•	•	•																										5
Polypodium vulgare	•	•	•		•	•									•	•		•	•				•													5
Polystichum munitum			•			•	•																													10
Blechnum spicant	•																																			3
Athyrium felix-femina																		•																		1
Maianthemum dilatatum	•	•	•		•	•	•	•		•	•	•	•	•	•	•					•	•														1
Boschniakia hookerii	•	•		•									•	•																						16
Goodyeara oblongifolia		•									•																									5
Number of Species:	11	8	6	5	5	10	7	6	5	4	5	5	5	3	3	3	3	2	2	2	2	2	1	1	1	1	0	0	0	0	0	0	0	0	0	2

Pacific yew islands

Shore pine/Douglas fir islands

FIGURE 5.3. Distribution of forest species on 35 midsized islands (1460 < $Area_{FL}$ > 2920 m^2). Sixteen species occur on these islands, with 2–34 islands/species and 0–11 species/island.

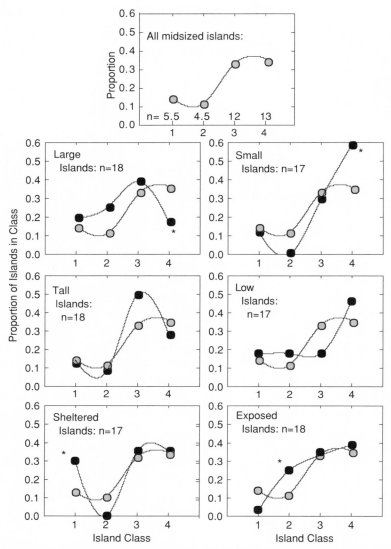

FIGURE 5.4. Thirty-five midsized islands are divided into four classes (proportions in upper graph), with class 1: Pacific yew islands, class 2: shore pine/Douglas fir islands; class 3: other species-rich islands and Class 4: species-poor islands with <2 forest species. Species-poor islands tend to be smaller (second-row graphs) and species-rich islands taller (third-row graphs) than other midsized islands. Pacific yew islands are sheltered in the inner Sound, while shore pine/Douglas fir islands tend to be more exposed with fewer nearby islands (lower graphs). Asterisks denote statistical significance from expected values, with $p < 0.05$.

common. For the subset of the midsized islands classified as sheltered, ANG_{100} averages $90.9°$ ± 70.0 SD and ANG_{100N} 4.3 ± 2.7 SD; for the exposed islands ANG_{100} averages $50.5°$ ± 59.3 SD and ANG_{100N} 2.03 ± 2.36 SD (the numbers are about twice as high for the sheltered islands, and significantly different by t-test; $p < 0.05$). Thus, yew islands are generally large and sheltered in the interior of the Sound, while pine-fir islands are smaller, low, and exposed toward to the outer Sound. Class 3 islands lack these indicator species and are large, tall islands of no particular exposure, and class 4 islands are small and low. Understory shrubs are sparse on islands of these sizes, but most occurrences (4/5) are on yew islands, where also the largely

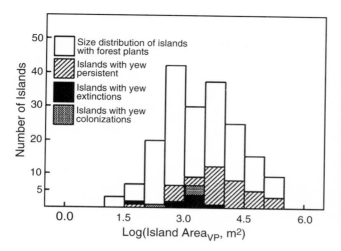

FIGURE 5.5. Distribution of Pacific yew (*Taxus brevifolius*) on Barkley Sound islands (hatched bars) over islands of different areas, relative to the distribution of islands with forest plant species (open bars). The solid bars represent islands where yew was recorded 1981–1982, on which it died by or prior to 1985. Stippled bars are yew colonization events.

arboreal fern *Polypodium glycyrrhiza* is especially common (EXP 1.7 and 8.3 on yew, nonyew islands; OBS 5, 5; $\chi^2 = 7.2, p < 0.05$). Of the remaining species, one only (*Maianthemum dilatatum*) is widely distributed, and it appears to be insensitive to the factors that determine presence or absence in yew, fir, and pine.

It is hardly surprising that small islands are inhospitable to forest vegetation in general and to tree species in particular. With decreasing area, island substrates become rockier with less soil buildup, and also they intercept less precipitation and lack the soil to retain what little is intercepted. Small islands are lower in elevation, with a reduced capacity to shelter leeward and northern slopes from wind and summer sun, and support only low vegetation with a limited capacity for shading substrates and ameliorating climatic extremes. Pacific yew is a subcanopy species and rarely grows beyond at least the partial shade of taller spruce, cedar, and hemlock; it thrives only on the larger, cooler, damper, and shadier of the midsized islands. The distribution of yews that were persistent on islands for the duration of the study, relative to the islands' vascular plant area $Log(AREA_{VP})$, is shown in Fig. 5.5 (hatched bars).

Yews did poorly on small islands in the decade preceding and up to 1985, which was a dry period in general (ave. annual precipitation 2682 mm), 10% less than the years of record since, 11% less than in preceding years of record. This period also includes the extreme drought year of 1985, when just 54% (1665 mm) of the long-term average annual precipitation fell. Yews can look unhealthy, moribund, or dying for years before they are effectively dead, but live yews recorded as single individuals on a number of small- and midsized islands at the beginning of the study period (1981) were classified as dead shortly thereafter, between 1982 and 1985 (nine islands; Fig. 5.5 solid bars). Three of these islands are classified as small, tall, and exposed by the criteria explained above, so a posteriori they are poor candidates to support yews permanently. Four other islands are large but low and/or exposed, but two are large, tall, and sheltered (A_{11}, A_{13}) and seem suitable for yew persistence. Four islands were colonized during the period of the study; on two (B_3, C_{11}) yews recolonized after previous extinction post-1981, in 1989 and 2003, respectively, being absent in the intervening years. Two islands were colonized de novo (C_8, D_{29}), the latter being a particularly small island [$\log(AREA_{VP}) = 2.45$ m^2], greater in size than only one other island with a persistent yew.

Earlier it was shown that Douglas fir is a tree species with a much more limited distribution on E-group islands in the center of Barkley Sound than on A–D islands (Fig. 3.8, lower left). The species avoids shady sites and favors sunny islands, those toward the Sound's periphery, close to mainland coasts and away from the Broken Group's fog banks. It and shore pine are most frequently found on south-facing island shores that are steep and rocky; however, Douglas fir has a broader distribution than the pine, and dominates the forest vegetation of A8: "San Jose NW," a large, low, and isolated island in the Trevor Channel that lacks other trees except for two small spruce. Pine and fir, like yew, have rather subdued colonization-extinction (C/E) dynamics. Over the course of the study, colonization and extinction events for Douglas fir totaled three and two, respectively (C/E = 3/2), for shore pine C/E = 3/3, all on small islands with $\log(AREA_{VP}) < 3.75$. All of the colonizations and extinctions in Douglas fir occurred on tall islands (ELEV > 4.8 m), all in shore pine on low islands (ELEV < 4.8 m); over species and between C- and E-events, the islands were equally divided between sheltered ($n = 5$) and exposed ($n = 6$).

On one small island, A3, Douglas fir survived the 1980s drought period but died back from 2.8 m (1985) to 1.8 m (1986), although it had regrown to 5.5 m in height by 2003. On the same island, a 1.2 m tall spruce died in 1985, presumably a consequence of the severe drought. Spruce, cedar, and hemlock all become increasingly sporadic in incidence when island size decreases, and their tenuous hold on small islands is reflected in both higher extinction rates and a slowed growth of new colonists. The heights of new tree colonists, invariably on small islands (as all larger islands have permanent resident spruce, cedar, and hemlock), were measured periodically, and these data are shown in Fig. 5.6. In the figure, the islands from which height data were obtained are divided into two size classes, $\log(AREA_{VP}) \leq 2.7$ m^2 and $2.7 < \log(AREA_{VP}) < 4$ m^2; new colonists on the smaller islands grew slowly (Fig. 5.6, left-hand column), sometimes even shrinking in height, and some dying within the study period. On larger islands (Fig. 5.6, right-hand column), the saplings generally grew more rapidly, although less convincingly so in spruce, and none died. I calculated regression coefficients b for sapling growth rates from the equation Tree Ht = $a + b *$ YEAR, for all $n = 13$ individuals that have four to nine data points (and no justification for nonlinear relationships). The coefficients vary in magnitude from 0.008 to 0.277, indicating growth from 0.8 to nearly 28 cm/y, and the range in R^2 values for these regressions is 0.86–0.99. In Fig. 5.7, the growth rate coefficients b are shown to be strongly related to island size, $\log(AREA_{VP})$, and there are no apparent differences among species. Both linear and nonlinear models are statistically significant, with more explanatory power in the latter ($R^2 = 0.84$, vs. 0.62 for the linear model).

Summing numbers of permanently occupied and newly colonized islands in the two size categories of Fig. 5.6, for the three tree species spruce, cedar, and hemlock collectively, extinctions occurred in 8 of 79 instances on smaller islands (10.1%), but in only 6 of 184 (3.3%) instances on larger islands. Clearly sapling longevity, as well as growth rate, increases with island size. Notably, extinction rates computed in a similar fashion for the two "dry-island" trees, Douglas fir and shore pine, are twice as high as for spruce-cedar-hemlock, namely, 22.2% on the smallest islands and 6.7% on those somewhat larger. Somewhat anomalously, while showing a preference for warm, dry slopes, the fir and pine do not persist as well as other forest trees in the sparse and shallow cracks of very small islands, though they persist well enough on larger islands with appropriate habitat.

Ideally, assembly rules should be stated as algorithms that use abiotic and biotic factors to predict accurately on which islands a given species, or assemblage of species, will occur. With the forest species discussed in this section, the "rules" are rather qualitative and far from the ideal. Area dominates species richness in forest plants, and increasing island area with deeper soils generates taller woody vegetation and the shaded, temperature- and humidity-ameliorated conditions conducive to additional subcanopy and forest-floor plants. Species are nested, with species falling out sequentially as island area is reduced, but "assemble all forest species on the larger islands" is a rather weak rule. No interspecific effects were discerned even on small islands with limited accommodations (Table 5.2), other than a shared preference for certain islands and avoidance of others as the dominant trees track in

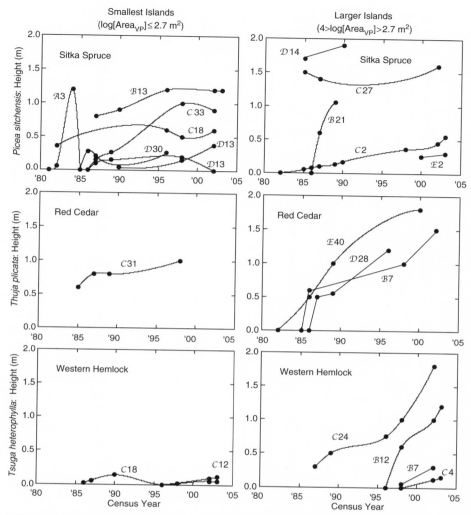

FIGURE 5.6. Measured growth rates of three common trees, spruce, cedar, and hemlock, recorded as new or recent colonists on small islands in Barkley Sound. The island of occurrence is noted on the graphs. The islands are divided into two size classes, smallest (left-hand column) and larger (right-hand column). Note that growth rates are generally higher on larger islands, but less obviously so for the spruce.

unison similarly suitable conditions for survival. Consideration of midsized islands with a limited range in area did aid in distinguishing other island features, such as elevation and sheltered versus exposed position, leading to a classification of yew islands in contrast to pine-fir islands based on these features. This constitutes a somewhat stronger rule for distinguishing among types of midsized islands: "look for yews on larger, taller, and more sheltered islands, firs and pines on somewhat smaller, lower, and more exposed islands."

SHORELINE HABITATS

Shoreline plants are quite different in most ways from forest plants; they are generally small, often tiny and inconspicuous, herbaceous rather than woody, and relatively short-lived (some are annual species). Their preferred milieu on the island shores affords ready access to the surrounding ocean and to dispersal opportunities via marine flotation ("hydrochory") among suitable living sites. This is quite different from the inhospitable barrier

NESTEDNESS AND ASSEMBLY RULES   85

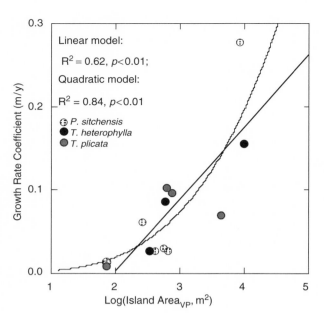

FIGURE 5.7. Growth rates (GR) on small islands of the three common trees, spruce, cedar and hemlock, are estimated by linear regression from data in Fig. 5.5. The trees grow faster on larger islands (abscissa: vascular plant area A_{VP}). The quadratic model (GR = $0.0017A^{1.44}$) gives an improved fit over the linear model (GR = $-0.17 + 0.087$ A), but both are shown here.

the ocean constitutes for the dispersing propagules of forest species; the ocean isolates the forest patches but, in contrast, connects the shoreline habitats.

The roster of shoreline species is given in Fig. 5.8, and at 44 species it is similar in length to that of the forest species. Nearly all of the annual species in the island flora are included here, and amount to 11/44 species (25%). A few of these species will be recognized as cosmopolitan weeds (e.g., *Stachys mexicana*, *Poa annua*, *Chenopodium album*, *Matricaria matricaroides*), some of these are aliens as also are others of more specialized shoreline habits (e.g., *Cakile* spp., *Sagina crassicaulis*). However, most are native plants specialized to shoreline environments (i.e., not found elsewhere), and the more broadly distributed and weedier species are included in the list because I find them only in shoreline habitats on the Barkley Sound islands. Taxonomically, they are a mixed bag, with nine grass species (Poaceae), four each in Brassicaceae (mustards) and Caryophyllaceae (pinks, chickweeds), and three species of Astercaeae (daisies) predominating. Note in the figure that several genera have different species representative of different shoreline habitats (*Festuca*, *Potentilla*, *Plantago*, *Puccinellia*, *Spergularia*).

There are other shoreline species that occur on mainland shores around Barkley Sound but have not yet been found on the islands. These include species of quiet, muddy bays and inlets around Bamfield, such as the composite *Jaumea caurina*, an owl-clover *Orthocarpus castillejoides*, and the sand-spurry *Spergularia canadensis*. Species common on broad, sandy mainland beaches are particularly depauperate on the islands, where this habitat has a particularly limited representation. Of common occurrence on Pachena and Keeha Beaches near Bamfield are the sedge *Carex macrocephala*, beach morning-glory *Convovulus soldanella*, dune tansy *Tanacetum bipinnatum*, and silver burweed *Ambrosia chamissonis*, but all are absent from the islands (though Bell and Harcombe [1973] report what was apparently a single record of the last-mentioned *Ambrosia* from Æ49: Clarke).

The shoreline plants are conveniently, and with relatively little ambiguity, divided into four groups reflecting their habitat preferences for cobbly, sandy, muddy. or rocky beaches. The last category is much the largest, and the plants in it, given the prevalence of rocky shorelines in the Sound, are the commoner and most widely distributed of the shoreline flora. Shoreline plants are further characterized by their position high

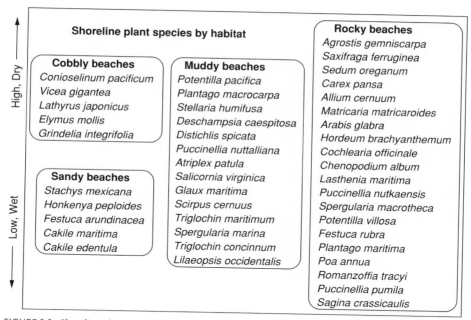

FIGURE 5.8. Shoreline plant species on Barkley Sound islands, segregated by typical shoreline habitat, and by height relative to the intertidal, from high/dry upper beach sites (top species) to low/wet lower beach sites (bottom species).

or low on the shoreline, and each is characteristically restricted to specific zones along the gradient from wet and often wave-splashed sites low on the shoreline to higher and drier sites up beyond the direct influence of the tidal cycle. In sheltered, muddy coves and estuaries, for example, the tiny umbel *Lilaeopsis occidentalis* spends a good proportion of its time immersed in seawater, as do the several species listed above it in Fig. 5.8 (e.g., *Triglochin concinum, Spergularia marina*). Indeed, another species of sheltered estuaries in the vicinity of Bamfield, the composite *Jaumea carnosa*, blooms undersea with yellow daisylike flowers, but I have not found it in any of the island coves. The lower species in each habitat category would be regularly flushed with seawater, while those at the top of the lists grow in high and dry sites. On rocky beaches, the upper four species listed are of somewhat limited occurrence and are to be found preferentially on the warmest and driest of steep, south-facing cliffs. The truly ubiquitous species of the Sound's shorelines are those rocky beach species such as the grasses *Hordeum brachyanthemum* and *Festuca rubra*,

the cinquefoil *Potentilla villosa*, and the sea plantain *Plantago maritima*, which occur on almost every island with vascular plants.

Like the forest plants, the shoreline species are significantly nested when islands are ranked by numbers of shoreline species and species by number of islands occupied (Wilcoxon rank-sum test; Cody 1999). Thus, species-poor islands generally support a subset of the shoreline species of species-rich islands. The data, however, are not as tidy as those for forest species, because there are a number of shoreline species that appear to follow a refuging lifestyle and occur sporadically on small and often isolated islets, while being absent from larger islands. These "waifs" or "tramps" include species such as *Cochlearia officinale, Matricaria matricaroides, Lasthenia minor* (ex. *Baeria maritima*), and *Puccinellia nuttalliana*. Clearly, island area is much less of a dominant factor for shoreline than for forest species, and even small islands can provide plenty of quite adequate coastline to support these small, herbaceous plants.

In Fig. 5.9, 44 shoreline species and 213 islands are ranked with island-rich species at the

	E76	E77	C36	A24	E44	A14	A2	E49	A1	C6	B12	D11	E54	E42	E68	A3	D36	C3	A7	A8	E58	D12	E45	A9
Potentilla villosa	●	●	●	●	●	●	●	●	●	●	●	●	●	●	●	●	●	●	●	●	●	●	●	●
Plantago maritima	●	●	●	●	●	●	●	●	●	●	●	●	●	●	●	●	●	●	●	●	●	●	●	●
Festuca rubra	●	●	●	●	●	●	●	●	●	●	●	●	●	●	●	●	●	●	●	●	●	●	●	●
Hordeum brachyanthemum	●	●	●	●	●	●	●	●	●	●	●	●	●	●	●	●	●	●	●	●	●	●	●	●
Sagina crassicaulis	●	●	●	●	●	●	●	●	●	●	●	●	·	●	·	●	●	●	●	·	●	●	●	●
Allium cernuum	●	●	●	●	·	●	●	●	●	·	●		●	●	●	●	●	●	●	●	●			·
Conioselinum pacificum	●	●	●	●	●	●	●	●	●	●	●	●	●	●	●	●	●	●	●	●	●	●		
Puccinellia pumila	●	●	●	●	●	●	●	●	●	●	●	●	●	●	●	●	●	●	●	●	●	●	●	
Elymus mol	●	●	●	●	●	●	●	●	●	●	●	●	●	●	●	●	●		●	●	●	·		●
Vicea gigantea	●	●	●	●	●	●	●	●	●			●	●	●	●	●	●		●	●	●		·	
Arabis glabra	●											·					●							
Salicornia virginica	●	●		●	●		●		●		●	●	●		·		●		●				●	
Honkenya peploides			●		●	·	●		·	●	·	·	●	·	●				●	●	●	·		
Deschampsia caespitosa	●	●	●	●	●	●		●	·	·	·	●	●	·	●	·	●		·	●				
Lathyrus japonicus	●	●	●		●		●		·		·	·	●	●	●		·		●	·				
Atriplex patula	●	●	●		●	●			·				●	·			·			·		·		
Poa annua		●			·		·	●		·			·						·		·			
Carex pansa	●		·	●				·					●		·				·		●			
Triglochin maritimum	●	●		●	·		●	·	●		·	·	·		·	·	●		·	●			·	
Sedum spathulifolium			●			●			·	·		●	·	●	●	·								●
Potentilla pacifica	●	●			●	·	●	●	●	·	·		●	·		●	·			·				
Romanzoffia tracyi				·	·			·															·	
Glaux maritima	●	●	●		·	●	●	·	●			●			·		●		·	●	●			
Distichlis spicata	●	●	·	●	●	●				●	●		·					●		·	●			
Puccinellia nuttalliana	●			●		●		·						●		·	●						·	
Grindelia integrifolia				●	·		●		●				·			●	·		·				●	
Triglochin concinnum					·	●					·	·		·		●	·		·		●			
Plantago macrocarpa		●			·	●	●		●								·		·	·			●	
Spergularia marina	●	●			·	·	●								·		·				●			
Cakile edentula				●				·		●		·			●									
Cochlearia oficinale					●			●																
Scirpus cernuus	●	●	·		·	·				·	·												·	
Spergularia macrotheca																								
Matricaria matricaroides																								
Chenopodium album					●		·	●		·														
Stellaria humifusa		●	·		●																			
Saxifraga ferruginea			●						●															
Agrostis gemniscarpa			●						●															
Baeria maritima																								
Stachys mexicana					●			●																
Puccinellia nutkaensis																								
Lilaeopsis occidentalis	●																							
Cakile maritima					·																			
Festuca arundinacea																								
	24.0	24.0	22.0	22.0	21.0	20.6	20.2	20.0	18.6	16.4	16.2	16.2	16.0	14.8	14.8	14.4	14.0	13.9	13.3	13.2	13.0	12.8	12.3	11.7

FIGURE 5.9. Roster of shoreline species. This species-by-sites matrix depicts the distributions of 44 species of shoreline plants over 213 islands, with species (rows) ranked top to bottom from most to fewest island occurrences, and islands (columns) ranked left to right from most to fewest records of the shoreline species. Island incidence over time is reflected by dot size within the matrix, with the three dot sizes, small to large, representing presence J_T in censuses <one third, between one third and two thirds, and >two thirds of the time. Note that the five commonest species have occurred on >80% of the islands over 75% of the time; the "Count" summary figure in the right-hand column of the last panel reflects both spatial and temporal incidence. Thus a handful of species are widely-distributed and occur on nearly all islands down to those of the very smallest sizes, whereas 90% of the listed shoreline species are much more sparsely, and patchily, distributed. Note that only the very large islands (e.g. those in the first panel of the matrix) support more than a quarter of the 44 shoreline species. The diversity of shoreline habitats (i.e. rocky, cobbly, gravelly, muddy or sandy substrates) seems limited on all but the largest islands, and there are likely also dispersal limitations in these plants that reduce island incidences (see text).

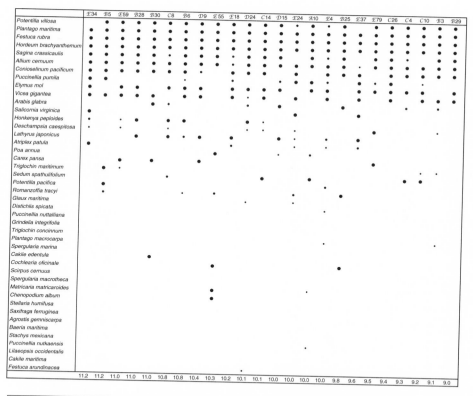
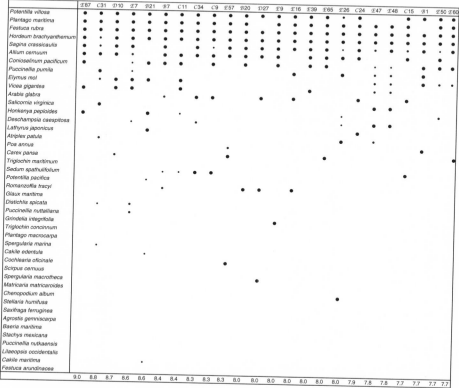

FIGURE 5.9. Roster of shoreline species (*continued*).

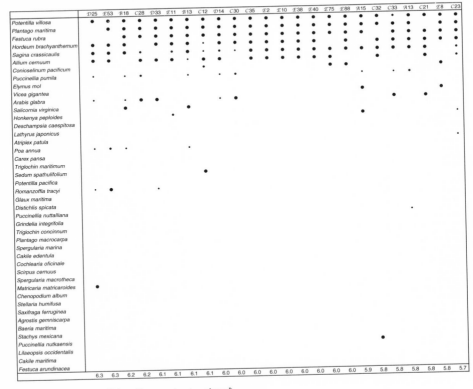

FIGURE 5.9. Roster of shoreline species (*continued*).

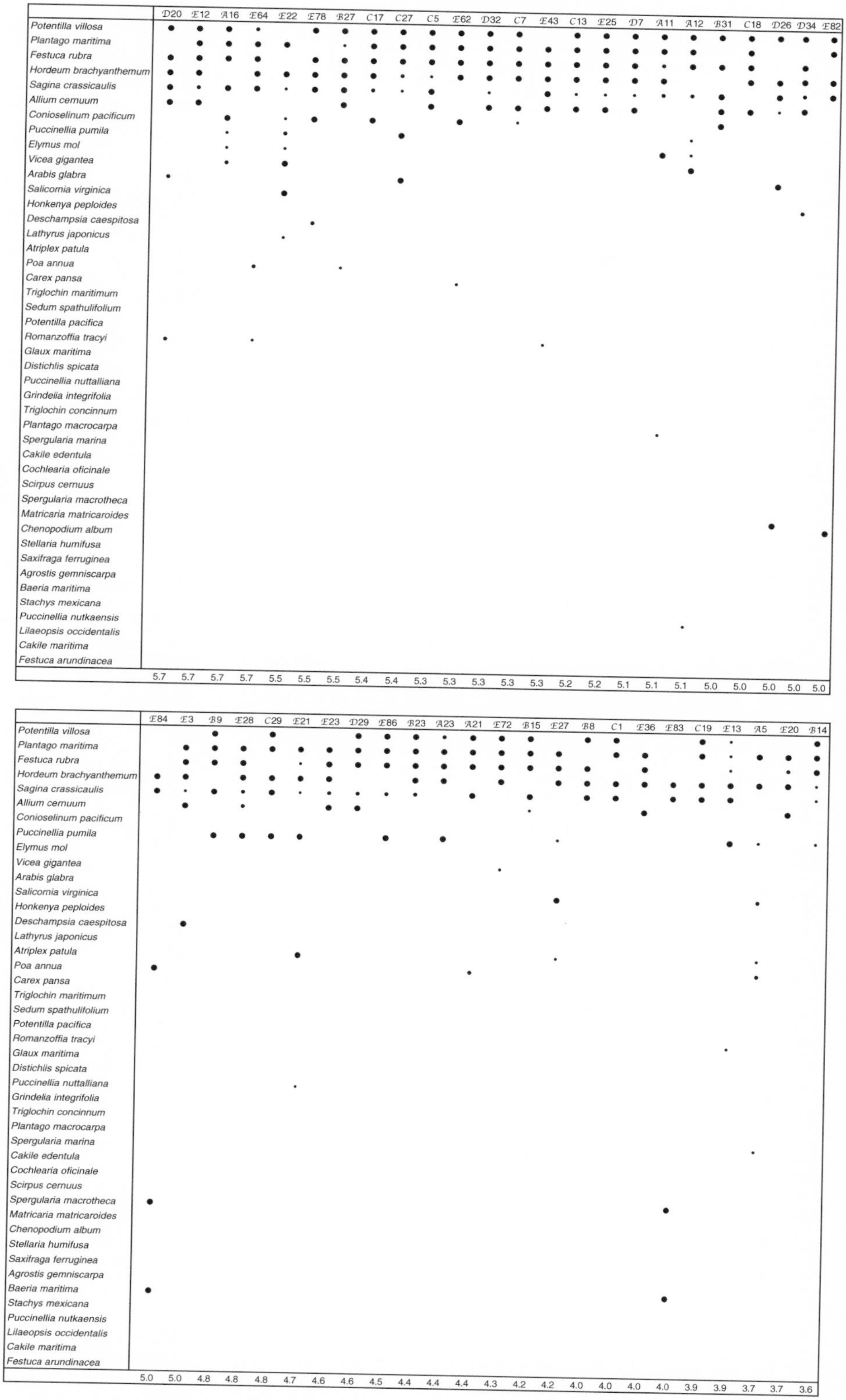

FIGURE 5.9. Roster of shoreline species (*continued*).

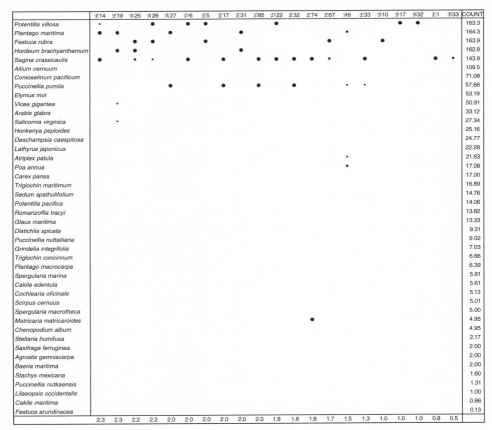

FIGURE 5.9. Roster of shoreline species (*continued*).

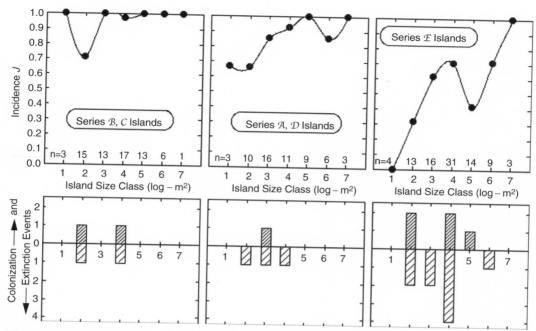

FIGURE 5.10. Incidence J of a shoreline obligate, fuzzy cinquefoil, *Potentilla villosa* (upper), on islands in B, C series (left), A, D series (center), and E series (right), as a function of island size (abscissa), showing progressively lower incidence in more sheltered islands. By chi-squared tests, series-E islands differ in incidence from both B–C and A–D islands. Sample sizes of island numbers are shown. Below are the recorded numbers of colonization (upper bars) and extinction (lower bars) events on these islands over the 22-year study.

top and species-rich islands to the left. Species presence is recorded with dot symbols in three sizes, with large dots indicating species present in ≥2/3 of censuses, small dots meaning present in ≤1/3 of the censuses, and medium-sized dots representing presence in the median tercile of temporal incidence. The top five, most widely distributed species occurred at some time or another on >80% of the islands (i.e., the proportion of entries nonblank), and on these islands occurred >75% of the time (i.e., for *Potentilla villosa*, row J_T values sum to 163.3, and 163.3/213 = 77%). At the other extreme, eight shoreline species occurred on average (over time) on two or fewer islands. and one, the sandy beach grass *Festuca arundinacea*, was found on just one occasion on one island ($D24$: "Big Stud"). Conversely, the most species-rich islands, $E76$ Turtle and $E77$ Willis, supported an average of 24 shoreline species at each census, while the lowest score, for $B33$: "Seppings Intermed. Rk," had a single species, *Sagina crassicaulis*, in one of two census years.

The top five species in Fig. 5.9, which occur on ≥2/3 of all islands, differ in several respects, including morphology, dynamics and longevity. Incidence functions for two of these, *Sagina crassicaulis* and *Plantago maritima*, were shown earlier in Fig. 3.7, where the rapid turnover of the former contrasted with the much slower dynamics of the latter. The incidence of a third species, *Potentilla villosa*, which occupies montane rocks and talus slopes inland as well as these rocky shorelines, is shown in Fig. 5.10. The data are subdivided by island groups; incidence is highest on series B–C islands, intermediate on A–D islands, and lowest on E islands (significantly so, by chi-squared tests). These differences seem to correlate with a preference in this plant for high, exposed rocks on the outer edge of the Sound, and it is less common on the lower, sheltered, foggy. and often shaded coastlines of the E-group islands (on

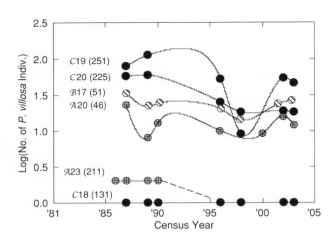

Figure 5.11. Population size trends in *Potentilla villosa* on some small islets in Barkley Sound. The species was present (but uncounted) on B17 since 1985, on C18–20 and A20 since 1982. The 1998 dip in numbers follows the record wet year of 1997. On A23 the population, size 2, went extinct between 1990 and 1996. The numbers following island codes are island areas (AREA$_{FL}$) in m^2.

several of which, even some of the larger islands, it has remained unrecorded). Lack of suitable habitat, rather than dispersal limitations, is indicated by the colonization and extinction numbers. In general, the species has slow turnover dynamics, with a total of 8 colonization and 14 extinction events recorded over the course of the study. Most of these events occurred in the Æ-group islands, where apparently persistence of *P. villosa* is more tenuous. However, the species appears to be long-lived and persistent over time even at low population sizes. Numbers of individuals were counted in successive censuses on several small islets, and these data are shown in Fig. 5.11. One islet maintained a "population" of a single individual which persisted for 20 years (C18), verifiably the very same individual at least from 1987 on! On A23, two individuals persisted for 5 years before they both disappeared. Note the general similarities in population trend data in the figure, especially between A20 and B17 (3.76 km apart across the Trevor Channel). These and other *P. villosa* populations took a dip after the record wet year of 1997 (>4 m annual precipitation), reinforcing the notion that the species prefers sunny, dry rocks over shady, damp sites.

Pearlwort, *Sagina crassicaulis*, is similarly widespread, but in contrast to *P. villosa*, it is, effectively at least, an annual species, with very high turnover and poor persistence. It has the distinction of being recorded from more islands (182; 85%) than any other species, though in an average year in occurs on a subset of just 68% of them. Of the 95 islands on which ≥7 censuses were conducted 1981–2003, the species has maintained a remarkably regular incidence over time, generally being present between 3/5 and 4/5 of the time (Fig. 5.12). Over the course of the study, 59 colonization events and 58 extinction events were recorded on this island set. These events occurred at higher rates on islands with more neighbors within a kilometer, as both colonization and extinction events are significantly correlated with ANG$_{KM}$ ($r = 0.162, 0.179$, respectively; $p < 0.05$). These turnovers in time occurred on about one-half of the islands (45/95; 47%), with 44 of the islands (46%) retaining apparently persistent populations and the remaining 6 islands perennially lacking the species. On most islands, *Sagina* populations are represented by a few scattered individuals, and on islands with persistent populations the species was noted to shift around the shorelines over time as some individuals died off and others became established.

Sagina crassicaulis fits well the classical perception of a tramp or refuging species, and it is a species that satisfies most closely the prerequisites of a metapopulation in a stochastic steady state. It occupies (a) a patchy habitat, with (b) a substantial risk of yearly extinction,

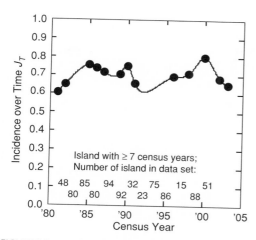

FIGURE 5.12. Pearlwort (*Sagina crassicaulis*) is a common shoreline plant that occurs on more islands than any other. On islands censused in ≥7 different years, incidence over time averages between 60% and 80%, as this annual species blinks in and out with high colonization and extinction rates.

but (c) good recolonization potential, and (d) has no really dominant source or "mainland" population (Hanski 1998). Estimates of true colonization and extinction rates, rather than simply the minimum number of events recorded, can be derived from censuses in adjacent years. Denoting colonization and extinction rates as c and e, and presence as 1, absence as 0, the transitions between adjacent years 1–1, 1–0, 0–1, and 0–0 will occur with frequencies $(1 - e) + ec$, $e(1 - c)$, c, and $(1 - c)$. Note that a rescue effect is built into these calculations, as a population perceived as persisting on an island (1–1) might actually persist $(1 - e)$ or might go extinct but then recolonize or get rescued (ec). Likewise, a perceived extinction (1–0) requires both extinction (e) and lack of recolonization $(1 - c)$. Estimates of true colonization and extinction rates are shown in Table 5.3, where 121 islands (third column) over seven size classes (first column) contribute transition data (columns 4–7) from censuses in adjacent years. No colonization events were observed in three island size classes (1, 6, and 7), and therefore no calculations for c, e are possible. But for the remainder, c and e are computed as indicated above (columns 8, 9), and the predicted incidence J of the species on islands of these size classes 2–5 is derived. Islands lacking the species will be colonized at a rate c, producing $(1 - J)c$ new island records (between adjacent years), a number that, in steady state, will balance the number of islands that are newly empty: $Je(1 - c)$. Thus $(1 - J)c = Je(1 - c)$ and $J = c/(c + e - ec)$; the incidence values computed from colonization and extinction data (Table 5.3, column 10) are not very different from the observed values.

There will presumably be abundant (but redundant) "colonization" of already occupied islands, and this is expected to be commonplace, especially on the smaller sample sizes of larger islands, although it goes unrecorded. Such unneeded and ineffectual "rescue" is especially hard to detect. If we make conservative guesses that the real colonization rates of size 1, 6, and 7 islands are 0.05, 0.3, 0.3, respectively, the predicted incidences for islands of these sizes would be 0.355, 0.863, and 1.0, similar to observed values. These values are added to Table 5.3 in parentheses. Multiplying the numbers of islands in each size class (column 2) by predicted incidences J (column 10) gives the average proportion of occupied islands in the set, a value of 0.69. Across years, then, 69% of the island set is expected to be occupied by *Sagina*, based on computed c, e values (and reasonable extrapolations from them); this is agreeably close to the average value of J_T in Fig. 5.12, namely, 70% ($\Sigma_i n_i J_{Ti}/\Sigma_i n_i$ over size classes i).

Given that *Sagina* incidence flutters around mean values and populations are winking on and off continuously and rapidly in both space and time, there might be neighborhood effects in the presence/absence patterns, and perhaps local synchronicity in colonization and extinction. Clearly synchronicity can be detected only between islands that share many of the same census years and moreover are not either permanently occupied or permanently empty. There are 27 islands which share ≥7 census years with *Sagina* incidence $0 < J_T < 1$. Islands A_2 and A_3, for example, were both censused in

TABLE 5.3
Colonization-Extinction Data for Pearlwort (Sagina crassicaulis) from Island Censuses in Adjacent Years

ISLAND SIZE	NO. ISLANDS TOTAL	CENSUSED IN ADJACENT YEARS	TRANSITIONS TO PRESENCE (=1) FROM ABSENCE (=0), AND VICE VERSA; OR NO TRANSITION				COLONIZATION RATE c*	EXTINCTION RATE e*	INCIDENCE J**	
			$1 \to 1$	$1 \to 0$	$0 \to 1$	$0 \to 0$			PREDICTED	OBSERVED
1	10	5	10	1	0	8	0	0.091	0 (0.355)	0.303
2	38	22	13	3	3	35	0.086	0.205	0.315	0.424
3	45	26	53	4	5	8	0.385	0.114	0.846	0.692
4	59	39	50	9	10	23	0.303	0.219	0.665	0.641
5	36	16	37	1	1	6	0.143	0.031	0.845	0.809
6	21	11	20	1	0	0	0	0	0 (0.863)	0.858
7	7	2	4	0	0	0	0	0	0 (1)	1

*Colonization rate c is derived from the proportion of empty islands that gained species: $(0 \to 1)/[(0 \to 1) + (0 \to 0)]$, and extinction rate e from the proportion of occupied islands that lost species: $e(1-c) = (1 \to 0)/[(1 \to 1)+(1 \to 0)]$.
**From the calculated values of c and e, the expected incidence is computed as $J = c/(c + e - ec)$, and compared to observed values.

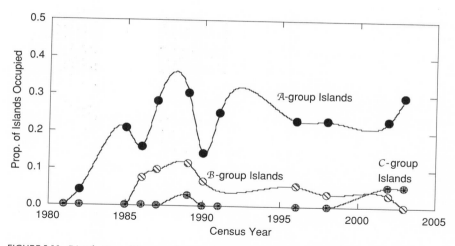

FIGURE 5.13. Distribution of the annual grass *Poa annua* over time. First appeared in A group 1982, found on ca. 25% of A islands after a decade, reached B group islands by 1986 but subsequently died out, and later (and more sporadic) occurrence in C group islands.

years 1981, 1982, 1985, 1986, 1987, 1989, 1996, 1998, 2002, and 2003. *Sagina* records for these years from A2 are 0-0-1-0-1-1-1-1-1-1 ($J_{T1} = 0.70$), and for A3 are 1-0-1-0-1-0-1-1-0-1 ($J_{T2} = 0.60$). Synchronicity was measured by the coincidences of the 1's and 0's; here there are six matches and four mismatches. The expected number of matches, if 1's and 0's are independent, is $10[J_{T1} * J_{T2} + (1 - J_{T1})(1 - J_{T2})]$, and the expected number of mismatches is 10 minus this quantity. Thus expected numbers are 5.4, 4.6, observed numbers are 7, 3, and a chi-squared measure of the disparity returns 1.03 (no significant synchronicity). I used positive chi-squared values where expected matches were greater than expected, negative values where expected matches were less than expected, and regressed the chi-squared values against the distance apart of the islands in the comparison. Here distance varies from a few meters to nearly 15 km, and overall the chi-squared values average 0.28 ± 1.47 SD, not significantly different from zero ($p > 0.05$). The regression tests the hypothesis that the relative synchronicity of censuses is a function of the island pairs' proximity, but although the coefficient of distance is negative, indicating lower chi-squared values between more isolated islands, the hypothesis is not confirmed ($p = 0.17$). It appears that even closely adjacent islands can be quite asynchronous in their incidences of *Sagina*, and that no local hot spots build up and attain temporary prominence within its density or distribution patterns.

The annual grass *Poa annua* is another shoreline species with a refuging island lifestyle (n.b., the species has a near worldwide distribution and is a well-known weed of damp and disturbed mainland habitats). This species was not found on any islands in 1981 but showed up the following year on a single tiny island, A6: "Hosie Rks d." In the next decade it reached 10 A-group islands, including 4 others nearby in the Hosie group and the 3 San Jose islands 1.8 km distant. It became extinct on some earlier islands while colonizing others and eventually seemed to stabilize at around 25% island occupancy (Fig. 5.13). I first recorded the species in the B-group islands in 1985, whence it spread to 10% of this group over a five-year period; but it never became widespread there and had petered out by 2003. Its arrival in the C-group islands was later still (1989), but it has remained sparse and sporadic there (Fig. 5.13). The species must have moved through the Barkley Sound islands in wavelike colonization episodes many times in the past, ebbing and flowing but nowhere gaining a firm foothold. The species was unrecorded by Bell and Harcombe (1993) in the Broken Group, but it is now found on *E*49: Clarke and a half-dozen of Clarke's near neighbors. Perhaps significantly, Clarke is the most frequently visited of the Broken Group islands by humans, and a popular camping locale.

Three other grasses of rocky shorelines, all perennials, are widely distributed on the islands and rank near the top in Fig. 5.9: red fescue *Festuca rubra*, meadow barley *Hordeum brachyanthemum*, and dwarf alkaligrass *Puccinellia pumila*. All three occur on both coasts of the continent, the fescue also in Europe, and the barley lives in montane habitats inland as well as in maritime sites. The first two are near ubiquitous on the drier rocks of the upper shoreline, while the last prefers lower sites near the tide line. However, on several small islets that lack *Festuca* and *Hordeum*, *Puccinellia* grows vigorously on high and dry rocks several meters above the tide line. All of these grasses are similar in size or biomass, and the two upper shoreline species are very similar in their erect growth forms, though young or nonflowering individuals can easily be told apart by the leaves—broad in the *Hordeum*, narrow and revolute in the *Festuca*. *Puccinellia pumila* is much more prostrate in growth form than either of these former two, though it reaches similar sizes. Most middle-sized and large islands support all three species (Fig. 5.9). If there are interactions amongst them over space, for example, with one or a pair of species excluding another, they are more likely to be manifest on the smallest islands, where space is more limited. However, even tiny islands of a few dozens of square meters of suitable habitat can sustain populations of hundreds of these grasses in their cracks and soil pockets.

To test for such interactions, I looked at the set of islands comprised of sizes 1 and 2 ($AREA_{FL} < 10^{2.5}$, or < 316 m^2). There are 48 islands in this combined category, and 10 of them have no grasses at all. Many of the remainder with grasses, however, are radically different one from another in grass species composition and density. Some have retained stable grass populations over the longer term, while others have undergone considerable changes. Several support an approximately 50:50 mix of *Festuca* and *Hordeum* (*A*23: "Nanette"; *C*18: "Ross E Adj S Rk"; *C*21: "Ross Mid Rks E"; *D*33: "Big Stud SW Rk"). Others are dominated by a single, common species but have been invaded in the last two decades, with varying success, by one or both of the other species. Some empty rocks were invaded within the study period too, and on some the new colonist expanded rapidly in population size, on others persisted but at very low numbers, and on yet other islands the colonists petered out. For example, on *C*22: "Ross Mid Rks W," *Hordeum* invaded in 1996 and had accrued >100 individuals by 2003; on *A*6: "Hosie Rks d," *Puccinellia* invaded (1 individual) in 2002 and persisted in 2003; on *C*20: "Ross E Tall Rk," where *Hordeum* was already abundant, *Festuca* invaded in 1996 and persisted with just 3 individuals to 2003; on *E*32: "Faber NE Rks S," *Puccinellia* increased from 5 individuals in 1986 to >100 in 2000 but decreased to extinction by 2003. Clearly, some of these events take place without any contribution from other, resident grasses, due presumably to the vagaries of small-island populations with variable conditions for persistence year to year, but in other cases the presence of common and persistent neighbors may have contributed to the fate of new colonists.

I next calculate and analyze the incidences of these grasses over both space and time. There are a total of 299 censuses over the period 1981–2003 on these small islands, with a maximum of 12 per island (two islands), and about one-half of the islands (23) were censused ≥ 7 times. There is a fair amount of turnover through time of the grasses, with C/E events recorded for *Festuca*, *Hordeum*, and *Puccinellia* numbering 7/7, 5/5, and 7/8, respectively. The joint incidences (in time and space) of the three grasses were measured in *Festuca-Hordeum-Puccinellia*, or F-H-P, at 0.411, 0.508, and 0.268. There are eight possible combinations of the three grass species: FHP, FHP′, FH′P, F′HP, FH′P′, F′HP′, F′H′P, and F′H′P′, where the "prime" designation indicates absence, and its lack means presence. If the species are distributed independently on these islands, with a narrow size range and a presumption of suitable habitat for all species, we can easily calculate the expected numbers of censuses with each of the eight grass combinations and compare them with the observations. In general, there are too few islands compared to expectations with all three grasses (FHP) or no grasses (F′H′P′), too few with just *Festuca* (FH′P′), too many with the two-species combination FHP′ and with the single species F′H′P ($\chi^2 = 40.24$, df = 7, $p < 0.001$).

This test is somewhat suspect, however, as the 299 censuses are not independent, and later censuses of the same island are more likely to retain similarities to earlier censuses despite the substantial degree of species turnover. Assigning the different species combinations to various of the 48 islands according to their predominance there over time (rather than using a given census as a sampling unit) is a more conservative protocol. When this is done, the various species combinations are seen to occur in frequencies not significantly different from those expected by chance, with no particularly favored combinations (chi-squared test; $p > 0.05$).

Given each island can be assigned to a particular species combination that predominates on it over time (even though, with species turnover, there is no absolute constancy in the occupant species), I investigate whether certain sorts of islands favor or disfavor certain grass combinations. The results are shown in Fig. 5.14, where the 48 islands are divided into three size categories, small, medium, large (based on $AREA_{1m}$ [area above the 1 m contour line] <40 m², 40–100 m², and >100 m²), and three elevation categories, low, moderate, tall (ELEV <2 m, 2–3.5 m, >3.5 m). All except 5 of the 48 islands fall into just five of the nine possible subdivisions, and for tidiness these 5 islands are assigned to one or another of five well-populated subdivisions closest to them. Six of the eight grass combinations are reasonably common, and these fall within island categories as indicated in Fig. 5.14. No tall or large islands are empty; *Festuca* is common on tall islands and coexists with *Hordeum* on both large-tall and "med-mod" (medium or intermediate area, moderate or intermediate elevation) islands. *Puccinellia* is common on small, low islands, some of which it shares with *Hordeum*; it occurs also on large islands of moderate height but tends to avoid medium-sized islands of moderate height that are dominated either by *Hordeum* alone or by *Hordeum* plus *Festuca*. *Hordeum* has the largest apparent niche, being represented on all five island types alone or with either *Festuca* or *Puccinellia*. Although there are no hard and fast boundaries between the different combinations, and the frequent colonization events attest to the broad range of conditions deemed suitable by each species, nevertheless there are some useful assembly rules represented in the figure.

Although neighboring islands were not especially similar, within years, in their incidence of *Sagina*, there do seem to be spatial consistencies over islands in the censuses of shoreline species in general. In Fig. 5.15 I show species numbers in three groups of islands in the $\mathcal{A}$ group: the North Hosie islands $\mathcal{A}$1–6, the San Jose islands $\mathcal{A}$8–10, and the Brady Beach islands $\mathcal{A}$15–22. The first mentioned are particularly rich in shoreline species, and the four larger islands lie well above the regression line describing the overall relation between numbers of shoreline species and island area ($AREA_{FL}$; Fig. 5.15, upper). Their shoreline species are closely nested, and species rank high there that are low ranked or absent on other islands. The San Jose islands are modestly species rich in shoreline species, and all lie with similar displacements above the regression. These islands also show nested shoreline floras, although they support the small-island tramps *Romanzoffia*, *Matricaria*, and *Puccinellia nutkaensis* that confound the nestedness picture at the bottom of the chart (Fig. 5.15, center right). The Brady Beach islands, all hugging the mainland

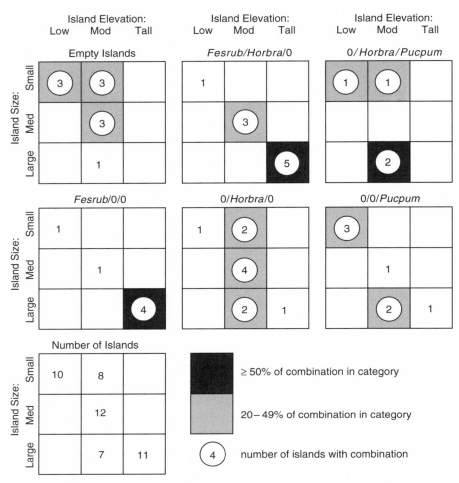

FIGURE 5.14. Distribution of various combinations of the three shoreline grasses *Festuca rubra*, *Hordeum brachyanthemum*, and *Puccinellia pumila*, over 48 small islands (of sizes 1, 2). Islands are subdivided by elevation ("low" <2 m, "med" >2, <3.5 m, and "tall" >3.5 m) and by size, using area above 1 m contour (Area 1): ("small" <40 m^2, "mod" >40, <100 m^2,"large" >100 m^2). Two uncommon combinations, Fesrub/O/Pucpum and Fesrub/ Horbra/Pucpum occur on two islands and are not shown. The common associations of island type and grass combination are shaded. See text for discussion.

shore south of Bamfield, are quite average in species richness (Fig. 5.15, lower left). Also nested, this shoreline florula shows a high ranking of *Sagina crassicaulis*, much lower of *Hordeum brachyanthemum*, and *Arabis glabra* appears here anew. Thus there are differences among the island groups in species composition and in the relative rankings of the species they share, but there are similarities within island groups in relative species richness, and a shared degree of nestedness that is in general strongly related to island area.

EDGE HABITATS

Edge habitats support by far the largest proportion of the flora of the Barkley Sound islands. In some respects this is a catch-all category, as I assign to it those species that are neither forest nor shoreline obligates. The list from edge habitats is comprised of some 161 plant species, including a few taller (*Alnus rubra*, *Pinus contorta*) and shorter (*Prunus emarginata*, *Pyrus malus*, *Rhamnus purshiana*) trees, many shrubs (*Arctostaphylos*, *Gaultheria*, *Vaccinium*, *Rubus*, *Rosa*,

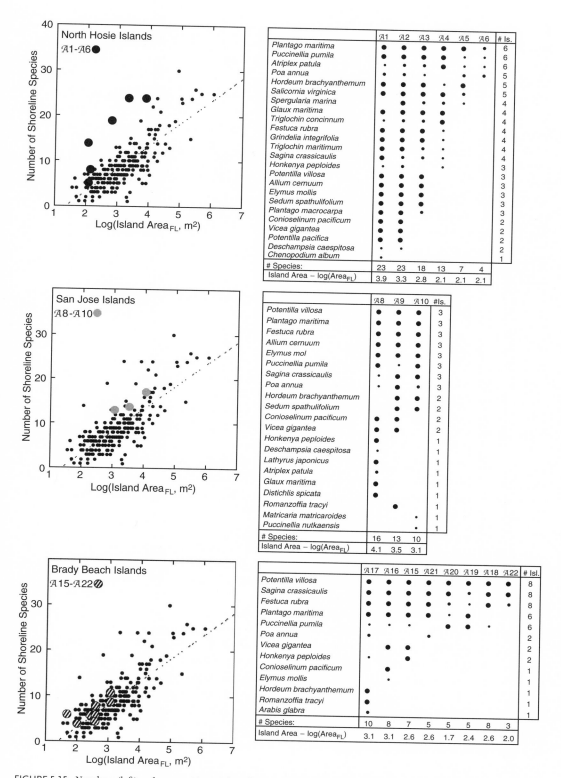

FIGURE 5.15. Numbers (left) and composition (right) of shoreline species on three different islands groups (top to bottom), showing similarities within groups of relative species richness (left) and species composition (right), but differences in these aspects between island groups. Large symbols at left show species numbers in the islands groups relative to those of all islands (small symbols).

Ribes, Cornus, Sambucus, Lonicera, Physocarpus, Holodiscus, Salix, Symphoricarpos, Crategus), and even more forbs, among which the genera *Achillea, Angelica, Aster, Castilleja, Epilobium, Fragaria, Galium, Hypochaeris, Mimulus,* and *Senecio* figure prominently. The roster also includes 18 genera of grasses, in terms of species numbers the largest familial representation, as well as 5 *Carex* and 3 *Juncus* species. Most of these edge species are widely distributed and can be seen along forest edges, river banks, and beside anthropogenic gaps and structures in disturbed sites on the mainland in general. There are a number of common aliens among the islands' edge flora, though most are native to the Barkley Sound region, and the vast majority of the forbs are perennials. Note that comparisons across species of incidence and colonization/extinction dynamics in this island system are informative despite the different origins of the taxa and the various lengths of time they have been part of the system; this seems especially tenable among plants of similar growth forms and with similar functional roles in the vegetation.

A matrix of ranked species (161) by incidence over ranked islands (217) by species numbers is both cumbersome and unattractive, and it will suffice to say that the edge species are not nested by the usual criteria. I have discussed previously (Cody 1999) the unpredictability of edge species composition among islands; variation in species composition from island to island is especially characteristic of the edge flora, and this is apparent even between neighboring islands of comparable sizes. The islands C1–C6 are neighbors of diverse areas ($AREA_{FL}$ = 1140–13,300 m^2) that collectively support 69 edge species, 49 of which have occurred on the largest island, C6. These six islands are ranked from fewest to most edge species, left to right across the abscissa of Fig. 5.16. The cumulative totals of edge species (i.e., cumulative over time, per island) are shown in the upper curve of Fig. 5.16 (upper); the total incidence of edge species ($\Sigma_j J_{Ti}$ over number of species i) returns lower numbers (lower curve of the same figure), because of the part-time incidence of some species. Each island shift, stepping from richer to poorer islands right to left, entails a reduced total species incidence because (a) species are lost right to left, while (b) some species are retained, and (c) new species are gained. These are the lower, center, and upper bars of the histograms in the lower part of Fig. 5.16. Note that "new" species are gained on smaller islands, namely, species that were absent on the larger islands, which constitute up to one-third of the total species' incidence on the poorer island. With perfect nestedness, there would be no species gain right to left, but this is not the case with edge species, which frequently occur on smaller islands even though they are absent on their larger neighbors.

Some of the most ubiquitous edge species are shrubs, and there are seven very common species with overall incidence on 217 islands of >0.4. These are *Lonicera involucrata, Gaultheria shallon, Vaccinium ovatum, Rubus spectabilis, Amelanchier alnifolia, Ribes divaricata,* and *Rosa nutkana*. They occur on between 99 (*Rosa nutkana*) and 168 (*Rubus spectabilis*) islands, and their incidences over time, on all islands, are, respectively, 0.687. 0.673, 0.652, 0.612, 0.457, 0.42, and 0.423. The incidence of each species over island size classes in shown in Fig. 5.17, where the mean overall number of these shrubs is shown as a smoothly declining function of island size (in the right-hand graph). The largest islands average between six and seven species, the smallest two or fewer.

These shrubs are all relatively similar in growth form and biomass, with larger individuals generally 1–2 m in height. Further, all are berry producers, with fruits variously purple to red and around a centimeter in diameter, and very likely rely on birds for dispersal of the seeds within the fruits (see chapter 7). Therefore, we might look to these shrubs, with their rather similar ecologies, for evidence of interspecific interactions, positive or negative, that affect their incidences. Selecting $SIZE = 4$ islands ($n = 59$), the incidences of the seven

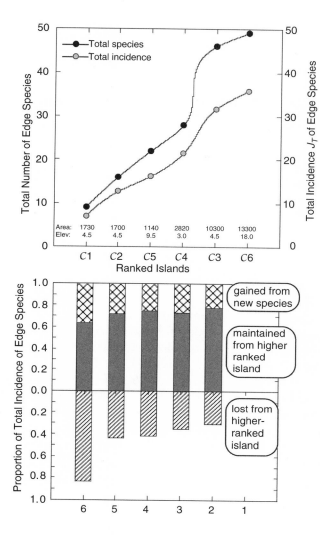

FIGURE 5.16. Turnover of edge species among neighboring islands $C1$–$C6$. Islands are ranked (above) from fewest to most species and lowest to highest total incidence, generally corresponding to island area and elevation. Some species are lost in the transitions from larger to smaller islands (lower), but new edge species are gained in the transitions, and these constitute around one-third of the smallest islands' total edge species (lower left).

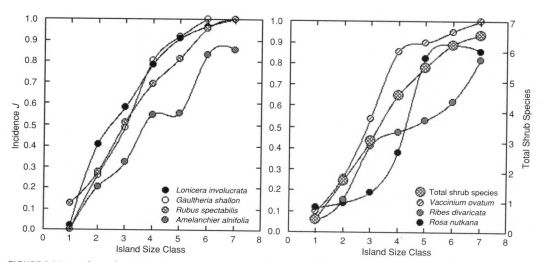

FIGURE 5.17. Incidence functions of the seven common edge shrubs with joint incidence over all islands >0.4; several have very similar incidence functions. The sum of the incidences, labeled "total shrub species," ascends smoothly from 0.41 to 6.53 with increasing island size class (right, larger symbols).

TABLE 5.4
Observed versus Expected Shrub Coexistence on Size 4 Islands

NO. SPECIES	NO. ISLANDS		CHI SQUARED*
	OBSERVED	EXPECTED	
6 or 7	22	14.01	4.56
5	12	18.63	2.36
4	10	17.23	3.03
3	8	6.55	0.32
0–2	7	2.58	7.57
Total	59	59	17.84

*$p \approx 0.001$ that observed and expected numbers match.

species (ordered as above) are 0.783, 0.856, 0.802, 0.692, 0.548, 0.381, and 0.477, respectively. If the shrubs are distributed over islands independently, the incidences can be used as before to calculate the numbers of islands with all seven, with six, five, and so forth down to one or no shrubs. Table 5.4 compares observed to expected island numbers, and $p \approx 0.001$ that the numbers match (chi-squared test). The mismatch is caused by there being too many islands with many (6 or 7) or very few (≤ 3) shrub species, and too few islands with intermediate numbers of shrubs (4 or 5). A possible explanation for this result is that some $SIZE = 4$ islands are particularly conducive to shrubs while others are unsuitable, and that interactions among shrub species are not strong enough to regulate species numbers on the preferred islands. It is also conceivable that the presence of one or another species enhances the likelihood that others will occupy the island, perhaps via some modification of soil, shade, or temperature conditions. But note that this result is the opposite of what we might expect if there were negative interactions among shrubs, such as competition for space and its resources, that might limit or regulate the numbers of species per island. If that were happening, the island categories overrepresented in the figure would have had smaller than expected values, not larger, and the categories underrepresented would have had larger than expected values.

Each shrub species has its island incidence somewhat differently affected by factors such as island size, isolation, and elevation. This is shown in Fig. 5.18, where incidence is shown for each species as a function of these three variables. The island sample in the figure is all islands with $1 < \log(\text{AREA}_{\text{VP}}) < 4$, for an $n = 176$, a range of distance from the mainland $0 < D_{\text{MLD}} < 10$ km, and a range of elevation $0 < \log(\text{ELEV}) < 1.8$ m. Note that, while all species have higher incidence on larger islands (i.e., those with larger $\log(\text{AREA}_{\text{VP}})$ values), some shrubs are not affected by isolation distance (*Gaultheria shallon, Lonicera involucrata, Rubus spectabilis*), whereas others are strongly (*Amelanchier alnifolia*) to mildly (*Vaccinium ovatum, Rosa nutkana*) affected by this variable. The seventh species (*Ribes divaricata*) is commonest at intermediate isolation values. In some shrubs, incidence increases sharply with increased elevation on smaller islands (*Gaultheria shallon, Vaccinium ovatum, Lonicera involucrata*), whereas others are unresponsive to this variable (*Rosa nutkana*), absent only from the lowest islands (*Rubus spectabilis*), or apparently prefer low islands (*Amelanchier alnifolia, Ribes divaricata*).

One way to look for possible interspecific interactions, direct or indirect, is to chart "incidence

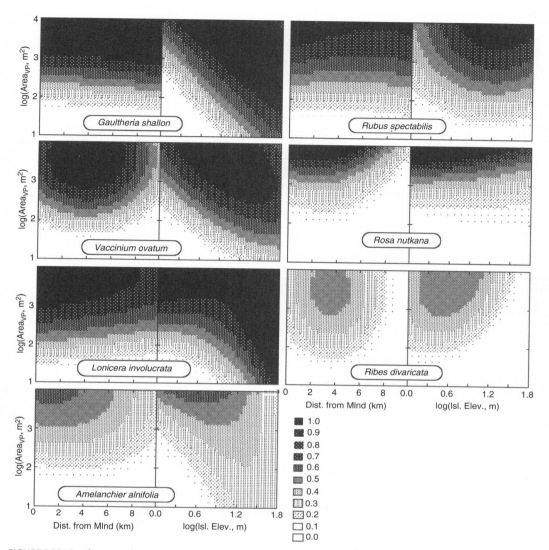

FIGURE 5.18. Incidences J of seven common shrub species on 176 small- to medium-sized islands, with respect to island AREA $_{VP}$ (ordinate), distance from mainland (km), and elevation (log -m; joint abscissas). See text for discussion.

shifts" on different island subsets with different properties and different species' prevalences. In Fig. 5.19 I select various island subsets on which each of the seven shrub species, respectively, reaches incidence $J = 1$ and plot how incidence shifts or changes among the other species on these various island subsets. The data set for this figure is that of islands with $1.8 < \log(\text{AREA}_{VP}) < 2.4$, a range chosen to give intermediate incidences for most of the shrubs and a sample size of $n = 49$. On these islands,

the incidences of the seven species (ordered as previously; subscript "A" for "all") are J_A = 0.591, 0.461, 0.495, 0.478, 0.344, 0.566, and 0.162, respectively.

The incidence shifts, with only a single exception, are to higher values on island sets where particular species score $J = 1$. That is, on those islands where species X is always present, $J(X) = 1$, incidence in the other species increases above their average: $J(Y)$, $J(Z)$, and so forth are also higher. The single exception, a

NESTEDNESS AND ASSEMBLY RULES 105

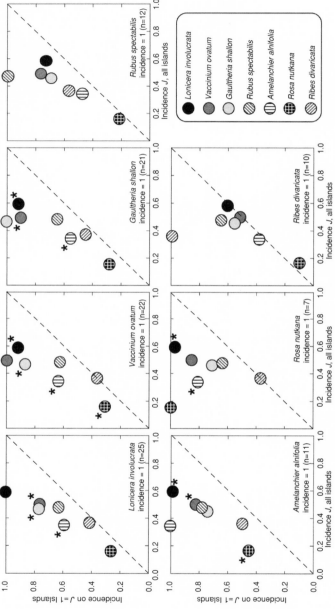

FIGURE 5.19. Seven common shrubs (see index, lower right above) have incidences on $n = 49$ islands with $1.8 < \log(AREA_{vp}) < 2.4$ as represented along the abscissa. Various subsets of these islands are selected on which each species, respectively, reaches incidence $J = 1$. The incidence shifts of all species on these island subsets (sample sizes given) are represented on the ordinate. Except for reduced incidence of *Rosa* on $J(Ribes) = 1$ islands, lower right, all incidence shifts are positive, indicating higher prevalence of nearly all species on islands particularly conducive ($J = 1$) to any one of them. Asterisks indicate statistically significant shifts of incidence. Note the relative strength and synchronicity of incidence shifts in the quartet *Lonicera-Gaultheria-Vaccinium-Amelanchier*, the reciprocity of responses in *Amelanchier-Rosa*, and the lack of significant incidence shifts of species to $J(Ribes) = 1$ and $J(Rubus) = 1$ islands. See text for further discussion.

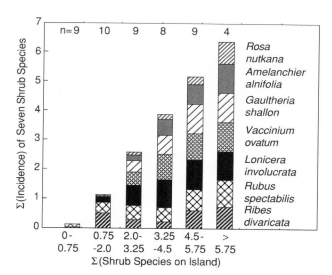

FIGURE 5.20. Composition of island shrub floras, measured by incidence (ordinate), on islands with increasing numbers of shrubs (left to right, abscissa). Note the predominance of *Rubus* and *Ribes* on species-poor islands, and the later and successive addition of *Lonicera*, *Vaccinium*, *Gaultheria*, *Amelanchier*, and *Rosa* on increasingly species-rich islands. The figure indicates the assembly rules for these shrubs on islands of increasing shrub species richness. Island sample size is 49 islands of $1.8 < \log(\text{AREA}_{VP}) < 2.4$.

slight drop of *Rosa* incidence below the 45° line on islands where $J(Ribes) = 1$, is statistically insignificant ($p > 0.05$ by chi-squared test; see Fig. 5.19, lower right). Each incidence shift can be similarly tested for statistical significance, using $J_A * N$, $(1 - J_A)N$ as expected values, $J_1 * N$, $(1 - J_1)N$ as observed values, where J_1 is the new (shifted) incidence on islands with $J = 1$ for some particular species and numbering N for the island subset. Thus, $J_A(Rosa)$ is 0.162, but on the 11 islands where $J(Amelanchier) = 1$, this is increased to $J_1(Rosa) = 0.455$. Expected numbers of islands overall with and without *Rosa* are 1.78, 9.22, and observed numbers on 11 islands that are *Amelanchier* strongholds are 5.01, 5.99, for a chi-squared value of 6.99 (df = 1, $p < 0.01$). Evaluating each incidence shift in a similar manner, there is strong reciprocity in chi-squared values between species pairs; that is, X's shift to higher incidence where $J(Y) = 1$ is very similar to Y's shift where $J(X) = 1$ ($r = 0.95$, $p < 0.001$, $n = 21$ paired values). Four species, *Vaccinium*, *Gaultheria*, *Lonicera*, and *Amelanchier*, shift incidence significantly and reciprocally (in 11/12 possible cases) on islands where each in turn has $J = 1$, and thus constitute a quartet with coordinated and similar, positive responses to islands. In contrast, high incidence in either *Rubus* or *Ribes* generates no significant incidence shifts in the other shrub species, and *Rosa* is intermediate. *Rosa* "pulls up" both *Lonicera* and *Amelanchier*, an effect reciprocated only by *Amelanchier*, and *Vaccinium* pulls up *Rosa* but without reciprocal effects.

These islands were selected because they are quite marginal for shrubs by dint of small size, and most larger islands support all or most of these shrub species (see Fig. 5.17). The buildup or assembly of shrub species richness on these small islands is shown in Fig. 5.20, in which different species have disparate patterns. Most one-species islands are *Rubus* or *Ribes* islands; thereafter, with increasing species numbers, *Rubus* incidence increases at a steady rate, peaking at $J = 0.89$, while *Ribes* incidence drops beyond two-species islands, then recovers somewhat to peak at 0.75 on the richest islands. *Lonicera*, *Vaccinium*, and *Gaultheria* are added and rise in incidence in that order with increasing richness, while *Amelanchier* and *Rosa* debut later, on islands with >2 species, with the former increasing rapidly in incidence to $J = 1$ and the later increasing slowly in incidence to $J = 0.75$. The figure is, in a sense, an assembly algorithm for the expanding edge shrub community.

There are two surprising aspects to the edge shrub distributions: there are diverse responses of shrub species to abiotic factors such as island area, isolation, and elevation, and there are essentially no detectable negative (biotic)

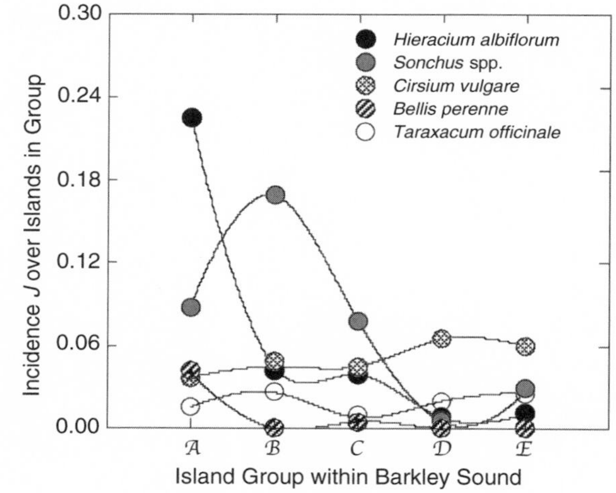

FIGURE 5.21. Incidences over island group (abscissa) of five rarer edge species in the family Asteraceae. Incidences are level across groups in *Cirsium* and *Taraxacum*, but higher in the $\mathcal{A}$ group in *Bellis* and *Hieracium*, high in the $\mathcal{B}$ group, and low in the $\mathcal{D}$, $\mathcal{E}$ groups in *Sonchus* species.

impacts of one species on another. The latter likely illustrates, more than anything else, methodological insufficiencies, in that the detection of potential negative impacts of one species on another is unrealistic with low-resolution incidence information and will require data on numbers or densities to resolve. The former abrogates overall similarities in growth form and dispersal syndromes and is presumably owing to interspecific differences in germination and survivorship related to morphological and physiological traits, which at this time remain unknown.

Numerically, the 161 edge species are dominated by perennial forbs, amongst which the family Asteraceae (or "Compositae") is prominent. While outnumbered by the grasses (Poaceae: 36 spp.), the "comps" (18 species in all; except for two rare subshrubs *Gnaphalium purpureum* and *Erophyllum lanatum*, all are herbaceous forbs), by dint of their size, visibility, and overall density, are more conspicuous. Large islands with higher disturbance levels, such as $\mathcal{E}$49: Clarke, support many weedy comps as well as other nonweedy natives, with 10 composite species in all recorded there with a combined incidence $\Sigma J = 7.32$. This total is matched on $\mathcal{A}$24: Cape Beale, with its lighthouse facility and attendant habitat modifications.

Amongst the composite forbs, many are wind dispersed, or anemochorous, and show distributional differences amongst the island groups. In Fig. 3.9, where incidence was compared between closer ($\mathcal{A}$–$\mathcal{D}$) and more distant ($\mathcal{E}$) islands, *Achillea millefolium* and *Anaphalis margaritacea*, a zoochore and an anemochore, showed no obvious effect of island isolation. Figure 5.21 shows the incidence of five species of rarer composites over the five island isolation groups, and there is a significant island-group effect in two taxa, *Hieracium albiflorum* and *Sonchus* spp. (ANOVA, $p < 0.001$ and $p = 0.026$, respectively). Here the taxon "*Sonchus* spp." represents a combination of three species *S. asper*, *S. oleraceus*, and *S. arvensis*, grouped together for the purposes of illustration, and a boosted sample size (all have similar growth form, morphology, all are quite rare, and they almost never co-occur on an island). Collectively, they are considerably more common on $\mathcal{B}$-group islands and rarest on $\mathcal{D}$- and $\mathcal{E}$-group islands, while *Hieracium* is far more common on the least-isolated $\mathcal{A}$-group islands than elsewhere. All but the common lawn daisy *Bellis perenne*, which lacks a pappus, are anemochores; the thistle *Cirsium vulgare* has a particularly large, fluffy, and near-spherical dispersal propagule with the seed centrally embedded, and it does especially well in reaching the more distant islands.

Incidence over island groups is shown for six common species of composite species in Fig. 5.22. Three of these, *Aster, Hypochaeris*, and

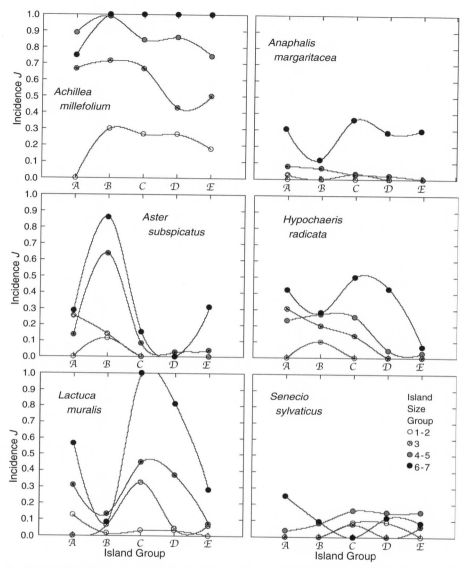

FIGURE 5.22. Distributions of six edge species in the family Asteraceae, by island groups (𝒜–ℰ) reflecting position in Barkley Sound, and by island size groups. Note the sensitivity of incidence to island group in *Aster*, *Hypochaeris*, and *Lactuca*, and the relative insensitivity to island area in *Anaphalis* and *Senecio*.

Lactuca, show significant island group effects on incidence (ANOVA; $p < 0.001$ in all cases); all three are anemochores, although dubiously so in *Aster* with its severely reduced pappus. *Senecio* is also an anemochore but it shows no island group effect, and neither do the two uppermost species (*Achillea*, *Anaphalis*). Islands are segregated by size in this figure, and the general effect of island size is to increase incidence (see *Achillea*, upper left, for a clear example of this pattern). However, island size influences incidence in different ways in different species. The smallest islands (size 1–2) are occupied by *Aster* and *Hypochaeris* (middle row) on ℬ-group islands only. With incidence curves falling to the right, $SIZE = 3$ islands are occupied in the 𝒜 and ℬ groups (*Aster*), and in the 𝒜, ℬ, and C groups (*Hypochaeris*), but not on the more isolated islands in either species. *Senecio sylvaticus* shows yet another pattern, with incidence being highest, except for the largest *A*-group islands, on *C–E* islands and in these groups being high on even the very small islands (Fig. 5.22, lower right).

Incidence is examined more closely in Fig. 5.23, where the independent variable is the species richness of the edge forb flora (E_F) of the islands. I divided the edge flora into trees and shrubs (E_S), graminoids (Poaceae, Cyperaceae, Juncaceae: E_G) and forbs (E_F: many families), with ranges in ΣJ from 0 to 20.4, 0 to 23, and 0 to 32 respectively. The best predictor of incidences of the common composites is consistently forb diversity E_F (Kendall's rank correlation coefficient τ, $p < 0.05$ for all six species), which in turn is most closely determined by island area occupied by vascular plants: $\log[\Sigma J(E_F)] = -0.327 + 0.328 * \log(AREA_{VP})$, $R^2 = 0.49$, $p < 0.001$ and $n = 167$ islands with $E_F > 0$. As total E_F incidence accounts for more explained variance in composite incidences, by a factor of about 1.5, than does $\log(AREA_{VP})$, it was chosen as the abscissa in Fig. 5.23.

Logistic regression curves are fitted to the incidence data of Fig. 5.23, but these models are only modestly successful, as measured by R^2 values (see figure). It seems that the main reason for the low R^2 values for the logistic models is the relative unpredictability of the occurrences of the species. Different islands, with similar edge forb diversities and over a narrow range of this variable on the abscissa, may record incidences of 0, or 1, or something intermediate. That is, for each of the species represented in the figure, $J(0)$ values overlap $J(1)$ values along the abscissa. This overlap zone constitutes a hysteresis, a zone of particularly unpredictable incidence. I took the $J(0)$ and $J(1)$ data for each composite species and used them to calculate 95% confidence limit on the lower limit for $J(1)$ and the upper limit for $J(0)$. For all six species, abscissa values for $J(1)$ were normally distributed (Lilliefors test for normality; $p < 0.05$), and subtracting 1.65*SD from the mean gave the lower 95% confidence limit. As abscissa values for $J(0)$ were skewed, with a left-hand tail, in all species except *Achillea*, I computed at upper 95% confidence limit by considering only abscissa values >0 (which gives distributions not significantly different from normal in all cases; Lilliefors test). The results define the hystereses drawn as stippled regions on the incidence functions of Fig. 5.23; these regions show over which ranges of values on the abscissa the incidence of the species in question is particularly unpredictable. Although the sample size ($n = 167$) is the same in all species, hysteresis zones vary in width with the extent of horizontal (abscissa) overlap between $J(0)$, $J(1)$ values and with data density in the overlap range; hysteresis zones would dwindle to insignificance if incidence transitioned smoothly from $J(0)$ to $J(1)$ with only intermediate values of incidence in the transition zone.

I identified hysteresis zones as those where biotic influences on incidence should be most apparent, that is, where the presence or absence of one species might affect the incidence of another. One approach is via ANOVA, with results shown in Table 5.5 (upper part). Here each hysteresis zone is identified for each species consecutively (column 1), with island sample sizes given in column 2, and significant effects are depicted in the body of the table. I tested for island group effects: does incidence vary among island categories *A–E*? It does in three of the six species (*Lactuca*, *Aster*, *Hypochaeris*), which all have significantly reduced incidences on the more isolated islands (see Fig. 5.21). Within the hysteresis zone, are there effects of edge forb diversity, such as higher incidence on more forb-rich islands? There are, in the same three species (as above), which all show significantly higher incidence as the total number of edge forb species increases (Table 5.5, column 3). Further, and more interestingly, *Achillea* incidence is negatively related to incidence in *Lactuca*, *Senecio*, and *Anaphalis*, accounting for 30–40% of its variance with the first two species (Table 5.5, top row). Conforming with these results, *Lactuca* incidence is positively associated with the incidence of both *Anaphalis* and *Senecio*, and *Aster* incidence also varies positively with that of the latter species (Table 5.5, rows 3 and 4).

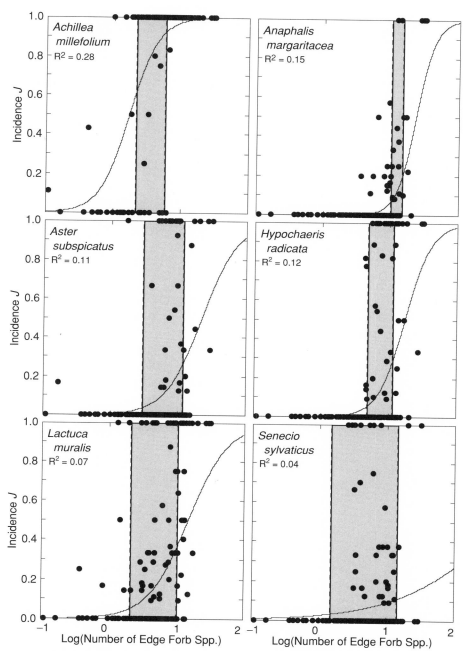

FIGURE 5.23. Incidence J of six edge forb species (ordinate) is plotted against the logarithm of the total number of edge forb species (abscissa) over all Barkley Sound islands. For each species, the distribution of values (along the abscissa) where $J = 0$ and $J = 1$ allow computation of an upper 95% confidence limit (CI) for $J = 0$ and a lower 95% CI for $J = 1$. The zone between the two CI limits defines a hysteresis, in which all values $0 < J < 1$ may be expected. Hysteresis zones are wider in species with patchier and less predictable distributions. Logistic regressions for incidence over Log(Number of Edge Forb Spp.) fit the data increasingly poorly (as indicated, upper left to lower right) and are least useful where the hystereses are broad. See text for discussion.

TABLE 5.5
Tests for Interspecific Effects on Incidence: Composite Forbs

SPECIES	n	ISLAND GROUP	LOG (NO. EDGE FORB SPECIES)	ACHILLEA MILLEFOLIUM	LACTUCA MURALIS	ASTER SUBSPICATUS	HYPOCHAERIS RADICATA	ANAPHALIS MARGARITACEA	SENECIO SYLVATICUS
ANOVA test results[a]									
Achillea millefolium	67	NS	NS	—	**p = 0.013, R² = 0.312**	NS	NS	**ρ = 0.049, R² = 0.090**	**p < 0.001, R² = 0.394**
Lactuca muralis	121	p < 0.001, R² = 0.266	p < 0.021, R² = 0.044	NS	—	NS	NS	p < 0.001, R² = 0.231	p = 0.001, R² = 0.251
Aster subspicatus	119	p < 0.001, R² = 0.270	p < 0.001, R² = 0.161	NS	NS	—	NS	NS	p = 0.015, R² = 0.213
Hypochaeris radicata	71	p = 0.040, R² = 0.139	p = 0.001, R² = 0.927	NS	NS	NS	—	NS	NS
Anaphalis margaritacea	19	NS	NS	NS	NS	NS	NS	—	NS
Senecio sylvaticus	154	NS	NS	NS	NS	NS	NS	NS	—
Analysis of joint influences[b]									
Achillea millefolium	81	—	—	—	ρ = −0.146	ρ = −0.154	r = −0.156	NT	ρ = −0.189
Lactuca muralis	81	—	—	**ρ = −0.352****	—	ρ = −0.173	ρ = −0.131	NT	ρ = 0.136
Aster subspicatus	81	—	—	**ρ = −0.245***	ρ = 0.090	—	**ρ = −0.258***	NT	ρ = 0.029
Hypochaeris radicata	74	—	—	ρ = −0.172	ρ = −0.215	ρ = −0.098	—	NT	**ρ = −0.240***
Anaphalis margaritacea	47	—	—	NT	ρ = −0.131	ρ = −0.079	ρ = 0.151	—	ρ = 0.221
Senecio sylvaticus	81	—	—	**ρ = −0.224***	ρ = 0.149	**ρ = −0.292****	**ρ = −0.236***	NT	—

[a] ANOVA tests (upper part of table) for independent influences on species' incidences within individual species' hysteresis zones (see Fig. 5.23). Boldface results are negative influences; others are categorical (e.g., Island Group) or positive (nonbold). NS = Nonsignificant.
[b] Residuals from species' logistic regression curves ($J * \log[E_F]$) are related (lower part of table) to other species' incidence (six right-hand columns) via Spearman's rank correlation coefficient ρ, within zones of overlapping hystereses ($n = 47–81$). Boldface values indicate statistical significance, level of significance noted by asterisks: **: $p < 0.01$, *: $p < 0.05$. NT = Not tested.

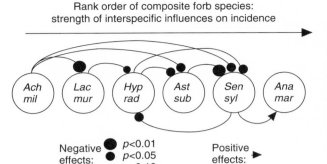

FIGURE 5.24. Composite forbs (species names abbreviated from Fig. 5.22) are ranked left to right in order of their capacity to reduce incidence in the other species (high in *Achillea*, low in *Anaphalis*). Eight negative effects, plus one near-significant positive effect were recorded; see text and Table 5.5 for derivation of this chart form of assembly rule.

A final step in the analysis of composite forb incidence considers the joint influences of the number of edge forb species (log E_F) together with the incidences of the other species. The results of the analyses are given in the lower part of Table 5.5, where Spearman rank correlation coefficients (i.e., nonparametric because of normality constraints) relate the residuals from the various species' logistic regression equations (of Fig. 5.23) to the incidences of other species within hysteresis zones. Note that the great majority of the correlation coefficients are negative, indicating that incidence in one species is reduced by higher incidence in the other. On the abscissa in Fig. 5.23, a region 0.45–0.80 with $n = 81$ islands encompasses much or almost all of the hysteresis zones of four of the six species in the figure (all except *Hypochaeris* and *Anaphalis*). In this zone, *Achillea* residual incidence is not significantly related to incidence in any other of the forbs (see Table 5.5). In contrast, *Lactuca*, *Aster*, and *Senecio* all have reduced incidence (i.e., lower residuals) where *Achillea* incidence is high. Further, there are reciprocal negative effects between *Hypochaeris* and *Senecio*, and additional significant negative effects by *Hypochaeris* on *Aster* and by *Aster* on *Senecio*. With the removal of forb species-richness effects, interspecific interactions shift somewhat but remain compelling. *Achillea* shows the greatest capacity to influence incidence in other forbs, while *Senecio* incidence shows the greatest susceptibility to be influenced by other species. This supports a view of *Achillea* as a dominant species in this assemblage, with *Senecio* and then *Aster* being the more subordinate. Indeed, a possible ranking of the whole species set by the dominance-subordinance clues in Table 5.5 might be *Achillea-Lactuca-Hypochaeris-Aster-Senecio*, as set out in Fig. 5.24. This scheme seems to constitute another sort of assembly rule, in which species higher up the rank order (i.e., to the left in Fig. 5.24) can negatively affect the incidence of one, several, or perhaps all of the species lower in rank order (to the right). The only indication of a balanced reciprocal effect here is found in the species pair *Hypochaeris-Senecio*, each of which reduces incidence in the other. Incidence rules of this type seem rather difficult to disentangle, but it is possible that experimental work on greenhouse or garden competition trials could facilitate the a priori prediction of such rules. If competitive superiority can be regarded as an incidence booster, it presumably has many components, including propagule production and dispersal, colonization, and germination success, and thereafter survival and persistence in the face of both abiotic and biotic adversity.

A final example is invoked regarding predictions of incidence from colonization and extinction events, this from two very similar species of small trees in the family Rosaceae, Pacific crabapple (*Malus fusca*) and bitter cherry (*Prunus emarginata*). Both are relatively common on the islands in Barkley Sound, both species thrive at the forest edge usually in partial shade, and both are similar in stature, reaching about 5 m

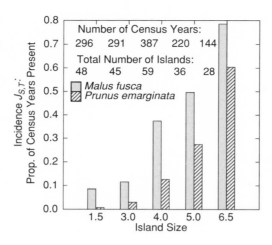

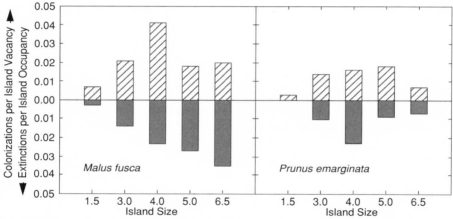

FIGURE 5.25. Distribution of two small trees (fam. Rosaceae) of edge habitats: *Malus fusca* and *Prunus emarginata*. *Malus* has higher incidence in general, but larger islands support both species with nearly equal frequencies. Colonization rates are higher in *Malus*, but *Prunus* persists better, especially on the larger islands.

tall. Further, the two species are similar in other respects: they produce small fruits of similar size (around 1 cm across), almost certainly zoochorous, and the leaves of the two species are similar in size and overall shape. Often we need to examine a (nonfruiting) tree carefully to determine species, as the distinguishing notched leaf edges of the crab apple are often not apparent at first glance. Thus, the two species seem ecologically very similar and might be considered prime candidates to show interspecific effects on each other's incidence.

Overall, *Malus* is more common than *Prunus*, especially on smaller islands, whereas on larger islands they occur at fairly similar frequencies. In Fig. 5.25 (upper) I plot incidence as the proportion of census years in which each species was recorded, as a function of five island size classes (sizes 1 and 2, and 5 and 6 are grouped). Note that the grand sum of census years, meaning the total censuses over islands × years, is 1338 (some lasting minutes, some days!); incidence here is the proportion of this total that the species was present. Over all islands and censuses, *Malus* averaged $J_{S,T}$ = 0.296, about twice the value for *Prunus* (0.142). The ratio between the two is approximately constant over island groups, with the exception of *E*-group islands, on which *Prunus* is only one-fourth as common as *Malus*. Using all of the methodological approaches employed above, perhaps more, I failed to substantiate the hypothesis that the incidence of one species influences the incidence of the other. It may be that the two species coexist with no direct effects because their colonization and persistence (= extinction resistance) dynamics differ. Possibly, like in the extreme cases of supertramp species versus "high-S" species, there is a trade-off between good colonization ability and good persistence ability; incidence might be relatively similar in species that compensate for weakness in one by strength in another.

I extracted from the presence records of each species those which followed an absence, and expressed them as a proportion of all absences. Likewise, I took the absence records and considered only the absences that followed a presence; this also was calculated as a proportion of all presences. These proportions are the ordinates of Fig. 5.25 (lower), showing how colonizations per island vacancy and extinctions per island occupancy vary in the two trees over island size. While *Malus* colonizes well, especially at intermediate island sizes, it persists increasingly poorly (with more extinctions per island occupancy) with increasing island size. *Prunus*, on the other hand, is a poorer colonist but is much better at persistence, especially on larger islands. Thus the relative similarity in incidence of the two species on larger islands is owed to good colonization in *Malus*, but more to good persistence in *Prunus*. Both species presumably cycle onto and off islands over time but do so with different limiting factors, independently, and with no apparent synchrony or disjunction in time. Clearly, we would need to learn how such factors as island size, position, area, elevation, microclimate, and so forth affect relative colonization and persistence rates of a host of potentially interacting species before their assembly rules could be constructed. There is a great deal of further research required, but this will surely be productive and exciting.

6

Species Turnover in Space and Time

COLONIZATION AND EXTINCTION, PERSISTENCE, AND TURNOVER

Through time, new plant species colonize and become established on islands in Barkley Sound continuously, just as old residents fall off the island roster as they become locally extinct. Since colonization and extinction rates are approximately equal over time (Fig. 3.4), an island's list of resident plant species remains about the same length, although the names change and species turnover is recorded as a function of time. Turnover between censuses taken in adjacent years is particularly informative, as it enables computation of actual colonization and extinction rates (see chapter 5, the discussion of shoreline plants), at least of reasonable estimations of them. Even these estimates may have inherent biases, since colonization rates are derived only from islands from which the species was previously absent. On these islands that are not permanently occupied, colonization rates are likely to be lower (because of their smaller size, greater isolation, less habitat suitability, etc.) than on islands with established resident populations of the species with high probability of persistence. These latter are islands on which colonization would be "redundant," where it could be detected only through marking and monitoring techniques that can identify any newcomers. Colonization that rescues a species that would otherwise have gone extinct is also not distinguished, nor measured.

Thus true colonization rates, if we want them to measure the arrival rates of new propagules on islands and their subsequent successful establishment, are likely higher than those measured here, even with the most comprehensive data sets, and this is worth keeping in mind. Note that, while we are comfortable in labeling the arrival and establishment of new individuals on an island as colonization, even though the species is already present, there is no equivalent on the extinction side. That is, "redundant extinction" does not make sense; there is an asymmetry in that individuals colonize but populations go extinct.

While colonization and extinction rates are approximately equal over fairly long time periods, over shorter time periods they may be unequal. For example, colonization events may occur in boom-and-bust spurts, possibly associated with favorable conditions for dispersal that occur only sporadically. This was noted in the Brady Beach islets of Fig. 4.1, where over periods of several years colonization events C exceeded extinction events E, and over other periods the reverse was true. The numbers of colonization and extinction events within years were negatively correlated on these islets, significantly so on four of the eight islands (A15, 16, 19, 20: $r = -0.69$, -0.86, -0.63, -0.58, respectively). Cross correlation between island pairs revealed that both colonization and extinction events occur independently amongst the islands (only one of 28 cross correlations, in both C and E, was significant). However, cross correlations within islands between C and E show that significantly more extinction events followed two to five years after higher numbers of colonization events were recorded, although gaps in the census series limit the power of these tests. For examples, cross correlations peaked after a two-year lag on A16, 19, 20, 22, at values of 0.37, 0.48, 0.32, and 0.26, respectively. Interestingly, significant cross correlation values were obtained from just one of the four larger Brady Beach islets (on A16), indicating perhaps a reduced degree of or a buffering in negative feedback between excess colonization and subsequent extinction on larger islands, where there are more space, species, and opportunities for persistence.

The periodicity of stretches of time with first high and then low colonization rates deserves closer attention. Perusal of Fig. 4.1 shows that colonization events exceeded extinction event in 1986 and again in 1998 and 2002, but extinctions outnumbered colonizations in 1991. The summers of low sunspot numbers are 1985–1986, 1997–1998, and of high sunspot numbers 1990–1991 and 2002–2002 (but no weather data exist for this last period). There are collectively 24 islands in the A–C groups that were censused in 1985 and 1986, in 1990 or 1991, and in 1998 (but not 1997). The 1986 C and E data are most relevant, since they follow a census year (1985), and therefore C and E are unambiguous one-year figures; in 1986, $C = 68$, and $E = 21$, a ratio >3:1 in favor of colonization over extinction events. Summers in both years were unusually warm and dry, but the 102 mm of summer precipitation in 1985 was a record low. Data from the other sunspot low or high years are not very useful, as they are not single-year data (they do not follow a census year), nor are the summer weather signals in those years as distinct as in 1985–1986. It does appear that years of sunspot lows may induce, via a stronger North Pacific High cell, especially warm and dry (and possibly windier) summers in the Sound, which, in turn, promote a powerful colonization pulse in the island plants (many of which are anemochores). More data than even I have are required to test critically this scenario.

Apropos of data limitations just mentioned, turnover between censuses made before and after longer, multiyear time intervals becomes increasingly difficult to evaluate. Repeated colonization and extinction events may have occurred in taxa that have apparently persisted, while new absentees reflect extinction after undetermined residence times and new residents represent colonization at a prior but unknown date. Over time and with a finite size of the pool of potential colonists, an increasing proportion of colonizations are actually recolonizations, representing returning species that were resident on the island during some previous time interval, rather than first-time residents. Barring autochthonous speciation (new species production within the island system) and without an influx of regional newcomers, such as newly introduced weeds or adventive range expanders from elsewhere, this proportion asymptotically approaches unity.

Thus far I have shown that colonization and extinction rates differ among taxa, even among those of similar dispersal strategies and growth forms; rates also differ among islands, with island size and isolation affecting particularly the former. In general, it seems that persistence times, population longevity, or extinction resistance must be a function of many variables that include individual plant longevity, island population sizes, and the island conditions of habitat extent and suitability. Persistence times will thus vary greatly and must be very different in the long-lived woody plants of stable island interiors relative to the short-lived herbaceous plants typical of edge habitats, sparsely and often patchily distributed in small enclaves. In Fig. 3.6, frequency distributions of persistence, the equivalent of incidence over time J_T, were plotted for several island groups and sizes; the quasi-U-shaped distributions indicate a good representation of long-term persistent plant species ($J_T = 1$). Less numerous but still conspicuous are short-lived residents ($J_T < 0.20$), and there is also a scattering of species with intermediate J_T values.

I used these three categories of persistence, $J_T = 1$, $1 > J_T > 0.20$, and $J_T < 0.20$, to classify the species list into three population longevity components: long-term residents **L**, midterm residents **M**, and short-term residents **S** (all three written in boldface), over the five island groups. On islands with $n > 7$ census years, sample sizes are 23 ($\mathcal{A}$), 16 ($\mathcal{B}$), 30 (C), and 19 ($\mathcal{D}$). To obtain comparable sample sizes in the $\mathcal{E}$ group, I relaxed the standard somewhat and included islands with $n > 5$ census years (31 islands). While the overall time span of the census data is similar (i.e., from 1981 or 1982 to 2002 or 2003), this equitability in samples sizes may be bought at the cost of a somewhat decreased resolution in the $\mathcal{E}$-group islands. This is because censuses more widely spaced in time allow more opportunities to miss short-term residents and confuse permanent or long-term residents with those that have been rescued by conspecific recolonizations (see above). I return to this point below.

Regressions of **L**, **M**, and **S** against island size (log[$AREA_{VP}$]) are highly significant on all island groups (highest p values 0.004, except for **M** species on the $\mathcal{E}$ group, where $p = 0.012$). For long-term (quasi-permanent) residents **L**, the R^2 values are similar to those of the all-species regressions of Fig. 4.11 (0.77–0.82). For midterm species **M** and short-term residents **S**, R^2 values are lower, averaging 0.59 and 0.53, respectively. In Fig. 6.1 the regression lines are summarized in additive fashion for the five island groups, where species in the **L**, **L** + **M**, and **L** + **M** + **S** categories are shown as functions of island areas. The proportions of species totals that comprise **L**, **M**, and **S** species differ among island groups. In the $\mathcal{E}$ group, the long-term residents **L** dominate the species totals over all island sizes. However, in the C and $\mathcal{D}$ groups, midterm residents are numerically dominant over a short range of the smallest islands (log $AREA_{VP} < 1.5$). In the $\mathcal{A}$ and $\mathcal{B}$ groups, first **S** species and then **M** species dominate the smallest and somewhat larger islands, until eventually a size is reached (log $AREA_{VP} = 1.55 - 1.75$) where the **L** species are numerically dominant. These transitions, the changes in the category of the dominant species at certain island sizes, are indicated by the vertical dashed lines in the graphs (Fig. 6.1).

The long-term residents, indicated by the **L** regression lines in Fig. 6.1, are all similar in intercept on the abscissa, but they differ somewhat in slopes among island groups and therefore differ in their rates of increasing species richness over area. Regression coefficients are tested among the island groups in Table 6.1 (upper), showing that **L** species in the $\mathcal{A}$ group are significantly elevated over $\mathcal{B}$- and $\mathcal{E}$-group numbers (these are reciprocally significant differences in regression coefficients). Amongst other pairwise comparisons between groups, there is a tendency for higher **L** species numbers in the $\mathcal{D}$-group islands over the $\mathcal{B}$- and $\mathcal{E}$- group islands, and similarly for higher numbers in $\mathcal{A}$-group islands over the C-group islands, and the latter in turn higher than the $\mathcal{E}$-group numbers. These tendencies are indicated by one-sided (nonreciprocal) significant differences in regression coefficients (for example, C-group data could conform with the $\mathcal{A}$-group regression coefficients [$p > 0.05$], but not vice versa). Thus the island groups can be rank ordered $\mathcal{A}-\mathcal{D}-C-\mathcal{E}-\mathcal{B}$ in richness of **L** species; this ranking appears roughly equivalent to their proximity to major land masses and inversely correlated to their exposure to maritime conditions, but I have found it hard

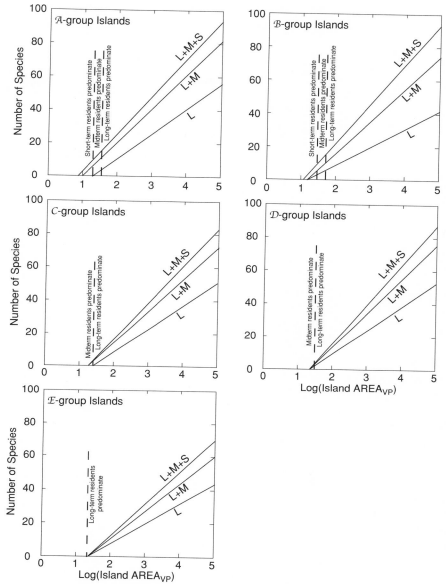

FIGURE 6.1. Numbers of plant species that are long-term (**L**; $J_T = 1$), midterm (**M**: L > 0.2), and short-term residents (**S**: $J_T \leq 0.2$) on islands in the five isolation groups 𝒜–ℰ, as a function of island size, log(AREA$_{VP}$). Long–term residents predominate over all island sizes in ℰ groups, but midterm residents dominate the smaller islands in 𝒞, 𝒟 groups, and shortterm "visitors" occupy the smallest islands in the 𝒜 and ℬ groups, where midterm species predominate on slightly larger islands. Regressions are based on islands with J_T evaluated over $n \geq 7$ census years (𝒜–𝒟) or $n \geq 5$ (ℰ); all are significant ($p < 0.015$).

to assess the latter in a more quantitative and useful fashion.

If regression lines for **L**, **M**, and **S** species all had a common intercept on the abscissa of Fig. 6.1, which is the case for the ℰ group and is approximately so for other groups, the proportions of these species categories would remain fixed over island size. Amongst four island groups, 𝒜, 𝒞, 𝒟, and ℰ, the proportions are remarkably similar, namely, **L** species comprise

TABLE 6.1
Differences among Regressions of Long- and Short-Term Resident Species Against Island Area

ISLAND ISOLATION GROUP	𝒜 GROUP	ℬ GROUP	𝒞 GROUP	𝒟 GROUP	ℰ GROUP
Long-Term Persistent Species ($J_T = 1$) vs. log(AREA$_{vp}$)					
𝒜 group	—	$p = 0.003$	NS	NS	$p < 0.001$
ℬ group	$p < 0.001$	—	NS	$p = 0.04$	NS
𝒞 group	$p = 0.002$	NS	—	NS	$p = 0.038$
𝒟 group	NS	NS	NS	—	NS
ℰ group	$p < 0.001$	NS	NS	$p = 0.025$	—
Short-Term Resident Species ($J_T < 0.20$) vs. log(AREA$_{vp}$)					
𝒜 group	—	$p < 0.001$	$p < 0.001$	$p < 0.001$	$p < 0.001$
ℬ group	NS	—	NS	NS	$p = 0.040$
𝒞 group	$p < 0.001$	$p < 0.001$	—	NS	NS
𝒟 group	$p < 0.001$	$p < 0.001$	NS	—	NS
ℰ group	$p < 0.001$	$p < 0.001$	NS	NS	—

Note: Regression coefficients for the island groups listed in the first column are tested for differences from regression coefficients in the island groups listed in the remaining columns. NS = nonsignificant.

0.60–0.62 of the species total, **M** species 0.23–0.27, and **S** species 0.12–0.15. Proportions differ in the ℬ group, however, with **L**, **M**, and **S** constituting on average proportions 0.48, 0.34, and 0.18 of the species total (i.e., a greater representation of midterm residents **M**). Persistence times appear to be lowered here overall, and again it seems related, with ℬ-group islands located at the distal end of the Deer Group islands and facing the brunt of the Pacific Ocean storms, to exposure. The island at the outermost end of this group, ℬ25: "Bordelais S Rks," (AREA$_{VP}$ = 2660 m^2), serves to sharpen the point. From the sea the island is a very impressive jumble of massive rocks and looks quite plant free, but in fact it accommodates a smattering of crevice-hugging plants, their distribution heavily skewed to the side of the island facing the interior of the Sound. They collectively amount to 21 species (in an average year), and 25 species cumulatively. The time-averaged forest, edge, and shoreline species number 1, 7, and 13, compared to species numbers predicted from the regressions in Fig. 4.12 of 16.6, 24.5, and 11.4. Thus, total species numbers on this outlier amount to just 40% of what is expected of islands in more benign locations, and shifts from the norms in species persistence ratios seemingly apply throughout this island group (Fig. 6.1, upper right).

Differences among island groups are clearer in the category of short-term residents **S** (Table 6.1, lower), which are significantly more numerous on the 𝒜- and ℬ-group islands. As these are predominantly short-lived edge species, good colonists but poor persisters, they presumably benefit from the greater accessibility of islands of the 𝒜 and ℬ groups, with their low and modest isolation, respectively. Mean distance-to-mainland (D_{MLD}) is 0.74 and 2.47 km, respectively, in these islands, versus 5.72 km for ℰ-group islands where the proportions of **S** species are lowest. 𝒞- and 𝒟-group islands are intermediate in numbers of **S** species and also in isolation distances (means 3.48, 3.74 km, respectively). An indication of these sorts of differences was given earlier in Fig. 3.9.

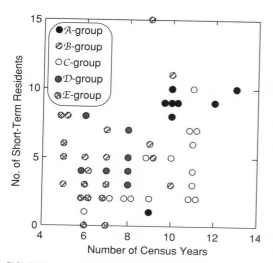

FIGURE 6.2. Number of short-term resident species on the five island groups as a function of the number of census years per island (*CYRS*). There are significant intergroup effects, but no significant effect of the number of census years (MANOVA: $p < 0.05$, groups; $p = 0.07$, CYRS).

It might reasonably be hypothesized that the low proportion of **S** species in the $\mathcal{E}$ group is a consequence not of greater isolation but of my using the criterion $n > 5$ census years rather than $n > 7$ census years over the more than two decades of monitoring. As each census can record "one-time" species whose occurrence is unique to that census year and which therefore will contribute to the **S** species total, the more census years the higher **S** might become. But this effect is countered by the potential for additional census years to reduce **S** numbers, by re-recording a species that had been found earlier, thus removing it from the ranks of **S** and adding it to the **M** species total. I plot the number of short-term residents against number of census years (*CYRS*) in Fig. 6.2, where an effect of island group is discernible, but no significant effect of number of census years (MANOVA: $p = 0.07$ that *CYRS* affects **S**). It appears that multiple censuses on an island produce new records of **S** species at about the same rate as they convert erstwhile **S** species to **M** species, and the two effects cancel each other.

Given that a majority of short-term resident plants are weedy edge species, most of which are common in disturbed sites on the mainland, a historical note on island disturbances is relevant here. **S** species are especially common on the $\mathcal{B}$-group islands, where the sample of islands with $n > 7$ census years is comprised of those in the vicinity of the Dodger Channel, which separates the large islands of Diana (to the northeast) from Haines and Edward King (to the southwest). The island $\mathcal{B}$12: "Haines SE" is a conspicuous outlier in Fig. 6.2 (see also Table 4.1), with **S** = 15 short-lived residents ($n = 9$ *CYRS*). This island is quite large (log[AREA$_{VP}$] = 4.0) but is nearly completely devoid of taller vegetation and has few native woody plants. It supports, besides a few short spruce and a couple of small, domestic *Prunus* and *Malus*, an inordinate number of weedy plants, many of which have also reached nearby islands. The island was a Huu-ay-aht dwelling site, Tlakakpikas, until the early 1950s, with a more substantial summer village, A-a-so-its, located nearby. Further, there was an early European settlement on adjacent Haines Island, with a sealing and whaling base, chapel, and trading post (*fide* Jane Peters, of the Bamfield Huu-ay-aht Information Center, and Dennis Morgan of the Bamfield Huu-ay-aht Community Forest, pers. comm.; also Scott 1970). Traces of the settlement are nearly eliminated on the larger island, but on smaller (and drier) "Haines SE," forest regrowth has been slow, and the legacy of disturbance survives there in an abundance of weeds, with spillover effects onto neighboring islands.

There is a fairly clear dichotomy in persistence times between the steadfast, long-term resident plants and the short-term residents that are more akin to temporary visitors. Figure 4.1 shows how, in the face of ongoing colonization and extinction events, cumulative species numbers on islands increase while the number of residents at any one time remains nearly constant. A simulation model, rather simplistic to be sure, illustrates some of the operative factors and processes in a two-component island flora

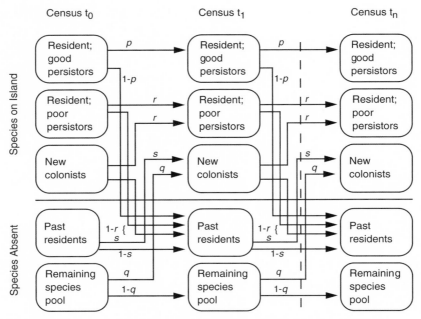

FIGURE 6.3. Categories of plant species on (upper) and off (lower) the island at any point in time. Island residents are divided into three components, long-time, good persistor residents, short-term residents with poor persistence, and new colonists. Species absent and candidates for invasion consist of past residents plus the remaining species pool. Probabilities p, q, r, and s describe the transitions of species among the component categories; p, q, and r are predetermined, while s is adjusted to maintain equilibrium species number.

with different dynamics for long- and short-term resident plants; it examines the parameters that affect turnover and cumulative species counts and the resultant trends (Fig. 6.3). I divide the original residents, censused at time t_o, into two components: (1) the long-term residents, accumulated over the longer time or even persistent since the island's creation, with good survival probability p; (2) the short-term residents, with poor persistence probability r ($r < p$). In any one year the residents are augmented by new colonists, which are either recolonists drawn from the pool of past residents (probability s) or are first-time colonists from the remaining pool of species candidates (probability q, $q < s$). The equilibrium species number $\hat{S}$ the size of the total species pool for the island (both larger on larger islands), and the probabilities p, q, and r are preset, with probability s adjusted to maintain a constant number of species on the island.

The results of 25 successive iterations of the model, a time period about equivalent to that of the island study, are shown in Fig. 6.4. The three parts of the figure apply to smaller islands with $\hat{S} = 10$ species (upper), intermediate sized islands with $\hat{S} = 20$ species (center), and larger islands with $\hat{S} = 30$ species (lower). The size of the species pool is changed, as would be appropriate for islands of different sizes, from 50 to 100 to 130, respectively (see Fig. 6.4). Transitional probability values also vary with island sizes, and two different sets are employed for each of the three island sizes; these are shown in the figure. Initial conditions are set with the "new colonist" box empty and island residents divided approximate 3:1 between good and poor persisters, and off-island species divided approximate 1:3 between past residents and the remaining species pool. The figure shows how cumulative species numbers increase over time, at faster rates where persistence values

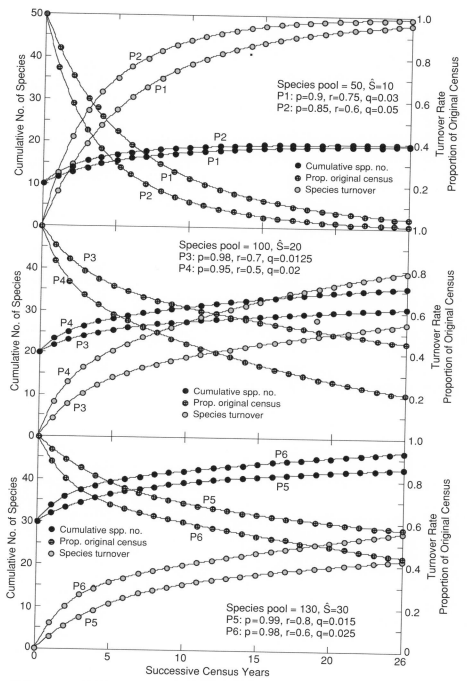

FIGURE 6.4. Simulation of cumulative species counts (left-hand ordinate), species turnover, and proportion of original species left on island (right-hand ordinate) in successive census years (abscissa). Three different scenarios are depicted, with equilibrium species number 10 and species pool 50 (upper), versus equilibrium number 20 and pool size 100 (center), and equivalent values of 30 and 130 in the lower figure. In each scenario, two alternative combinations of model parameters are given (e.g., P1, P2; P3, P4) with variations in p (probability of persistence for long-time residents), r (probability of persistence in new colonists and recolonists), and q (probability of a first-time colonist's arrival from the species pool). For each scenario and probability set, the model shows how cumulative species counts rise, how species turnover increases after the initial census, and how the proportion of original residents declines over time.

are lower and colonization rates are higher. Concomitantly, there is a turnover of plant species relative to the original census, and a parallel decrease in the proportion of the original census that remains on the island. Note that the rate of species accumulation slows over time, as more and more of the candidate species for island residence are recorded; likewise rates of species turnover slow, as island novelties accrue at decreasing rates in later censuses.

Part of the impetus for this simulation exercise came from the data in Fig. 4.1 and curiosity about what sorts of pool sizes and turnover rates would account for them. Note that on smaller islands with 5–10 species, the cumulative species number was about double that of the equilibrium species number in more than two decades, while on larger islands with 30–35 species, the cumulative count was around two-thirds higher than $\hat{S}$ in the same time period. The former increase in cumulative species is approximate that seen in the upper part of Fig. 6.4, while the latter is about that of the lower part of the same figure. This may indicate that the species pool sizes and the transition probabilities used in the model are not far from those operating on real islands. Thus on a 10-species island with species persistence probabilities between 0.9 for better and 0.75 for poorer persisters, and accessible to a total pool of 50 species, cumulative species number would reach 20 after two decades or so. A 30-species island accessible to a larger pool of 130 species would host an additional 13–17 species in the same time period, with good and poor persisters surviving with probabilities 0.98–0.99 and 0.6–0.8, respectively, and with recolonists and new colonists arriving annually at rates of around 20% and 2% of their off-island pool sizes, respectively.

All of this assumes a closed system, but in fact the species pools should be regarded as variable rather than fixed entities. We find one or two new alien weedy species each year, and some of the nonweedy native species are precariously rare and ultimately may disappear from the system. For example, we recorded the orchid *Spiranthes romanzoffiana* blooming on C15: "Ross E Cl. f'" in 1985, but it has not appeared since, neither at the original nor at any other location. There is a single *Acer macrophyllum* individual on A7: "S Hosie," and no others on any of the censused islands. *Gnaphalium purpureum* disappeared from C36: Sandford after 1987 (where it constituted the only island record). On this same island, the similarly restricted confamilial *Eriophyllum lanatum* persists, but represented by only a handful of individuals. The fern *Cystopteris fragilis* was located on the largest island surveyed, Æ76: Turtle, where a single individual was found in 2003; conceivably it may be generally distributed on other (uncensused) very large islands, although it was not recorded in Bell and Harcombe's (1973) survey. There are many other extremely rare species in the system, some perhaps relicts from pre-island times, and it is likely that more of them will be lost over time with little chance of reestablishment.

SPECIES TURNOVER IN TIME, AND ISLAND SIZE

Classical M/W theory predicts both high colonization and high extinction rates on small islands close to colonization sources, and therefore high species turnover (see Fig. 3.1). Here on the Barkley Sound islands, colonization rates are higher on less-isolated islands as predicted by the classical theory, but, because of target effects, these rates are higher also on larger islands (Fig. 3.5). But turnover rates T, the rates at which species identities change over time, are functions not only of colonization and extinction events, but also of the number of persistent species $\mathbf{L}$ on the island. Using the formula $T = [1 - Z(T_1 + T_2)/2T_1T_2]$, where Z = the number of shared species between islands with species totals T_1 and T_2, the maximum value attainable by T on islands with $\mathbf{L}$ shared (persistent) species and equilibrium species number $\hat{S}$ is $(1 - \mathbf{L}/\hat{S})$. For example, islands with 80% persistent species have a maximum turnover value of 0.2.

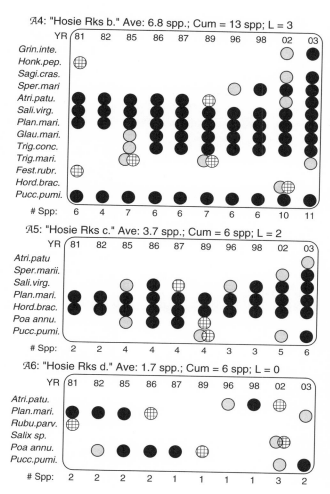

FIGURE 6.5. Census results from three small islands over a 22-year period, illustrating the many colonization and extinction events that occurred over the time period. Average species numbers on island, cumulative species numbers, and numbers of persistent species **L** are given. Solid symbols represent persisting species, shaded symbols are colonization events, cross-hatched symbols represent extinctions.

The dynamic aspect of island species lists is shown in Fig. 6.5 for three small islands of the Hosie group. The total number of species recorded on these three islands, which are only a few meters apart, is 17, but only three species, *Atriplex patula, Plantago maritima,* and *Puccinellia pumila,* occur on all three islands. Note that *Hordeum brachyanthemum* is a persistent grass on A5 and *Puccinellia pumila* is casual there; the reverse in true on A4, and no grasses are persistent on A6. The numbers of species in the average census declines from A4 to A5 to A6, at 6.8, 3.7, and 1.7 species, respectively. Cumulative species numbers are about double the average number of residents in the two larger islands, and about three times the average number on the smallest island; numbers of persistent species decline, top to bottom in the figure, from 3 to 2 to zero. The longevity of nonpersistent species varies considerably, with some lasting a decade or more and some with only single year "cameo" appearances. Several species recolonize the same island after going extinct earlier; examples are *Triglochin maritimum* on A4, *Puccinellia pumila* on A5, and *Plantago maritima* on A6. Note that there is some correspondence between events on different islands, such that *Poa annua* is present over a similar interval, early to late 1980s, on both A5 and A6, *Spergularia maritima* occurs from the late 1990s on A4 and in the early 2000s on A5, and *Puccinellia pumila* is present on both A5 and A6 in the early 2000s.

Species turnover is calculated for six Hosie islands, A1–6, in Fig. 6.6. These islands were censused between 1981 and 2003, a total of 9 times (A1, which lacks a 1981 census) or 10 times (all of the other five). Each pair of censuses is represented by a point in the six panels of this figure ($n = 45$ points for all but A1, which has $n = 36$). There is an unmistakable trend on all of the islands for species turnover to increase with time interval between censuses (Fig. 6.6). Further, turnover is much higher in absolute value on the smaller islands, although relative to the maximum values of T given the number of persistent species (at the horizontal indicated by T_{max}), turnover values are more similar. Note that turnover exceeds one-half of the maximum values of the largest islands after a time interval of about 4 years, but with decreasing island size this threshold, drawn on the figures, is passed after increasingly shorter time intervals (3, 2, and 1 year, respectively, in A4, A5, and A6).

It is also apparent that the largest turnover values between censuses are not always associated with the longest intercensus intervals; they are only on islands A3 and A4 (Fig. 6.6). Superimposed on the figures are boxed and connected points that represent turnover between the original census (1981 or 1982) and each successive census through time. The end points of these connected data points represent the highest turnover values in just two of the six islands. Clearly, species turnover involves colonization, extinction, and a good deal of recolonization, with the reappearance later in the time series of species that were initially present but absent over some intervening interval. New invaders are prone to high extinction rates but also enjoy high recolonization rates, such that island floras can, over time, deviate from and then cycle back to a species composition similar to that with which they started out.

The islands of intermediate size in the Hosie group, A2–5, which have 10 census years in common and 9 subsequent comparisons with the original census, show considerable synchronicity in species turnover. Table 6.2 (upper) gives Spearman correlation coefficients in year-specific turnover values among island pairs; all are positive, and several are at or approaching statistical significance (critical values for significance at the $p = 0.05$ and 0.10 levels are 0.70, 0.60, respectively). I examine this phenomenon of turnover synchronicity further with another island set, the Ross Islets (see Figs. 2.7, 4.5). Islands in the "Ross E(ast) Cl(uster)" include C7–12, "Ross E Cl a to f," which were censused in the same 11 years over a 22-year interval, giving 10 comparisons of later censuses with the original 1981 survey. Similarities in species numbers trends were shown earlier in Fig. 4.5. However, there is no compelling similarity in the timing and extent of species turnover among these islands (Table 6.2, lower); although positive correlations outnumber negative correlations by 11 to 4, all but one lack statistical significance. Thus **S** species come and go with only approximate synchrony here, and this appears to be due to patterns of colonization sequences within the island group. Species apparently colonize one or another island in a local group such as the "Ross E Cl," then spread in subsequent years to neighboring islands, meanwhile becoming extinct on some islands that were occupied earlier. Such time delays clearly will diminish the extent of turnover synchrony.

I illustrate this phenomenon of range spread via local island hopping in Fig. 6.7, with four weedy, edge species: a grass, two composite forbs, and a plantain. Three of the species, all except the grass *Aira praecox*, have maintained permanent populations on one (*Lactuca muralis*, *Plantago lanceolatum*) or on three (*Hypochaeris radicata*) of the archipelago's islands, different islands in each case. In general, incidence over time has been higher on the larger islands ("Ross E Cl b, c, d and f"), lowest on the smallest islands ("Ross E Cl e, e′, e″, e‴"), and low also on the one island completely isolated even at low tide levels ("Ross E Cl a"). Range expansion over time was steady in *Aira* and *Plantago* (top and bottom in Fig. 6.7), and somewhat more hesitant in the two composites (center panels).

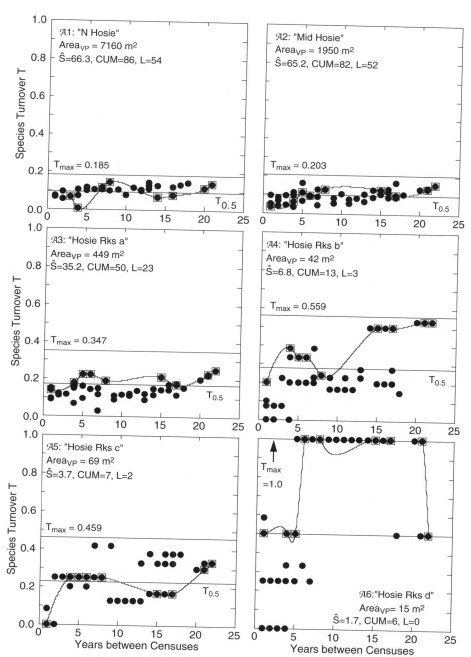

FIGURE 6.6. Species turnover (ordinate) as a function of the number of years between censuses (abscissa) on six islands of various sizes (areas given). Islands vary in equilibrium number of species $\hat{S}$, cumulative species number CUM, and the number of persistent species **L** (see above). Each pair of census years produces a data point, and the turnover between the original and subsequent censuses is shown by boxed and connected points. Given the number of persistent species **L**, the maximum value attainable by turnover, T_{max}, is indicated, as is also the value attained if nonpersistent species change by 50% between censuses. See text for discussion.

TABLE 6.2
Synchronicity of Species Turnover among Years

ISLANDS OF THE HOSIE GROUP ($n = 9$ INTERYEAR COMPARISONS)				
	A2	A3	A4	A5
A2	–	0.753[a]	0.633[a]	0.691[b]
A3		–	0.585	0.852[a]
A4			–	0.531
A5				–

ISLAND OF THE ROSS EAST CLUSTER ($n = 10$ INTERYEAR COMPARISONS)						
	C7	C8	C9	C10	C11	C12
C7	–	−0.098	0.275	0.492	0.343	0.443
C8		–	−0.196	−0.352	−0.138	0.107
C9			–	0.564[b]	0.122	0.363
C10				–	0.738[a]	0.488
C11					–	0.366
C12						–

[a] $p < 0.05$.
[b] $p < 10$.
Note: Spearman rank correlation coefficients among the species turnover rates of midsized islands in two groups. In each case, species turnover is assessed by later comparisons, within islands, to the initial census; correlation between these turnover values, among islands, is given in the table.

Summed incidences ($\Sigma_i J_i$) of the four species are highest on the islands "d," "b," "f," and "c" (2.75, 1.83, 1.75, and 1.58, respectively). Factors likely to affect range expansion are both the island area available for occupancy by edge forbs, and island location relative to other islands in the group that serve as sources or stepping stones. With this in mind, I conducted a multiple regression analysis with (1) distance from the center of the island cluster D_{center} (measured from the dot in the lower panel of Fig. 6.7), and (2) residuals from the regression of number of edge species versus log($AREA_{VP}$), an area-corrected surrogate for edge habitat extent (EA_{resid}). Both of these variables are significant contributors to the combined incidence of the four target species: $\Sigma_i J_i = 1.47 + 6.03 * EA_{resid} − 0.69 * D_{center}$; $R^2 = 0.53$, $p = 0.069$). Apparently both island edge habitat and island position within the archipelago, central versus peripheral, influence the residence and spread of these weedy species.

SPECIES TURNOVER IN SPACE

TURNOVER BY DISTANCE

Species turnover can be measured within islands over time, and also between islands over space. It might be hypothesized, for example, that nearby islands will share more plant species than islands that are far apart, and that species turnover will increase with the distance between the islands being compared. This phenomenon has been called "distance decay" (e.g., Nekola and White 1999) and will be sensitive to species' vagility and dispersal potential, to the barriers that impair dispersal, and also to changes in habitat suitability that possibly are correlated with distance. I tested this notion of distance decay by taking a subset of the plentiful

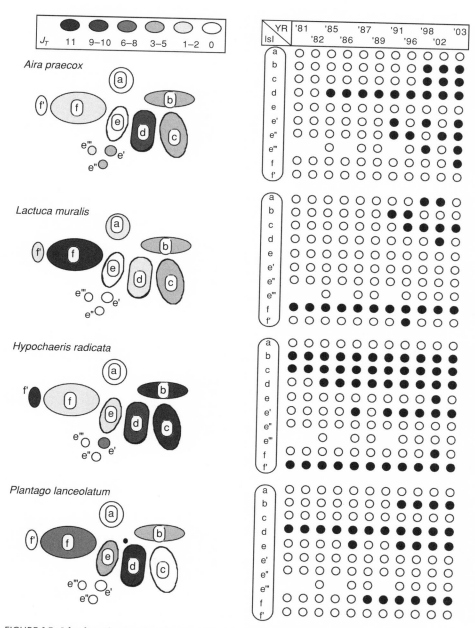

FIGURE 6.7. Islands in the Ross East Cluster group (C 7-15, C 35) are shown diagramatically on the left, with approximate relative positions and areas. Census results in 11 different years over a 22-year period are given in the matrices on the right for four weedy edge species: *Aira praecox* (early hairgrass), *Lactuca muralis* (wall lettuce), *Hypochaeris radicata* (hairy cat's-ear), and *Plantago lanceolatum* (narrow-leaf plantain). Distributions of these species are depicted on islands (left) using different degrees of shading to represent years of residence. Note that three of four species have persisted throughout the study on at least some islands, serving as sources of colonization of other islands in the archipelago. Species expand sporadically, but with some relation to island size and proximity.

size 4 islands (with AREA$_{FL}$ = 1000–3010 m^2), with the additional restrictions of ELEV > 4 m (to control for island elevation and its demonstrable influence on florula size), and with $n >$ 5 census years (to control for magnitude and temporal spread of census effort). This yields some 28 islands spread over groups A–E, with 5, 3, 11, 2, and 7 islands in each group, respectively. Collectively these islands support 149 plant species and an average species number of $\hat{S}$ = 28.5 ± 10.8 (in summed plant incidence over time, ΣJ_T). As a measure of species turnover between islands, I sum the absolute values of the differences in incidence between islands 1 and 2: $\Sigma_i(|J_{1i} - J_{2i}|)$, over all species i. This formula returns a value of zero if all plant species have the same incidences on the two islands, and a value equal to the sum of the two islands' incidence totals if no species are shared. For 14 rather standard islands taken from across the sample range, I compared each of the other islands in the 28-island set to these 14, evaluating species turnover T as a function of the distance apart of the islands.

Species turnover between A-, B-, C-, and D-group islands shows no significant relation, by the regression analyses, to island separation distances in 10 of 11 test islands (Fig. 6.8). The single exception is turnover from C10, which increases with distance from C10 (R^2 = 0.17; p = 0.031). In contrast, in all three E-group islands tested (E2, E9, E24), there is a significant distance decay, with species turnover increasing as island separation increases. Turnover explained by distance accounts for 28%, 44%, and 21% of its total variation in E2, E9, and E24. These three regression lines (and data points) are shown in the three panels of Fig. 6.8, on which regression lines for the A–D islands are superimposed, respectively. There are almost no systematic species changes with distance from the representative A–D islands, and there is therefore no spatially defined local character to the island floras in these groups. On the other hand, increasing turnover with distance from E-group islands means that there are locally defined characteristics of species composition in

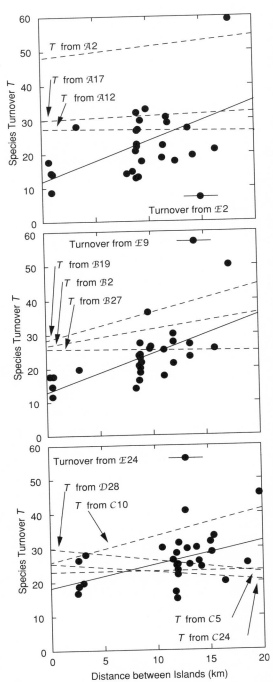

FIGURE 6.8. Regressions of species turnover T (ordinate) against distance apart of islands (abscissa) are significant for E-group islands (E2 [upper]: p = 0.004; E9 [middle]: p < 0.001; E24 [lower]: p = 0.016) but not for islands in other groups with the single exception of C10 (lower), with p = 0.031. See text for discussion.

ℰ-group islands that change gradually with increasing distance from focal islands.

I interpret these differences among island groups in spatial turnover or distance decay in the light of Fig. 6.1, which shows that lower proportions of short- and medium-term residents, **S** and **M** species, occur on ℰ-group islands, which are the most isolated in the Sound and have lower colonization rates by weedy edge species. Thus 𝒜- and ℬ-group islands especially, with many **S** and **M** species that by their nature are widely if erratically distributed, will show relatively high turnover with other islands close by, neighbors that generally support different subsamples of the **S/M** species. The mean turnover T amongst the 5 𝒜-group islands in the sample is 36.0 ± 11.7 SD ($n = 10$); the regression intercepts (Fig. 6.8) are also very high (neighboring islands have different species composition). Further, these 𝒜-group islands display relatively low turnover with distant (ℰ-group) islands, where the quotas of **S/M** species are much reduced and whose species lists will likely be subsets of those on the 𝒜- and ℬ-group islands. However, the ℰ-group islands are more likely to share their reduced quotas of **S/M** species with other islands locally (see Fig. 6.7). The mean turnover T amongst the seven ℰ-group islands in the sample is 16.7 ± 6.1 SD ($n = 21$), less than half that of the 𝒜-group islands, and again corresponding to the lower intercepts of the three ℰ-group regressions of Fig. 6.8.

I further illustrate differences in spatial turnover between island groups by choosing representative 𝒜- and ℰ-group islands and plotting the accumulation of new species on size 4 islands at increasing distances from these focal islands. In Fig. 6.9, species are gained with distance from the 𝒜12 and ℰ9 islands at quite different rates. The species gained are classified into **L**, **M**, and **S** species using a variation of the previous definition. This modification is needed because species that are **L** or **M** species on one island might be **M** or **S** species, respectively, on another island. I therefore treat species operationally in the 149-species list as "**L**" species if their summed incidence (across islands) is

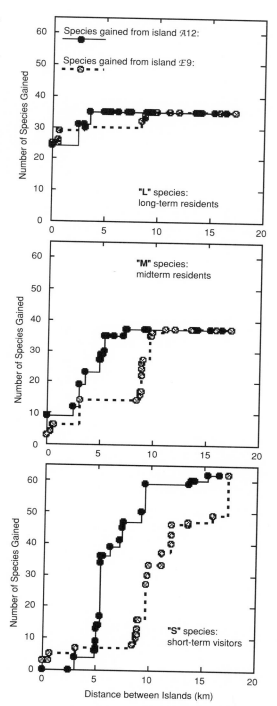

FIGURE 6.9. Among "size 4" islands, new species are gained (accumulated) faster in 𝒜-group representative 𝒜12 than in ℰ-group representative ℰ9. The difference is particularly obvious in M and S (midterm and short-term) species residents.

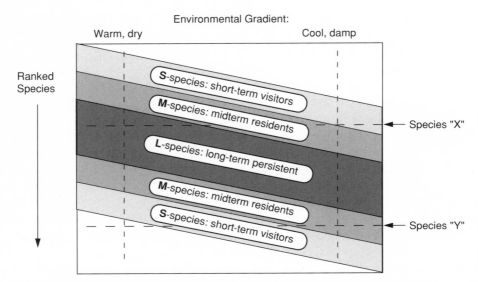

FIGURE 6.10. Putative distributions of longterm (**L**), midterm (**M**) and short-term (**S**) resident plant species on islands across an environmental gradient (abscissa). Species are ranked (ordinate) from those with strong preferences for warm, dry sites (top) to those with strong preferences for cool, wet sites (bottom). Species (horizontal rows) have different **L**, **M**, **S** designations in warm, dry versus cool, wet habitats, such that turnover of **L**, **M**, and **S** species occurs as one moves across the gradient left to right. Species "X" becomes less persistent left to right, whereas species "Y" becomes more persistent. Species that are long-term residents (**L**) at the warm, dry end of the gradient become successively mid-term (**M**) and then short-term (**S**) residents as environments become cool and damp to the right (see dashed lines).

≥ 7.0 ($n = 35$ species), as "**S**" species if it is <2.0 but >0.25 ($n = 62$ species) and as "**M**" species if intermediate ($n = 38$ species). This procedure captures the distinctions among species' residence times on islands in a similar fashion as before; the quotation marks signify the computational difference. In all three species groups, species are gained from A_{12} much more rapidly, that is, from closer islands, than they are from E_9. In general, E_9 requires contributions from islands some 5–10 km further distant to reach a comparable species gain to A_{12}, and thus the full complement of species within the "**L**," "**M**," and "**S**" categories is reached from A_{12} from much closer islands relative to E_9.

One last point about distance decay might be made about the possible contribution of habitat or environmental gradients to it. If there were such gradients across Barkley Sound or among islands within it, then one might expect that they would be distinguishable in terms of changing species' incidences, and changing rosters of **L**, **M**, and **S** species, as reflected in Fig. 6.10. There is little if any support for this scenario, at least not among the size 4 islands, among which the incidences of "**L**" species are unaffected by island position and ubiquitous, and "**S**" species incidences predominantly determined by island isolation.

TURNOVER BETWEEN ISLAND TWINS:
BIOGEOGRAPHIC LEDERBERG PLATES

Some islands form pairs of individual islands that are closely adjacent and similar in size and topography (see a stylistic representation of this in Fig. 6.11). These islands are termed "twins," and with shallow channel depths between the individual islands indicating that they recently comprised a single island, and are only more recently separated, they are rather like Siamese twins. Since their separation, the florulas of the island twins have had an opportunity to diverge as colonization and extinction events occur at least

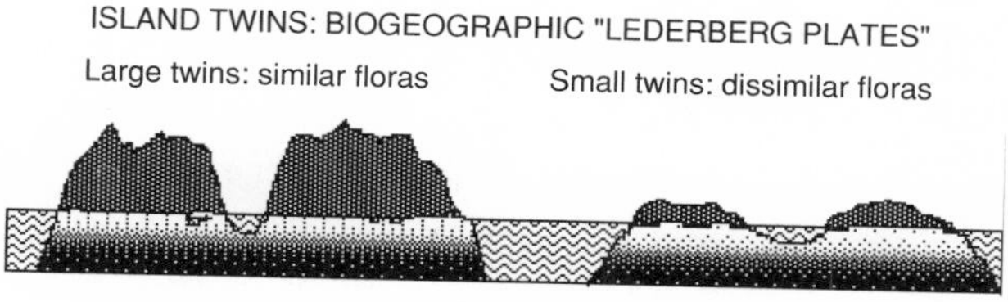

FIGURE 6.11. Island twins are similar in size and topography, closely adjacent in position, were recently conjoined, but are now separated by channels of <1 m in depth. Large twins retain similar floras but small twins, through the vagaries of colonization and extinction, have diverged widely in plant species composition.

partially independently on the two islands, and each can develop a discrete floristic makeup. I regard these island twins as akin to biogeographic Lederberg Plates, as they are templates that test for the random accretion of differences in florulas, just as the original plates tested for the random occurrence of mutations. There are 17 pairs of island twins, along with a triplet, which collectively span a range of areas over two orders of magnitude in size. They come from all five island groups, $\mathcal{A}$–$\mathcal{E}$, are almost all mutual near neighbors, and the larger exceeds the smaller by an average factor of just 1.15 ± 0.21 SD, and the taller is an average of 1.34 ± 0.47 SD times higher in elevation than the lower.

Species differences were computed between island twins as turnover T (computed as above) expressed as a percentage of the maximum turnover between an island pair, namely, $\Sigma_i J_{1i} + \Sigma_i J_{2i}$. The results are shown in Fig. 6.12, which plots percent turnover as a function of the mean area of the twin islands. There is a strong negative relationship, given by $\%T = 106.4 - 22.8 * \log \text{AREA}_{FL}$ ($R^2 = 0.37$, $p = 0.009$). Thus the florulas of twin islands have indeed diverged over time but are much more distinct between small island twins than larger twins. Between very small twins of around 100 m² in area, there is about a 2/3 turnover in plant species; for twins two orders of magnitude larger, the turnover drops to about 1/6 (see Fig. 6.12). Clearly, the vagaries of colonizing and surviving on small islands, with few persistent species, contribute to differences in species composition that are likely to lead to divergence in all but the most ubiquitous shoreline plants. Larger islands, on the other hand, would retain a greater proportion of their initially shared species and also provide more predictable targets for new colonists. There is no evidence from these data that species turnover differs between the eight $\mathcal{E}$-group twins and the nine twins on other island groups (although there is a slight tendency in that direction, with 6/9 of the $\mathcal{E}$-group twins lying below the regression line). Perhaps larger sample sizes would elucidate the expectation (see above) that turnover between $\mathcal{E}$-group twins would be less than between twins in the $\mathcal{A}$–$\mathcal{D}$ groups, but that would require relaxing the criteria for twin classification and perhaps introduce other sources of variation.

TURNOVER AND A POTENTIAL HISTORICAL LEGACY

The obverse of divergence in the florulas of twin islands is their potential for retaining species in common by dint of a recently shared past. Such historical legacies constitute a potential for islands with past connections to display reduced turnover. So we ask, are islands that have a recently shared contiguity, at times of lowered sea levels, likely to retain any evidence in their florulas of the recent connection, or has ongoing turnover through time erased its imprints? One approach to this question might employ the historical island phylogenies presented in chapter 2, together with information on the areal extent of past, conjoined islands. But there are two sorts of difficulties with this approach, one with the plants and another with the islands. Firstly, short-term visitors, the weedy edge species that comprise the bulk of the flora,

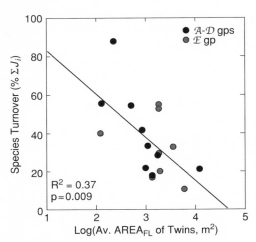

FIGURE 6.12. Species turnover between individual islands of island twins. Turnover is measured as the sum, over plant species i, of incidence differences between islands 1, 2 ($\Sigma_i |J_{1i} - J_{2i}|$) expressed as a percentage of the maximum possible difference between islands (i.e., no shared species, max. $= \Sigma_i J_{1i} + \Sigma_i J_{2i}$). Small twins are relatively different in species composition, large twins relatively similar. There are not significant differences in turnover between $\mathcal{A}$–$\mathcal{D}$ twins and $\mathcal{E}$-group twins.

are not at all likely to bear in their distributions any stamp of historical legacies lasting several millennia. On the other hand, the bulk of the long-term island residents are very broadly shared among islands throughout the Sound, and most occur on all islands of appropriate size; therefore they will contribute nothing to the question's resolution.

There are, secondly, other obfuscating factors that pertain to the island environments. Take, for example, a "beach satellite" islet such as a small rock stack separated from a much larger mother island only at high tide. The satellite might be home to a species that normally occurs on much larger islands, but the species may persist on the satellite for several reasons, which have already been discussed in relation to species richness (chapter 4). Firstly, the local environment (in terms of abiotic factors such as temperature and humidity) of the satellite will likely be influenced by its much larger neighbor and thus be atypical of small but more isolated islets. Secondly, because of close proximity to the large neighbor, it may be positioned to enjoy high colonization rates, such that unusual species occurrences on the satellite may be maintained by rescue effects from the nearby source of propagules. Thirdly, the satellite would have been a part of the mother island with a sea level drop of just a meter or so, and the oddities in its species list may be a signal of this historical legacy. Distinguishing among these possibilities, even with a comparatively long-term data set, requires both knowledge of dispersal dynamics and detailed measurements of microenvironmental factors; only a very limited resolution is possible here.

I discuss these options using a group of generally rare, but long-lived island species that show evidence of long-term persistence. These species were found on between 5 and 24 islands of the total 217 censused, and their distributions are shown in Fig. 6.13. The sizes of occupied islands are shown by the vectors at the bottom of each separate panel in the figure, which give the numbers of islands from the corresponding frequency distribution in the top left frame. In the top line of the figure are two manzanitas (*Arctostaphylos* spp.). The distribution of *A. columbianum* seems readily explained by habitat constraints; it occurs on only 6 islands at the extreme interior of the Sound, toward the northeast. These are the warmest and driest islands, and indeed this plant is quite restricted to the south-facing banks of these islands. The islands are all large with one exception, $\mathcal{D}$34, which is a beach satellite of the much larger $\mathcal{D}$24 "Big Stud." With a 1 m drop in sea level, $\mathcal{D}$34's present area of 228 m^2 would expand to 3341 m^2 to include $\mathcal{D}$24, 29, 32, 33, 34, and their immediate submarine surroundings. Unfortunately but not surprisingly, the health of the single individual of *A. columbianum* on $\mathcal{D}$34 is not good, and it has been looking rather poorly of late.

The bearberry (*A. uva-ursi*), has a distribution that is nearly the opposite of its congener, and many of the 24 islands it occupies are positioned southwesterly toward the cooler and foggier outer Sound. It also avoids most smaller islands; the two smallest on which it occurs, C15 and C16, with present areas of 953 m^2 and

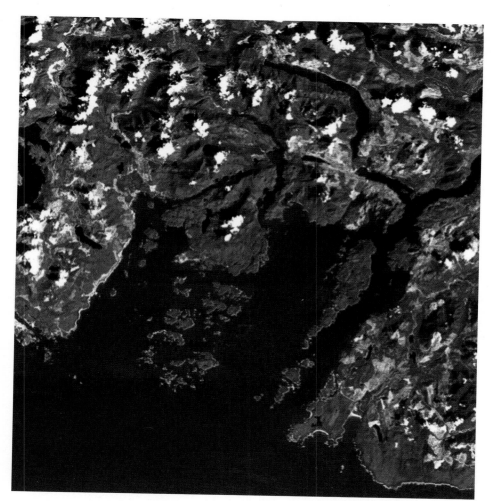

PLATE 1. A satellite view of Barkley Sound.

PLATE 2. View west across Barkley Sound from Pachena Dome on the Vancouver Island "mainland," with the open ocean off to the left (south). Across the Trevor Channel (foreground) are the Deer Group islands and beyond, across the Imperial Eagle Channel, lie the Broken Group islands situated in the middle of the Sound. This study is focused on some 200 islands of the many hundreds available.

PLATE 3. Transport to and among the Barkley Sound islands is facilitated by our inflatable Novurania, seen here anchored at Ohiat (*B3*) island. Helby Island lies in the midground (right), and in the far background (to the left, northeast) is the Vancouver Island mainland north of Bamfield.

PLATE 4. Nanat (*A11*) is a small (0.9 ha) but tall (21 m) island in the Trevor Channel and close to the mainland. Tall 40 m forest protects a damp interior that harbors near-black slugs and shades most of the coastline, precluding most herbaceous edge species. However, in one exposed site a tiny population of wall-lettuce *(Lactuca muralis)* has persisted for over two decades and led to studies of the evolution of reduced dispersal in the island anemochores.

PLATE 5. Middle Ross (*C4*) is a typical small- to mid-sized island, with area 2820 m^2 above the *Fucus* line, 351 m^2 of this area higher than the 2m contour, and with 970 m^2 occupied by vascular plants. It has averaged 37.75 plant species over 22 years, and a total of 46 species have been found there.

Clockwise from top left:

PLATE 6. The tiny island of Ross NNW (C5; 0.11 ha) is around 10 m tall and has averaged about 26 plant species over 22 years (34 cumulatively). It is the smallest island with a persistent population of the lady fern (*Athyrium felix-femina*) and is notable also for its populations of *Lactuca muralis* and *Senecio sylvaticus* that fluctuate widely in size (10X to 50X) between years.

PLATE 7. Poison Ivy (D7) measures 1960 m^2 above the *Fucus* line, but with its extensive low reefs, only 450 m^2 is occupied by vascular plants. The species count has varied from 19 to 23 species, 17 of which have been persistent over the study period (22 years). It is one of the smallest islands with surviving slugs, and also supports piggy-back plant (*Tolmiea menziesii*), which generally occurs only on much larger islands.

PLATE 8. Most of the islands' plant diversity is found at and near the shorelines, such as here on the east coast of the very large Sandford (C36: 290,000 m^2). In this muddy fresh-water seep are found such relative wetland rarities as *Plantago macrocarpa, Deschampsia caespitosa, Triglochin maritimum* and *Stellaria humifusa*, and the site was colonized by *Potentilla pacifica* in 1991.

PLATE 9. Many island shorelines are steep, dry and rocky, such as the west coast of Owen (E54) pictured here. Shorelines this steep support relatively few weedy edge species, but as Owen is large, with a cool and damp forested interior, it hosts such large-island rarities as *Aruncus sylvester, Streptopus amplexifolius,* and *Tolmiea menziesii*.

Clockwise from top left:

PLATE 10. Eleven species of ferns have been found on the islands, and three occur together here in a shallow cave on the north coast of Seppings (B6): *Polystichum munitum*, above; *Adiantum pedatum*, center left and lower right; *Polypodium scouleri*, e.g., center right. Seppings has six fern species in all; fern species decline very regularly with decreasing island size, with only those with the thickest fronds left on the smallest and driest islands.

PLATE 11. Licorice fern *(Polypodium glycyrrhiza)* is the only island fern commonly epiphytic (on mossy tree trunks). Though a long-lived perennial its fronds are annual, deciduous, and much thinner than the perennial fronds of its congener, leatherleaf fern *(P. scouleri)*. As a consequence, *P. glycyrrhiza* is absent from smaller islands that are drier and largely devoid of trees, on which *P. scouleri* flourishes.

PLATE 12. Dispersing propagules are very much at the mercy of chance, and new colonists may land in rather unconventional sites. This fireweed *Epilobium angustifolium*, an anemochore, became established on a driftwood log but nevertheless has managed to reach reproductive maturity.

PLATE 13. Some island colonists become established in rather inhospitable places (Tim Cody commiserates). This elderberry *(Sambucus racemosus)* has red bird-dispersed berries that are deposited haphazardly in open rocky sites where long-term survival is unlikely. Elderberry is a very successful colonist but a poor persister, with some 46 colonization events and 49 extinction events observed on 96 oft-censused islands.

Clockwise from top left:

PLATE 14. One of the most ubiquitous of the shoreline plants is fuzzy cinquefoil (*Potentilla villosa*), with an amphiberingian distribution (East Asia to North America). Perennial with individual longevity up to two decades, the species persists best on the dry rocks of exposed islands, has lowered incidence on sheltered islands with shaded coasts, and suffers population declines in wet years.

PLATE 15. Yarrow (*Achillea millefolium*) makes the top ten list of the most widely distributed island species, and occurs on 80% of the 152 islands greater than 1/20 ha. Its fruits mature in late summer and early fall, when they are likely dispersed on the feet of migrating shorebirds; hence there is no reduction of *Achillea* incidence with increasing island isolation.

PLATE 16. The classic model for a wind-dispersed plant (anemochore) is a herbaceous dandelion-type plant such as this cat's ear (*Hypochaeris radicata*). However, weedy anemochores are less prevalent on more isolated islands, indicating a dispersal limitation. *H. radicata* occurs on 21% of islands <4 km from the mainland (n=112), but on <3% of islands >4 km away (n=105).

Clockwise from top left:

PLATE 17. Wall-lettuce *(Lactuca [Mycelis] muralis)* is a widely distributed anemochore with good colonization rates and moderate persistence rates. As the years of establishment of many island populations was known, evolution in the morphology of the dispersal unit, pappus P and achene A, could be related to population age. Strong selection on islands for reduced dispersal ability decreases the ratio P/A (measured as volumes) and is detectable after just a decade of island residence.

PLATE 18. Having propagules that drift on ocean currents is a convenient dispersal mode of some of the plants along the island shorelines, the "floaters." Pea family plants have pods that seem nicely pre-adapted to dispersal via drift floating; the beach pea *(Lathyrus japonicus)* is a good example. Ocean currents, however, may make some islands hard for floaters to reach.

PLATE 19. Many of the shrub species at the forest edge are characterized by having conspicuous berries eaten and distributed by birds and some mammals (via zoochory). Included in this group are huckleberries and relatives (*Gaultheria, Vaccinium*; family Ericaceae), but most are in the family Rosaceae (e.g., *Amelanchier, Crategus, Malus, Prunus, Rosa, Rubus, Sorbus*). Here red berried mountain-ash *(Sorbus sitchensis)* grows with purple-berried salal (*Gaultheria shallon*) at an island edge.

Clockwise from top left:

PLATE 20. The largest plant family on the Barkley Sound islands is the grasses (Poaceae), with about 60 species. Two grass species are particularly prevalent along rocky shorelines, the meadow barley (*Hordeum brachyanthemum*, shown here) and red fescue (*Festuca rubra*). On medium-sized and larger islands both species are present and common, but on small to tiny islands one or the other species becomes absent or very rare; the two seem to be close competitors.

PLATE 21. Most sedges are wetland species, including many bog specialists in the genus *Carex* (the largest genus in British Columbia). *Carex pansa* is generally considered to be rare in the coastal regions of the province, but it is quite common on the islands of Barkley Sound (especially in the central *E*-group), where it is typical of dry rocks along island coastlines.

PLATE 22. Several of the largest islands have bogs, the smallest of which is this 0.2 ha bog on Turtle (*E*76). It supports 25 species of specialist bog plants, like this water-lily *(Nuphar polysepalum)*, and is not obviously species-poor. Turtle Bog is the only known island location for several of these bog specialists, such as bur-reed *(Sparganium emersum)*.

Clockwise from top left:

PLATE 23. Bunchberry *(Cornus canadensis)* is a routine component of mainland bogs and occurs also in all three island bogs surveyed—on Nettle, Effingham, and Turtle (E76) islands. Though just a few centimeters tall, its red berries, likely indicating zoochory, render it very conspicuous in mid to late summer.

PLATE 24. Banana slugs *(Ariolimax columbianum)* are found on virtually all islands over a half-hectare in size, but on smaller islands slug color patterns diverge from the usual (variably black-spotted). On small, damp islands the slugs become almost totally black, but on small, dry islands, most slugs are unspotted, as here on the 1/10 ha Fleming E Rk. (C24). Spotting patterns likely affect thermoregulation and water loss, and vary with island temperatures and relative humidity.

PLATE 25. Prolific mucus production in banana slugs serves as an aid in transportation, possibly also in moisture retention, and additionally as an anti-predator device. Birds such as thrushes show acute discomfort after attacking slugs, but a Northwestern salamander *(Ambystoma gracile)* had no problem ingesting a slug of nearly its own body size on Clarke (E49).

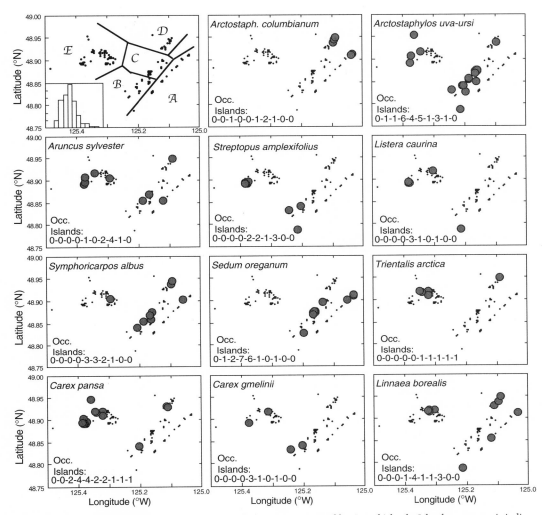

FIGURE 6.13. Distributions of some rare but long-lived plant species on Barkley Sound islands. Island occurrence is indicated by the hatched symbol, and island numbers and size by the vector at the bottom, which relates to the frequency distribution at top left. These are candidate species for possibly distinguishing among habitat factors, spatial position, and historical legacies in determining distribution; See text for discussion.

115 m², respectively, would join together with C5, C7–16, and C35 (on 6 of which it also occurs), to constitute a large 43,540 m² island, with just a 1 m drop in sea level.

In the second row of Fig. 6.13, goat's beard (*Aruncus sylvester*) and the lily *Streptopus amplexifolius* are recorded on large islands toward the outer edge of the Sound, with one exception. The former occurs on the very large $D36$: Weld (AREA$_{FL}$ 220,559 m²), far to the Sound's interior, and indicating, I believe, that very large islands can compensate for interior position by large size. Islands such as Weld can provide cool and moist habitats that to some extent mimic conditions on the outer Sound islands, if they are of sufficient size. Cool, damp islands of the outer Sound are where the nonwoody, tuberous-rooted *Streptopus* is found. The species is very likely to be long-lived, as I have found that individuals of its relative *Smilacina racemosa* in the Sierra Nevada to live for at least a half-century. (N.b., these geophytes can be aged by tracking annual increments to their subterranean tuberous rhizomes.) However, I do not see any evidence for historical legacies in *Streptopus*, only a preference for the outermost islands of the Sound. Another rare monocot is the orchid *Listera caurina* (twayblade), which is also undoubtedly long-lived; I have found this plant on just five islands, on one of which (*E*19) we were unable to find it after the first discovery. Three of its other islands are grouped off the northwest corner of *E*49: Clarke, and are *E*58: "Big Pigot," *E*59: "Little Pig E," and *E*60: "Little Pig W." These three islands plus nine others would join Clarke to form the 385,870 m^2 paleo-island "Clarke-Pigots" at sea levels lowered by 1 m. However, while the species persisted on the larger *E*58 through its last census, in 2000, the two individuals found on *E*59 in 1986 and 1989 were not relocatable in 2000, and the several initially discovered on *E*60 could not be found after 1986. Conceivably the species can persist subterranially in a dormant stage (as can other geophytic orchids) without the need to emerge each and every year. The species appears to have suffered from the dry summers of the mid- to late 1980s, and it will be interesting to see whether or not it stages a recovery.

Snowberry (*Symphoricarpos albus*), and stonecrop (*Sedum oreganum*), in the third row of the figure, are virtually absent from the *E*-group islands. The former is an edge shrub of open sites, the latter an herbaceous succulent that prefers sunny and rocky banks. Snowberry was found as a single individual on *E*18: Onion (AREA$_{FL}$ = 46,200 m^2) in 1982 but disappeared thereafter; another (single) individual was found growing on a log on Onion's north coast in 1989, but it too failed to persist. The species occurs mainly on larger and more open islands, although it certainly colonizes others that subsequently prove unsuitable (as above). It persists on *B*1: Wizard, which is a midsized island (6290 m^2) but low in elevation (3.4 m tall) and with very open vegetation. The stonecrop is absent from *D*- and *E*-group islands, curiously so in the case of the former. Seven of the *Sedum*'s 18 islands are those in the *C*-group islands co-occupied by *A. uva-ursi* (see above), islands that were united into a single, large island when sea levels were 1 m lower. Lastly in the third row is starflower (*Trientalis arctica*), rather widely distributed on the boggy and mossy forest floors of the mainland but extremely restricted on the islands. My five records (*D*36, *E*44, and *E*76–78) and one additional record in Bell and Harcombe (1973) on Effingham Island suggest the species is extant only on some of the very largest islands. The listed five are 220,559, 85,000, 1,146,110, 474,559, and 10,892 m^2 in size, and Effingham is even larger. The last mentioned, *E*78: "Willis W End," is the smallest, eight times smaller than the next larger, but *E*78 would join neighboring *E*77: Willis at sea levels 1 m lower. The occurrence of starflower on conspicuously smaller *E*78 seems to be a probable legacy of its recent past, when it was a part of an expanded 571,640 m^2 paleo-Willis.

Two sedges are shown in the last row of Fig. 6.13, with *Carex pansa* the more widely distributed species (on 18 islands, including 1 *B*- and 2 *D*-group islands, but the remainder all in the *E* group). It is found on relatively dry and rocky shorelines. Most of *C. pansa*'s islands are very large, but 5 are appreciably below one hectare in area. Three of these would join the expanded Clarke-Pigot paleo-island (see above), but given the preferred habitat of the species, and the questionable idea of historical legacies for shoreline plants, it seems inappropriate to invoke that sort of explanation. Indeed, another small island on which *C. pansa* occurs is *E*57: Puffin (AREA$_{FL}$ = 2340 m^2), which is a deepwater island and has had no recent land-bridge connections. The second sedge shown, *C. gmelinii*, is much more restricted in distribution and habitatwise occurs on gravelly and open edge sites. I have found it on just four islands (it was unrecorded by Bell and Harcombe [1973] as was *C. pansa*), in two areas with no apparent positional or environmental similarities. It occurs on *B*28: Leach (AREA$_{FL}$ 28,400 m^2) and its neighbor *B*30: Folger (ca. 180,000 m^2), two islands that have a long legacy (see Fig. 2.8) of shared isolation, since sea levels 50 m below present, and mutual isolation just since sea levels of 2 m below present. The other two occupied islands, $E47$ and $E48$, each 5900 m² in area, are twins off the south coast of the larger Chalk (uncensused) and until very recently would have been joined together, with Chalk and other neighbors, in a 176,790 m² paleo-island. In both *C. gmelinii* locations there is a suggestion of historical legacy; predicting and subsequently confirming the species' presence on Chalk Island would bolster the case, and this is high on my to-do list.

Lastly the twinflower (*Linnaea borealis*) is shown in Fig. 6.13 (lower right), with a broad distribution across the Sound. It occurs on the three largest of the A-group islands, three of the four largest D-group islands (including the twins $D9$: "Geer N" and $D10$: "Geer S"), and on four E-group islands (plus another three recorded by Bell and Harcombe [1973]). All of these are quite large islands (the Geer twins in combination are 33,865 m² at −1 m sea level), except the two islands off the south side of $E44$: Walsh, namely, $E45$: "Walsh WSW" ($AREA_{FL}$ 9220 m²) and $E46$: "Walsh SSW" ($AREA_{FL}$ 2250 m²). These two smaller islands support thriving *Linnaea* populations, as does neighboring Walsh, and they share with Walsh a common ancestry, being all part of an extended 147,740 m² "Greater Walsh" at sea levels 1 m lower. Given that *Linnaea*'s wide distribution argues for its broad habitat and environmental plasticity, and recognizing a strong association with large islands, there seems to be some basis for arguing that the current *Linnaea* distribution retains elements of historical legacies.

7

Dispersal Syndromes, Incidence, and Dynamics

DISPERSAL IN PLANTS: OPTIONS

The older literature on plant dispersal is largely descriptive and phenomenological (Ridley 1930; van der Pijl 1982), but it provides a broad background to the natural history and the remarkable diversity of the means by which plant propagules manage to reach new habitat. The review of the ecology of seed dispersal by Howe and Smallwood (1982) set the stage for more recent experimental and analytical approaches, which have developed markedly since that time. Resurgent interest in dispersal questions has produced a variety of promising approaches, and two edited volumes provide good coverage of new theory and new technique and give case histories of a range of detailed ecological studies (Clobert et al. 2001; Bullock et al. 2002). Amongst the more plant-oriented studies of dispersal, the role of frugivorous vertebrates in seed dissemination has received the most attention; recent compilations are those of Estrada and Fleming (1986) and Levy et al. (2002).

While adult plants are relatively sessile and limited to a local exploration of space via vegetative growth, crabgrass-type rhizomes, or strawberry–type runners, other life stages are remarkably mobile. Like many marine invertebrates but unlike most terrestrial animals, plants have two levels of dispersal, one of which (the zygote propagule) is relevant to the colonization of islands. Both gamete and zygote dispersal have implications for the genetic structure and contiguity of populations. Male gametes disperse via pollen grains, and species that are wind pollinated, such as oaks and pines, may have a much greater dispersal potential with their pollen than with their heavy seeds or fruits. But while pollen grains are very mobile, their longevity in air seems to be a matter of hours or less, UV radiation and desiccation being limiting. Pollen dispersal distances in forest trees seem limited in practicality to a few hundred meters (Smouse and Sork 2004); certainly the 36 km gap between the Sierra Nevada and the Inyo Mountains, southern California, has proved too great for pine pollen to bridge (Nathan et al. 2002). However great or narrow the range of pollen transport, seeds or fruits that are dispersed by wind or by sea currents have a lot of potential to reach far distant establishment sites. Assessing the extent of dispersal via the genetic turnover among populations (F_{ST} values) has been facilitated by recent advances of molecular techniques and is likely to become an increasingly important tool in dispersal studies (Rousset 2001).

The fact that plants have colonized the most remote patches of land on the globe readily attests to their colonization abilities. A single archipelago, or a lone and isolated oceanic island, may house a diverse flora representing colonists originating from a variety of different sources and using various means of transportation, as described by Joseph Hooker for the Galapagos Islands and discussed in chapter 1. Carlquist (1974) undertook detailed analyses of the origins of the Hawaiian flora, and also that of other Pacific Ocean islands. Riding air currents ("anemochory") and sea currents ("hydrochory"), and using the most vagile vertebrates, namely, birds ("ornithochory"), were all popular and successful transportation options. Remote islands are, however, not equally accessible to the species in potential source floras, and insular floras reflect variation in the efficacy of the various plant dispersal modes for long-distance, transoceanic transport. The floras of especially remote islands have been labeled "disharmonic," reflecting an absence or scarcity of some taxa expected on the basis of habitat suitability, and overrepresentation of others; the result is a species mix unlike that of typical mainland floras. Carlquist (1974) documents many examples of disharmony in the plants of isolated islands; he views disharmony as a condition predisposing island floras to adaptive radiation and does not exempt depauperate continental islands from its consequences.

I described earlier (see Fig. 3.10) some indications of potential dispersal limitation on island incidence in Barkley Sound. There were no effects of distance to mainland in four shoreline species, two weedy composites, a forest fern, and an orchid (Fig. 3.10, top row). In other species, congruence in incidence versus distance relationships indicated strong effects of island position, and an implication of habitat constraints or abiotic factors in their shapes (Fig. 3.10, center row). In four other plant species (Fig. 3.10, bottom row), isolation distance was a prominent associate of incidence, but with the exception of the grass *Elymus glaucus*, the relations are not monotonic, and factors besides isolation distance per se appear to be involved. Dispersal limitation is further examined in Fig. 7.1 for four plant species. Early hairgrass (*Aira praecox*; Poaceae) is a weedy, annual alien of dry, rocky edge habitats; its lemmas are awned, and seeds are likely dispersed in bird feathers. Small-flowered alumroot (*Heuchera micrantha*; Saxifragaceae) is a native, long-lived forb of partially shaded, steep banks; its seeds are small spined and contained in beaked capsules, and it also seems designed to attach to vertebrates, most likely birds, to effect dispersal. Oceanspray (*Holodiscus discolor*; Rosaceae) is a forest-edge shrub with small and hairy achenes (2 mm long) that are likely wind dispersed under dry conditions. False azalea (*Menziesia ferruginea*) is a forest understory ericad shrub, but unlike most of its confamilial blueberries and huckleberries, it is dry fruited (as implied by its other common name, "fool's huckleberry"). Its capsules might conceivably appeal to fruit-eating birds, but I should think it must depend on other, more limited dispersal means. Species' incidence is shown as dependent on isolation in Fig. 7.1, with island size effects partially controlled. *Aira* is predominantly a small-island species, and a subset of islands in *SIZE* groups 3 and 4 ($n = 104$) is considered. *Heuchera* is prevalent on midsized islands ($SIZE = 4, 5$; $n = 95$), and the two shrubs are typically confined to larger islands ($SIZE > 4$; $n = 64$).

Both *Aira* and *Heuchera* are distributed across all island groups (*A–E*), and in both incidence falls relatively linearly and at comparable rates with distance from mainland D_{MLD} (Fig. 7.1, upper; D_{MLD} coefficients –0.079, –0.092; $R^2 = 0.20, 0.17$, respectively; $p < 0.001$). *Holodiscus* (Fig. 7.1, lower left) was not found on any of my censused *E*-group islands, although Bell and Harcombe (1973) found it on two islands (Weibe, Dicebox) in that group; the regression

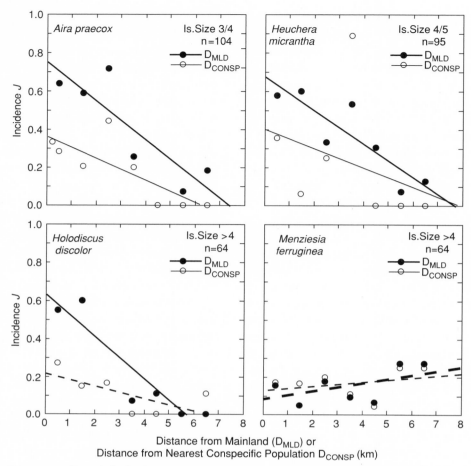

FIGURE 7.1. Effects of island isolation (abscissa) on incidence J (ordinate). Thick regression lines and filled symbols: isolation measured as distance from mainland (D_{MLD}, km); thin regression lines and open symbols: isolation measured as distance from nearest known conspecific population ($D_{CONSP.}$, km). Dashed lines indicate nonsignificant regression statistics. See text.

of its incidence against D_{MLD} is also significant, with similar statistics (coeff. −0.079; R^2 = 0.211; p < 0.001). In *Menziesia*, however, there is no significant distance effect (p = 0.097; R^2 = 0.04), and as indicated earlier other factors control its incidence. Of these four species, only *Menziesia* is likely to be relictual in distribution from times of lowered sea levels when the islands formed part of an extended Vancouver Island mainland. The other three would very likely not be represented in continuous forest habitat and, therefore, would require dispersal to reach newly formed islands.

It is readily apparent that colonist sources are not restricted to those on the Vancouver Island mainland, but sources of potential colonists for empty islands must include extant island populations that could act as secondary, if not primary, points of origin. I include an analysis of this effect in Fig. 7.1, where incidence is plotted as a function of distance from known, extant conspecific populations (D_{CONSP}). As not all Barkley Sound islands were censused, there is some resultant ambiguity in these data. But given the broad distribution of the censused islands, errors from incomplete coverage hopefully will not completely override any patterns that may exist. Neither *Aira* nor *Heuchera* occur on islands >4 km from known and potential source populations on other islands, even though there are vacant islands that are apparently suitable. These absences strengthen the implications of dispersal limitation. In *Aira*, multiple regression with both D_{MLD} and D_{CONSP} as independent variables shows D_{CONSP} as a marginally significant contributor to incidence (coeff. 0.072; $p = 0.066$) and R^2 is marginally improved from 0.20 to 0.23. In *Heuchera*, D_{CONSP} is a significant addition (coeff. 0.087; $p < 0.01$), and R^2 increases from 0.17 to 0.24.

Although *Holodiscus* incidence falls with distance from conspecific populations (Fig. 7.1, lower left; coeff. 0.013, NS), this variable does not add significantly to D_{MLD} in explanatory power. It would appear that plumed achenes are at a disadvantage in reaching the foggy and damp Æ-group islands, even though suitable habitat appears plentiful there. Bell and Harcombe reported that the species is surprisingly abundant on Dicebox Island, though it has not managed to spread from this one, fortuitous foothold. Lastly, incidence in *Menziesia* is unrelated to distance from extant conspecifics, as might be expected of a species with a relictual distribution. Note that even in the Æ-group islands, with their forest-friendly microclimates, *Menziesia* is not widely distributed, and it is represented on only about one-quarter of the larger islands. There seems, then, to be a potential for rescue effects in *Menziesia* populations, but rescue apparently happens rarely, since it has been ineffectual in maintaining the original (pre-insular) distribution. This again speaks to a dispersal limitation in the species.

FERNS

Ferns, with their light, windborne spores, clearly have tremendous dispersal potential. These dustlike specks average around 30–50 µm in size, about the same size as pollen grains, though those of wet forests species are generally somewhat larger (W. H. Wagner, in Carlquist 1974). Fern spores in the process of aerial travel have been captured from aircraft trawls in the upper atmosphere and clearly have the potential to reach places far from their points of origin. Spore longevity also is impressive, ranging from several weeks to over a century (Lloyd and Klekowski 1976), and spores are produced in vast numbers from sori beneath the fronds of mature (sporophyte) ferns, on the order of 10^5 per frond. Imagine a hectare or so of forest ferns at routine densities producing around 10^{10} spores each year! The power of fern dispersal is attested by their success in reaching, not once but repeatedly, isolated land specks such as the Hawaiian and the Juan Fernandez Islands in the North and South Pacific, respectively; along with many other remote oceanic islands, these have well developed and diverse fern floras. It has been estimated that over 100 fern taxa have independently colonized the Hawaiian Islands, where, with adaptive radiation and autochthonous speciation, there are now nearly 200 native species, two-thirds of them endemic.

Another line of evidence for the vagility of ferns comes from the extremely wide geographic ranges of both genera and species. Bracken *Pteridium aquilinum* occurs in suitable habitat worldwide, a feature that undoubtedly prompted Lawton's (1984) geographically comprehensive studies of bracken's fauna of insect herbivores (with an assessment that there is little evidence for convergence in size or organization of these herbivore communities). This is one of the 10 fern species of Barkley Sound listed in Table 7.1, and its distribution encompasses Hawaii, New Zealand. and Taiwan (as well as Eurasia, Australia, and North and South America). Of the nine genera in the table, all occur in Taiwan, eight in South Africa, seven in Hawaii, and six in New Zealand. Of the listed fern species, *Asplenium trichomanes* and *Cystopteris fragilis* occur also in South Africa and Taiwan, besides Barkley Sound. The latter species was represented by material from Tasmania and illustrated as "*Cystopteris tasmanica*" in Sir Wm. J. Hooker's (i.e., J. D. Hooker's *père*) 1854 volume *A Century of Ferns*, but the taxon was relegated to synonymy with *C. fragilis* by Bentham (1878).

TABLE 7.1
Fern Species of Barkley Sound

			ISLAND SIZE (LOG[AREA$_{FL}$]) AT INCIDENCE $J = 0.20$			
SPECIES	HABITAT CHARACTERISTICS	LEAF SPECIFIC WEIGHT (g/m^2)	ALL ISLANDS	A–D ISLANDS	E ISLANDS	P ISLANDS[a]
Adiantum pedatum	Forest interior; wet, shaded rocks	12.3 ± 2.3	4.57	5.01	4.71	5.68
Asplenium trichomanes	Forest interior; shaded rocks	—				
Athyrium felix-femina	Forest interior	27.5 ± 8.6	4.44	4.48	4.1	4.96
Blechnum spicant	Forest interior	60.4 ± 12	3.72	3.96	3.56	4.31
Cystopteris fragilis	Wet forest interior: 1 record (E76: Turtle)	—				
Dryopteris austriaca	Forest interior	39.8 ± 11	4.53	4.93	4.56	5.18
Dryopteris felix-mas	Forest interior; 1 record (E7: "S Hosie")	—				
Polypodium glycorrhiza	Chiefly forest and epiphytic on spruce	81.3 ± 30	2.87	2.54	3.2	2.98
Polypodium scouleri	Forest edge and exposed rocks	209 ± 49	2.25	2.24	2.23	2.31
Polystichum munitum	Forest interior to edge	106 ± 23	2.81	2.91	3.49	3.11
Pteridium aquilinum	Edge, sunny banks; southern exposure	66.1 ± 13	4.48	4.87	5.15	5.03

[a] P islands are those where the species is permanently resident.

Overall, then, ferns appear to be the classic plant taxon in which there are no dispersal limitations, and thus distribution will be determined solely by the characteristics of the sporophyte habitat. And for habitat characteristics, island size appears to be an appropriate surrogate, as suggested by the incidence functions in Fig. 7.2. Most of the commoner species, those that occupy a broader range of island sizes, display smoothly declining incidence with decreasing island size, and in Fig. 7.2 these curves are well fit by logistic regressions. The figure shows incidence curves for eight fern species; not represented are *Cystopteris fragilis*, which we found only on E76: Turtle, the largest island surveyed, and *Aspenium trichomanes*, with two island occurrences (A11: Nanat and E49: Clarke, medium and very large sized, respectively). Neither of these rarer species was reported by Bell and Harcombe (1973), although they did find a grape fern, *Botryichium virginianum*, on a single E-group island, Hand (unsurveyed by us), a species we encountered only in mainland bogs. Note that some of the incidence variation on larger islands in Fig. 7.2 may be attributable to low sample sizes in

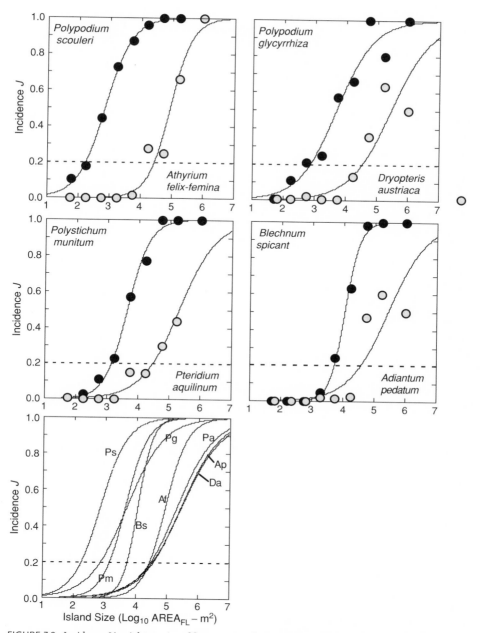

FIGURE 7.2. Incidence J in eight species of ferns, across all Barkley Sound islands. Logistic regression equations are fitted to the incidence data, which in the commoner species increase smoothly from 0 to 1 with increasing island size. Note that incidence falls off at different island sizes in different fern species. Number of islands in nine size categories are 10, 38, 44, 59, 36, 16, 5, 5, and 2.

those ranges. The numbers of islands per size group (in half order-of-magnitude increments) is 36–59 between $\log_{10}(\text{AREA}_{FL})$ two to four, but there are just five islands in each of the ranges 4.5–5 and 5–5.5, and only two islands with sizes $\log_{10}(\text{AREA}_{FL}) > 5.5$. *Pteridium aquilinum* is ab-sent from both of the largest islands, but this datum is not shown in the figure.

The lower left panel of Fig. 7.2 superimposes all eight of the fern incidence functions. While the three more restricted species, *Pteridium aquilinum*, *Adiantum pedatum*, and *Dryopteris austriaca*, are similar in incidence, the other five differ in that incidence drops at different rates with decreasing island size, and the island size at which incidence drops off also varies among species. The dashed line in the figure, drawn at incidence $J = 0.20$, intersects the incidence functions of different species sequentially at different island sizes, and these values can be used as a threshold index of critical island size (see below).

With dispersal limitations potentially excluded as a factor in fern distribution, I next show fern incidence with islands subdivided into the $\mathcal{A}$–$\mathcal{D}$ groups, nearer to the Vancouver Island mainland and more peripheral within Barkley Sound, and the centrally located $\mathcal{E}$ group. The results of this island split are shown in Figure 7.3. Four ferns of shady forest interiors, *Blechnum*, *Athyrium*, *Dryopteris*, and *Adiantum*, all appear to be more widely distributed on the $\mathcal{E}$-group islands. I interpret this to be a consequence of the generally lower maximum temperatures and higher relative humidities on $\mathcal{E}$-group islands. Bracken (*Pteridum aquilinum*) is of generally low incidence but shows a tendency to be commoner on the drier and sunnier $\mathcal{A}$–$\mathcal{D}$ islands. The three fern species with the broadest incidences are the two *Polypodium* species plus *Polystichum*, and they are not unequivocally higher in incidence on $\mathcal{E}$-group islands, unlike the forest interior species. *Polypodium scouleri* is commoner on $\mathcal{E}$-group islands except for the very small islands, and *Polystichum munitum* is commoner on the $\mathcal{A}$–$\mathcal{D}$ islands possibly excepting the largest islands. The largely epiphytic *Polypodium glycyrrhiza* shows a definite incidence split, being commoner on larger islands in the $\mathcal{E}$ group but commoner on smaller islands in the $\mathcal{A}$–$\mathcal{D}$ group. Given that *Polystichum munitum* and *Polypodium glycyrrhiza* are quite generally distributed on smaller and drier islands and obviously tolerant of the conditions prevalent on such islands, it might appear that their surprising scarcity in small $\mathcal{E}$-group islands might be attributable to a dispersal limitation. However, small islands are small targets for dispersing propagules, and the islands in the center of the Sound may simply receive fewer colonists than those closer to the edge, and closer to the large sources of spore production in the mainland forests.

It should be noted again (see Fig. 3.3) that incidence functions are a static representation of dynamic processes, and that, over time, some islands in a size class will lose a particular species, others will gain it, while on other islands the species may be perceived as a permanent resident. As the fern incidence functions of Figs. 7.2 and 7.3 are computed not by presence-absence but by average incidence over time, J_T, they include records of island incidence bolstered by transient populations; only on a proportion of the islands per size class are the ferns permanently resident (i.e., for the duration of this study). Figure 7.4 illustrates, then, another sort of incidence gap; in this figure, fern incidences are split by populations that are permanent residents **L** versus those that are transients, either short-term residents (**S**) or midterm residents (**M**). In some species, the difference in incidence attributable to the transients is proportionately similar over island sizes, and the curves, with and without the transients, are approximately congruent. This is the case with *Polypodium glycyrrhiza* and *Polystichum munitum*. In species typical of wetter and shadier forest, including *Blechnum spicant*, *Athyrium felix-femina*, *Adiantum pedantum*, and marginally with *Dryopteris austriaca*, the incidence gap is lower on the curve, indicating that transients are more prevalent on smaller islands (which are drier and less shaded). In two species associated with open and sunnier sites, *Polypodium scouleri* and *Pteridium aquilinum*, the incidence shift bulges higher on the curve, indicating more transients are recorded on larger islands where open, sunny sites are in general scarcer. The interpretation of fern distributions dominated by habitat factors and suitable conditions for persistence, rather than by dispersal limitations, becomes the more tenable in the light of Fig. 7.4.

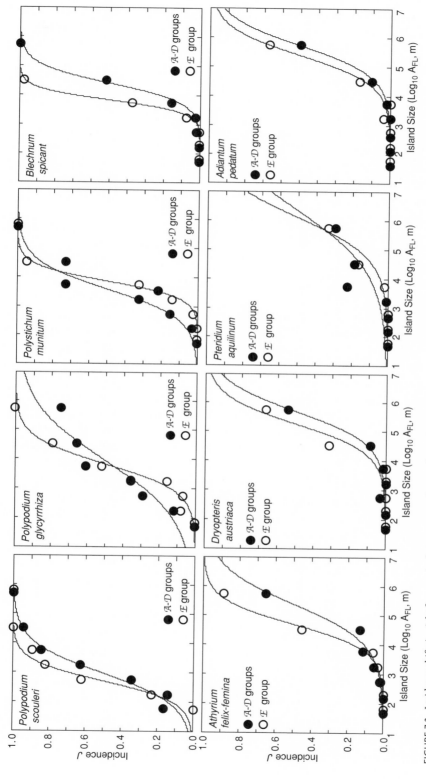

FIGURE 7.3. Incidence shifts in eight fern species between near-shore and peripheral Barkley Sound islands 𝐴-𝐷 groups and central Barkley Sound 𝐸-group islands. Note the general tendency for most fern species to be more widely distributed on smaller islands in the center of Barkley Sound. Numbers of islands in the seven size categories for 𝐴-𝐷 islands are 6-25-29-28-22-12-4, and for the 𝐸 islands are 4-13-15-31-14-9-3.

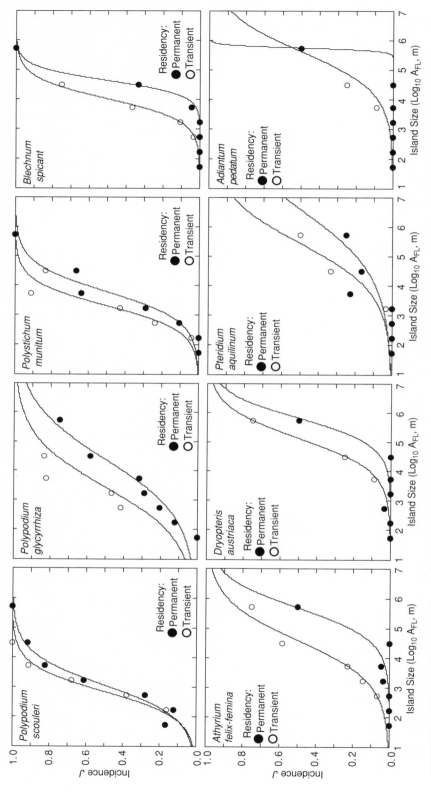

FIGURE 7.4. Incidence gap between populations that are permanent residents on islands (solid symbols) and populations that are transient (short-term S or midterm M) populations. Ferns are transients on islands that are considerably smaller than those on which they maintain permanent populations. Sample sizes for island size groups as in Fig. 7.2.

Yet dispersal limitations are not entirely excluded by patterns in the distribution data. As we expect small A–D islands to be less suitable for most ferns than small E islands, higher fern incidence on small A–D islands than on E islands (Fig. 7.3; see above discussion) is likely maintained by rescue effects and higher colonization rates. Other examples bolstering this point are provided by islands such as $E37$: "Exotica," a small ($AREA_{FL}$ = 2190 m^2), low (2.8 m) and open, cobbly island with a patch of low shrubs at its apex. It is, however, a mere 240 m east of the long shoreline of $E76$: Turtle, which supports some six fern species. If $E37$ were to support a resident fern, it would be most likely *Polypodium scouleri*, but we have never recorded this species there. But four fern species have showed up as transients on $E37$: a single individual of *Athyrium felix-femina* was present in 1989 and 1990, absent in 2000, but a single individual was back in 2003. That year it shared a damp log with a single *Blechnum spicant*. Two other one-census records on this island are of *Polypodium glycyrrhiza* (one individual) found in 1986 and a single *Polystichum munitum* in 1990. $E37$ seems well positioned to receive transient ferns from its large and close neighbor, Turtle, but they do not persist well at all. A similarly sized island is $A2$: "Mid Hosie" ($AREA_{FL}$ = 2200 m^2), located midway between much larger, fern-rich islands to the north ($A1$: "North Hosie;" $AREA_{FL}$ = 7920 m^2) and $A7$: "South Hosie;" $AREA_{FL}$ = 42,900 m^2) and about 100 m from each. Mid Hosie has two permanently resident ferns, *Polystichum munitum* and *Polypodium glycyrrhiza*, but between 1981 and 2003 there were single-year records of *Polypodium scouleri* (1982), *Blechnum spicant* (1996), and *Athyrium felix-femina*. The last-mentioned colonized in 2002 and was still present, eye-high on a wet, driftwood log on the north coast, in 2003. $B3$: Ohiat is a larger island, and with $AREA_{FL}$ = 5170 m^2, the predicted incidences (based on all fern data from A–D islands) of *Polypodium scouleri*, *P. glycyrrhiza*, and *Polystichum munitum* are 0.84, 0.61, and 0.73, respectively. All are in fact permanent residents on $B3$: Ohiat, as also is *Pteridium aquilinum* (predicted J_T = 0.23). For the large-island ferns *Adiantum pedatum*, *Blechnum spicant*, and *Dryopteris austriaca*, expected incidences on an island of this size are 0.024, 0.16, and 0.014, respectively, and all three have been recorded on Ohiat. *Adiantum pedatum* colonized in 1996, persisted in 1998, and two individuals were present in 2002. *Blechnum spicant* was present in 1982 (absent 1981, 1985) and present again (two individuals) in 1989 (absent 1987, 1990). *Dryopteris austriaca* was absent in 1981, colonized in 1982, and persisted until 1985, but was gone in 1986. Fern access to Ohiat may be facilitated by its position 320 m west of the 0.8 × 10^6 m^2 Helby and 680 m north of the 2.5 × 10^6 m^2 Diana, both copious sources of various fern spores.

The tolerance in different fern species for the drier and more open conditions on small islands is clearly related to their varied morphologies. All four of the species to the left in the panels in Fig. 7.2 (represented by solid symbols) have once-divided fronds. These consist of a rachis from which pinnae spread laterally; the pinnae have broad attachment to the rachis in all but *Polystichum*, which is the only species of the four whose pinnae have serrated (versus entire) edges. The other four species have twice-divided or more complexly subdivided fronds, becoming especially broad and feathery in *Adiantum* and *Dryopteris*. They are also conspicuously different in the thickness of their photosynthetically active surfaces; field collectors know that if *Adiantum pedatum* fronds are not pressed immediately after collection they shrink and deform like wet candyfloss. I have measured the "leaf specific weight" (LSW, in g/m^2) of most fern species on the islands; these values ranged from 12.3 g/m^2 (*Adiantum pedatum*) to 210 g/m^2 (*Polypodium scouleri*) and thus vary among species by a factor of 17. Higher LSW is presumably advantageous in desiccation resistance, perhaps also in resisting other aspects of exposure such as wind and high temperatures. Resistance to the higher maximum temperatures and lower relative humidities of

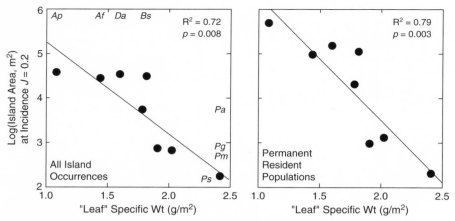

FIGURE 7.5. Fern species vary in "leaf" specific weight (abscissa; log scale), and fern species with lower mass per "leaf" area (to the left) fall to low incidence values sooner on smaller islands (ordinate: log[island area] at which incidence falls to 0.20). Species are abbreviated by genus and species letters at the margins of the left-hand panel. Both regressions are highly significant (see p values); the regression that considers only the incidence of permanent residents (right) has a somewhat higher predictive value relative to incidences that include transient populations (left).

smaller islands relative to larger islands (Fig. 2.13) likely will be related to intra- and interspecific differences in LSW.

Figure 7.5 shows that LSW is effective in predicting the lower limits of occupied islands, where fern incidence falls to 0.20 ($J_{0.2}$). When all occupied islands are considered in computing incidence (Fig. 7.5, left), critical island size at $J_{0.2}$ decreases by two orders of magnitude for an order of magnitude increase in LSW. Setting aside the transients and repeating the analysis, islands that support permanent resident populations (Fig. 7.5, right) decrease in size at $J_{0.2}$ by nearly three orders of magnitude per order of magnitude shift in LSW. Clearly, conditions for the maintenance of persistent populations are more stringent than those for temporary occupancy and are met across the board on correspondingly larger islands.

Most of these fern species are relatively conservative in habitat preferences, and all but one show no obvious morphological variability. The glaring exception is leather-leaf fern (*Polypodium scouleri*), which is found in a wide range of growing sites, from sunny and exposed rocks to shaded and damp banks, though we have never met with it in forest interiors. It also displays an astounding range of morphological variability. These ferns are long-lived perennials with persistent fronds, each of many year's duration. An individual fern may have many (i.e., ≫10) fronds, at all stages of development; individual fronds mature sequentially, die after maturing, and at any one time there are usually several mature fronds and many that are immature. Here I consider a collection of $n = 129$ mature fronds, that is, fronds bearing mature sori, taken from 10 $\mathcal{A}$–$\mathcal{D}$ islands and 17 $\mathcal{E}$-group islands. Individual fronds vary in length, from rhizome to tip, from <3 cm to around 65 cm. The fronds have variable numbers of pinnae, from a single, small terminal lobe without lateral pinnae to subopposite, paired pinnae that number up to around 40. On long fronds, the longest pinnae reach around 10 cm in length, but they are shorter on shorter fronds. Mature fronds bear sori beneath the pinnae, which are arranged in parallel rows on each side on the pinna's midvein. Sori may be restricted to the terminal lobe and the more distal pinnae or may be more generally distributed over two-thirds or more of the pinnae, ranked along the rachis. Tiny, mature fronds with a single terminal lobe a couple of centimeters long have a just a few sori, maybe 2 or 3, whereas a large frond with three dozen pinnae may have over 600

FIGURE 7.6. Variation in frond morphology in the coast polypodium, *Polypodium scouleri;* all of the illustrated fronds are mature and bear ripe sori. The fronds vary from about 2 cm in length (upper left), with two or three sori, to around 35 cm in length, with up to 40 pinnae and over 600 sori.

sori. Although I have not measured them, the sori all appear be the same size, rusty-colored and fuzzy, circular "blobs" around 4 mm in diameter. Frond thickness, as measured by LSW, is also highly variable. In this sample, LSW ranges from 112 to 507 g/m², and averages 261 ± 81 SD. Some of this variation in frond morphology is shown pictorally in Fig. 7.6.

Frond morphology in *Polypodium scouleri* seems to be a developmental response to conditions at the growing site. This species is a good example of the "niche variation hypothesis" presented by van Valen (1965) 40 years ago, in which broader niches, covering a wider range of biotic or abiotic conditions, may be better exploited by species with greater morphological variability. Frond variations are not island specific and are not expected to be, since large islands do provide some sunny and open sites for the fern, and some small islands have damp and shady nooks. The variations are almost certainly microsite rather than genetically determined, although the capacity to generate morphological variation may be more easily accomplished in polyploids such as *P. scouleri.* The different conditions at the edges of species' ranges, for example, are exploited often by polyploidy versions of species that are diploid in the range center. An example of this is the composite shrub *Alvordia,* in which endemics *A. brandegeei* and *A. fruticosa* of the Cape Region of Baja California have chromosome numbers $n = 15, 30$, but the polyploid *A. glomerata* ($n = 60$) extends north into the center of the peninsula and onto islands in the Sea of Cortés (Cody et al. 2002).

I have come to regard the morphological variation in leather-leaf ferns as a sort of bioassay of the abiotic characteristics of the growth site, as the shortest and thickest mature fronds (heaviest per frond area) are restricted to warm and open sites and the largest and thinnest fronds restricted to the shadier and cooler locations. Fronds grow and mature, presumably, in response to feedback from the growing site conditions. Decisions must be innately programmed as to whether to continue growing before maturing, to grow by increasing frond area

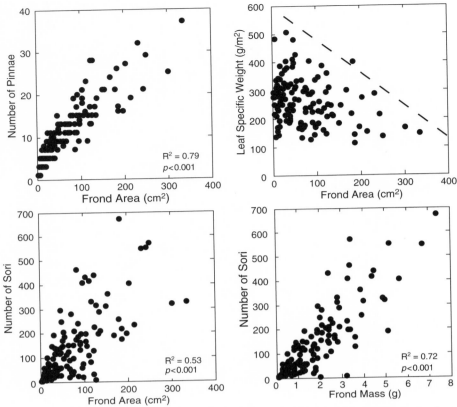

FIGURE 7.7. Morphological variation in the fern *Polypodium scouleri*. Numbers of pinnae (top left) vary from 1 to nearly 40, increasing with frond area. Leaf specific weight (top right) varies nearly fivefold, with small fronds showing a wide range of mass/area, but larger fronds increasingly restricted to lower mass/area. Numbers of sori per frond vary from very few to nearly 700 (lower panels) and increase with frond area, but track frond mass more closely.

(by adding pinnae or increasing their lengths), increasing frond mass, or both, or else mature sooner at smaller frond area or mass. Lacking data on frond age, however, it is by no means apparent that small, mature fronds are produced faster than larger fronds; the reverse might easily be true.

Morphological variation in *Polypodium scouleri* is shown graphically in Fig. 7.7. Numbers of pinnae are closely associated with frond area (upper left). While LSW varies widely in low-area fronds (upper right), high-area fronds are increasingly restricted to low LSW values. Regarding reproductive effort and the production of sori as the main goal of fern fronds, numbers of sori are rather loosely associated with frond area ($R^2 = 0.53$; lower left), but more tightly with frond mass ($R^2 = 0.72$; lower right). Frond area adds nothing to the regression's predictive value once frond mass is factored in, although numbers of pinnae do contribute a low but significant additional 0.03 to the R^2 value of the multiple regression. There is clearly much remaining to be learned about this fern's reproductive strategies over its wide habitat range, and it should certainly reward further research efforts.

FRUITING SHRUBS

A common strategy in Barkley Sound plants, especially the shrubs, is dispersal via fruits or berries attractive to vertebrates, and birds are most commonly the vectors of the fruits and the seeds within them. This mechanism is called zoochory; in berries where transport involves passage through the gut of the berry-eaters, it is endozoochory, to distinguish it from the epizoochoric transport of burs, barbs, stickers, and the like in fur or feathers. Among the three major sorts of plant-animal interactions that currently receive a great deal of attention from ecologists—herbivory, pollination, and seed dispersal—the latter has produced a particularly copious literature. Fruits and their vertebrate symbionts are one of the better-studied aspects of seed dispersal, and there are several recent syntheses (Estrada and Fleming 1986; Herrera and Pellmyr 2002; Levy et al. 2002).

Table 7.2 lists most of the conspicuous zoochorous plants in Barkley Sound; three families dominate the plant list: Caprifoliaceae, Ericaceae, and Rosaceae, while birds dominate the roster of dispersal agents. Note that five berry-producing species on the list are alien, including three "escaped" garden ornamentals: a holly, *Ilex, Cotoneaster*, and a rowan, *Sorbus aucuparia*. None of these aliens is widely distributed on the islands, and they are all more common on islands closer to the mainland (Table 7.2). In the genus *Rubus* also are two weedy, alien species, evergreen blackberry (*R. laciniatus*) and Himalayan blackberry (*R. discolor*), and these likewise have extremely limited island distributions. The remainder of the list is comprised of native plants.

Among plants with non-bird-dispersed fruits and seeds, I have included skunk-cabbage (*Lysichitum americanum*), as bears wreak havoc on the pulpy spathes when they mature in summer and presumably disperse the seeds. The achlorophyllous ground-cone *Boschniakia hookeri* (first described by Wm. Hooker in his *Flora Boreali Americana* [1838, 1840]), is parasitic on the roots of salal (*Gaultheria shallon*). It belongs to a sister taxon with various Scrophulariaceae that includes Barkley Sound hemiparasites (chlorophyllous root parasites) including *Castilleja miniata* (Indian paintbrush) and the related owl-clover *Orthocarpus castillejoides* (de Pamphilis et al. 1997). *Boschniakia*'s dry-capsuled

and multiseeded stems are apparently a major summer food source for bears (*Ursus horribilis, U. americanus*; MacHutchon 1989), which are its most likely seed dispersers. Bears and other mammals such as coyotes are probably the major dispersers of the mealy-fruited manzanitas *Arctostaphylos columbianum* and *A. uva-ursi*, as the latter's name suggests. Similar service may be provided for the coastal strawberry (*Fragaria chiloensis*) by deer, as the plant is avidly consumed by these browsers (in Pojar and MacKinnon 1994).

Apart from these exceptions, most zoochorous plant fruits in the table conform to the "syndrome of bird diaspores" of van der Pijl (1982), in that they are soft and juicy, conspicuous by dint of color and display, and around a centimeter or so in diameter. Many if not most of these are also eaten by mammals such as bears (e.g., *Lonicera involucrata*, which is sometimes called bearberry honeysuckle), but I assume that sea crossings are effected nearly exclusively by birds. Note that the *Rubus* berries are larger (Table 7.2), but these soft fruits easily disassemble into smaller bites. There are four common thrushes locally available to consume these fruits, including two typical of forest interiors (Hermit Thrush [*Catharus guttatus*], Varied Thrush [*Ixoreus naevius*]) and two more characteristic of the forest edge and of more open, brushier sites (American Robin [*Turdus migratorius*], Swainson's Thrush [*Catharus ustulatus*]). In addition, Cedar Waxwing (*Bombycilla cedrorum*) and Northwestern Crow (*Corvus caurinus*) are part-time frugivores and common in edge habitats, while Steller's Jay (*Cyanositta stelleri*) frequents forests.

There are a few forest and bog species in Table 7.2, a few small trees and herbaceous species, but the majority of the berry-producing plants (>50% by species, and much more by overall berry biomass) are shrubs of edge habitats. Berry color spans the range from yellow-orange (salmonberry [*Rubus spectabilis*]) to many red-, purple-, and black-fruited species; late-season fruits tend to darker colors, possibly related to fall changes in foliage color from green to

TABLE 7.2
Barkley Sound Plants with Seed Dispersal by Animals

PLANT SPECIES, BY FAMILY	GROWTH FORM	HABITAT	FRUIT COLOR	FRUIT SIZE	DISPERSAL AGENT	INCIDENCE ON ISLAND GROUPS				
						A	B	C	D	E
Taxaceae										
Taxus brevifolia (Pacific yew)	Tree	Forest	Red	±10 mm	Birds	0.33	0.15	0.31	0.23	0.29
Aquifoliaceae										
Ilex aquifolium (English holly)	Garden escape		Red	8–10 mm	Birds	0.11	0.00	0.06	0.00	0.00
Araceae										
Lysichitum americanum (Skunk-cabbage)	Herb	Bogs	Green	Spathe	?Bears	0.04	0.00	0.00	0.00	0.02
Caprifoliaceae										
Lonicera hispidula (Hairy honeysuckle)	Shrub	Edge	Black	±10 mm	Birds	0.04	0.00	0.03	0.00	0.00
Lonicera involucrata (Black twinberry)	Shrub	Edge	Black	±10 mm	Birds	0.74	0.82	0.78	0.73	0.73
Sambucus racemosa (Red elderberry)	Shrub	Edge	Red	5–6 mm	Birds	0.74	0.36	0.31	0.70	0.10
Symphoricarpos albus (Snowberry)	Shrub	Edge	White	8–12 mm	Birds	0.04	0.09	0.06	0.07	0.01
Cornaceae										
Cornus canadensis (Bunchberry)	Herb	Bogs	Red	6–8 mm	Birds	0.00	0.00	0.03	0.00	0.01
Cornus stolonifera (Red-osier dogwood)	Shrub	Edge	Whitish	6–9 mm	Birds	0.00	0.00	0.00	0.00	0.01
Empetraceae										
Empetrum nigrum (Crowberry)	Shrub	Bogs	Purplish	4–5 mm	Birds	0.00	0.00	0.00	0.00	0.01

Ericaceae										
Arctostaphylos columbianum (Hairy manzanita)	Shrub	Edge	Reddish	6–8 mm	Mammals	0.07	0.00	0.00	0.13	0.00
Arctostaphylos uva-ursi (Bearberry)	Shrub	Edge	Red	7–10 mm	Mammals	0.04	0.24	0.31	0.03	0.04
Gaultheria shallon (Salal)	Shrub	Edge	Purplish	6–10 mm	Birds	0.63	0.76	0.67	0.67	0.67
Oxycoccus oxycoccus (Bog cranberry)	Shrub	Bogs	Reddish	5–10 mm	Birds	0.00	0.00	0.00	0.00	0.02
Vaccinium ovalifolium (Blueberry)	Shrub	Forest	Purplish	6–9 mm	Birds	0.04	0.00	0.08	0.13	0.08
Vaccinium ovatum (Evergreen huckleberry)	Shrub	Edge	Purplish	5–9 mm	Birds	0.59	0.79	0.78	0.70	0.68
Vaccinium parvifolium (Red huckleberry)	Shrub	Forest	Red	6–9 mm	Birds	0.59	0.45	0.67	0.60	0.56
Grossulariaceae										
Ribes bracteosum (Stink currant)	Shrub	Edge	Blackish	±10 mm	Birds	0.19	0.00	0.00	0.10	0.02
Ribes divaricatum (Wild gooseberry)	Shrub	Edge	Purplish	±10 mm	Birds	0.41	0.58	0.72	0.73	0.43
Orobanchaceae										
Boschniakia hookeri (Groundcone)	Parasite	Forest	Yellow	±12 mm	Bears, ?Slugs	0.37	0.18	0.25	0.30	0.33
Rhamnaceae										
Rhamnus purshiana (Cascara)	Small tree	Edge	Purplish	6–9 mm	Birds	0.22	0.00	0.03	0.07	0.22
Rosaceae										
Amelanchier alnifolia (Saskatoonberry)	Shrub	Edge	Purplish	10–14 mm	Birds	0.56	0.52	0.64	0.63	0.49
Cotoneaster sp.	Garden escape					0.11	0.00	0.00	0.07	0.03
Crataegus douglasii (Hawthorn)	Shrub	Edge	Blackish	±10 mm	Birds	0.15	0.03	0.06	0.07	0.00

(*Continued*)

TABLE 7.2 (*Continued*)

PLANT SPECIES, BY FAMILY	GROWTH FORM	HABITAT	FRUIT COLOR	FRUIT SIZE	DISPERSAL AGENT	INCIDENCE ON ISLAND GROUPS				
						A	*B*	*C*	*D*	*E*
Fragaria chiloensis (Coast strawberry)	Herb	Shore	Red	±15 mm	?Deer	0.63	0.73	0.81	0.73	0.66
Prunus emarginata (Bitter cherry)	Small tree	Edge	Red	±10 mm	Birds	0.41	0.18	0.31	0.30	0.09
Pyrus (Malus) fusca (Crabapple)	Small tree	Edge	Reddish	10–15 mm	Birds	0.63	0.30	0.44	0.30	0.34
Rosa nutkana (Nootka rose)	Shrub	Edge	Reddish	to 20 mm	Birds	0.44	0.52	0.50	0.50	0.42
Rubus laciniatus (Evergreen blackberry)	Shrub	Edge	Black	10–15 mm	Birds	0.00	0.03	0.03	0.00	0.01
Rubus parviflorus (Thimbleberry)	Shrub	Edge	Red	10–15 mm	Birds	0.70	0.27	0.33	0.40	0.12
Rubus procerus; discolor (Himalayan blackberry)	Shrub	Edge	Black	±15 mm	Birds	0.07	0.03	0.03	0.00	0.00
Rubus spectabilis (Salmonberry)	Shrub	Edge	Yell.-red	12–15 mm	Birds	0.78	0.73	0.92	0.87	0.71
Rubus ursinus (Trailing blackberry)	Shrub	Edge	Black	±10 mm	Birds	0.22	0.09	0.08	0.27	0.02
Sorbus aucuparia (Rowan)	Garden escape		Red	±10 mm	Birds	0.07	0.00	0.00	0.00	0.00
Sorbus sitchensis (Sitka mountain-ash)	Shrub	Edge	Red	±10 mm	Birds	0.00	0.00	0.03	0.00	0.01
Liliaceae										
Maianthemum dilatatum (False lily-of-the-valley)	Herb	Forest	Reddish	5–6 mm	Birds	0.56	0.45	0.42	0.33	0.48
Streptopus amplexifolius (Twisted-stalk)	Herb	Forest	Red	10–12 mm	Birds	0.04	0.06	0.00	0.00	0.07

yellows, browns, and reds in deciduous leaves. Burns and Dalen (2002) found that fruits were harvested faster when displayed against foliage with stronger color contrasts.

Wintering, frugivorous birds in southern Spain showed strong preferences for the fruits of certain plant species (Herrera 1998), specifically those with high lipid or high soluble-carbohydrate content. Among the preferred fruits were those of *Lonicera implexa, Rubus ulmifolius, Pistacia lentiscus, P. terebinthus,* and *Phyllyrea latifolia*; the first two genera are prominent among the Barkley Sound fruit-producing shrubs. It seems that variation in fruit quality might well affect dispersal potential by frugivores, but no data on the relevant fruit characteristics from this area are available.

K. C. Burns conducted detailed studies at Bamfield on the berry-producing plants for his dissertation research. He noted that there is a general correspondence, over species in both taxa, between bird and berry abundances, at a wide geographic range of study sites (Burns 2002, 2004a). However, berry production at Bamfield peaks in August, when bird abundances are already declining after the breeding season. It may well be that the phenology of berry production is adapted more closely to the fall migration of the thrushes and waxwings (the crows and jays are resident), than to the presence of more stationary and largely territorial breeding birds. Clearly, the timing of fruit maturation, which varies considerably over species (i.e., May to October), may affect both their likely consumers and thence dispersal potential, and different plant species might avail themselves of both the spring and fall bird migration traffic.

Some fruiting plants are "mast" producers in that their large fruit crops mature synchronously; oaks (*Quercus*) and beeches (*Fagus*) are temperate tree examples, and there are many more examples in tropical forests. In temperate shrubs, fruit maturation is more commonly spread over extended time periods, both within individuals and between individuals within species. A further expansion of fruit availability

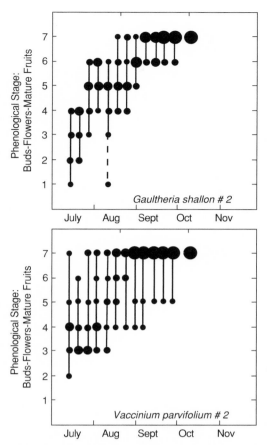

FIGURE 7.8. Spread of phenological stages within two individual shrubs, from weekly counts of buds (stage 1), flowers, fertilization, small and immature to large, mature fruits (stage 7). Progress through these stages to mature fruits is shown over time from early July to October.

over time is achieved by a displacement of fruiting times among species, such that some species have early-season maturation and others ripen later. These various patterns of berry production apply to the shrubs of Barkley Sound. Individual shrubs produce mature fruit sequentially over an extended time period, as buds and flowers move through the progression to maturity on offset schedules and fruit asynchronously. Figure 7.8 illustrates this for two individuals, of *Gaultheria shallon* and *Vaccinium parvifolium*. These shrubs were monitored over a four-month period in 1986 (with the assistance of Dr. Alan Burger at BMSC), and each produced newly ripened fruit over a two-month time period. The variation between individuals

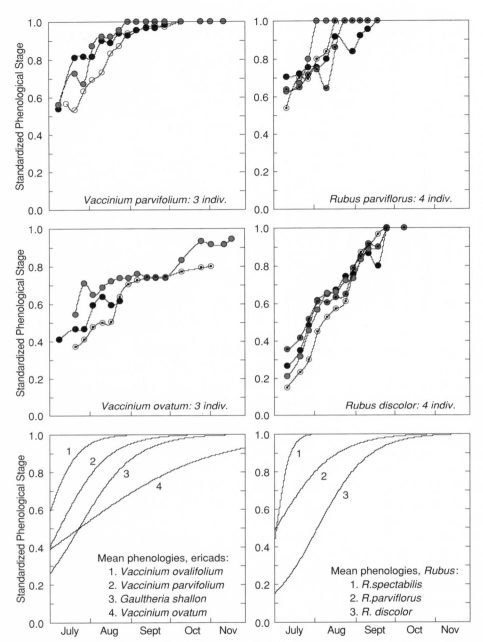

FIGURE 7.9. Phenological progress from buds to mature fruits (ordinate) over time (abscissa) in several individuals each of four shrub species (upper panels), and average species' phenologies (lower two panels) described by logistic regression models. Note variation within species among individuals (upper panels), and displacement in time among species (lower two panels).

within species is illustrated in Fig. 7.9 for two *Vaccinium* species and two *Rubus* species. Individual phenologies varied, within species, by several weeks, although alien *Rubus discolor* bushes were relatively closely synchronized.

Phenologies were averaged within species and the data fitted to logistic regression models (Fig. 7.9, bottom panels), to allow easy comparison among species and illustrate the rather dramatic differences among congeneric or closely

156 DISPERSAL SYNDROMES, INCIDENCE, DYNAMICS

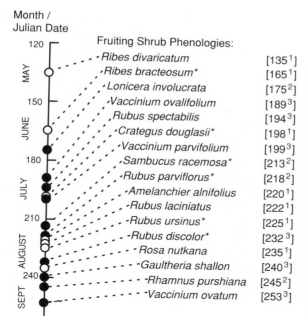

FIGURE 7.10. Phenologies of fruiting shrubs, in terms of mean date of first mature fruit (left-hand ordinate; Julian date in right-hand ordinate). There is a four-month spread of fruit maturation across species. Species represented by open symbols on ordinate and indicated by [¹] in right-hand column are approximations based on information in floras and field guides. In species with solid symbols, data are based on field monitoring studies, by Burns (2003; tagged [²]), and by Cody and Burger unpubl. 1986 data [³]. Species marked with asterisks are those underrepresented on more isolated $\mathcal{E}$-group islands.

related species in berry maturation rates and in the seasonality of their fruit production. Salmonberry (*Rubus spectabilis*), for example, is an early-season species that matures fast and is finished fruiting by about mid-July. Fruiting in the other two *Rubus* illustrated is staggered to sequentially later times, each peaking about three to four weeks after the earlier species in the series. The same sequential fruiting is seen in the four ericad species, with the forest *Vaccinium ovalifolium* the earliest and the chiefly edge species *Vaccinium ovatum* the latest, retaining mature fruits until well into late fall and early winter.

A broader picture of shrub fruiting phenologies is given in Fig. 7.10, which gives average dates for (first) fruit maturation for some 17 species. The data are compiled from our own studies (Cody and Burger unpubl. 1986 data) and from Burns (2003; 1998–1999 data) and are amplified by approximate dates taken from general literature sources on other species as indicated. Together, the shrubs supply mature fruits over at least a four-month period, spanning the spring migration period, the local breeding season, and the fall migration and extending (in the edge ericads; see above) into winter. These late fruits would be available to the year-round resident corvids but would largely miss the migrants and the summer visitors. Considering the local breeding season to last the two months from mid-June until mid-August, the six species that are substantially less common on the $\mathcal{E}$-group islands (indicated by asterisks) are all fruiting during this time period. This might suggest, rather weakly and unconfirmed statistically, that their midseason fruiting periods have limited the species' distribution within the Sound. After all, some midseason frugivores such as crows are often seen in passage between mainland and islands and presumably can transport seeds throughout their extended production periods.

An intriguing possibility is that there is a form of trade-off, of a type much beloved by ecologists, between dispersal ability and persistence ability, with incidence determined jointly by these two factors. I examined this notion in 13 species of the commoner fruiting shrubs by considering census records on the 96 islands that have ≥7 census years (to more accurately estimate persistence between censuses). Many of these shrub species were discussed in chapter 5 with respect to their patterns of joint coexistence on the islands; incidence functions over island area were given in Fig. 5.17 and related to isolation and island elevation in Fig. 5.18. The 96 islands that have been extensively recensused yield 758 "transitions" between adjacent censuses, for each of which absence (0) may be followed by absence (0→0) or presence (0→1), and each presence (1) can be followed by absence (1→0) or presence (1→1). These data are presented in the first four numerical columns of Table 7.3. Note that colonization events (0→1) are closely matched by extinction events (1→0), both overall (251 vs. 230) and within species. For each species, colonization rate c is estimated by the ratio of (0→1)/(0→0 + 0→1), and persistence rate $(1 − e)$ by the ratio (1→1)/(1→0 + 1→1), columns 5, 6. For the top eight commoner species in the table (with $J_{ALL} >$ 0.475; column 7), there is a significant negative relation between c and $(1 − e)$, as shown in Figure 7.11 (see regression statistics; Spearman rank correlation coefficient = −0.92, $p <$ 0.01). However, the remaining species (bottom five of Table 7.3) do not conform to the relation; clearly, low-incidence species have either low values of c, or of $(1 − e)$, or both, and will lie below the trade-off line determined by the high-incidence species.

Incidence on the islands is a function of both colonization and persistence rates, and a given level of incidence can be achieved by relatively good colonization and relatively poor persistence, or vice versa. *Gaultheria shallon* and *Rubus spectabilis*, for example, have very similar overall incidence on the islands (0.656 and 0.649, respectively; Table 7.3), but the former is the better persister and a poorer colonizer, the latter is the opposite. Since incidence J is given by the quotient $c/(c + e)$, the estimated c, e values can be used to generate an estimated incidence J^*, the second to last column in Table 7.3. The observed and predicted incidence values correspond closely. If persistence is enhanced by rescue effects, and apparent persistence is measured by $(1 − e + ec)$, an "adjusted e" is computed ("Adj e"; Table 7.3) to give an incidence J^{**} that is slightly higher than J^*. The nonlinear regression model is shown in Fig. 7.11; it fits parameters K_1 as a multiplier of c and K_2 as a multiplier of $(1 − e)$. K_2 is essentially unity, indicating that $(1 − e)$ is an accurate estimate of persistence, but K_1 has a fitted value of 0.834, indicating that c overestimates the true colonization rate. This may come about if early colonists have a much reduced persistence rate, which seems reasonable, but this, then, must more than offset a number of factors that lead to underestimation of colonization rates (rescue effects, redundant colonization, missed colonization events in the interval between censuses >2 years apart).

If island floras are considered to be comprised of well-established populations of some species along with more precarious populations of others, including recent immigrants, then a given species contributes to one or another of the **L**, **M**, or **S** components of an island's flora (see above). These components are maintained with differential reliance on colonization and persistence, and moreover, a species may change its **L**, **M**, or **S** designation from island to island, most obviously shifting down through this sequence as a function of decreasing island size. The notion of dividing the island plants into categories of more stable and persistent populations versus more transient populations with faster turnover dynamics was employed in the simulation model in chapter 6. Given the variability of colonization and persistence rates, across both species and islands, simply recording a species' presence on an island in a given size (or other) category, is clearly unsatisfactory. Because island populations may be either firmly established with negligible dynamics or basically transient, colonization dependent, and with poor persistence prospects, the distinction between these categories clearly needs to be made an integral part of island plant studies.

Several genera in Table 7.3 have both evergreen and deciduous member species (*Rhamnus, Rubus, Vaccinium*), but the listed species are all deciduous except for *V. ovatum* and *Gaultheria shallon*. These two species rank conspicuously at the top of the list in terms of computed incidence J^*. It seems that the

TABLE 7.3

Colonization, Persistence, and Extinction Rates in Berry Producing Shrubs

	TRANSITIONS BETWEEN SUCCESSIVE CENSUSES: UNOCCUPIED (0 →) OR OCCUPIED (1 →) ISLANDS MAY SUBSEQUENTLY BE UNOCCUPIED (→ 0) OR OCCUPIED (→ 1)				COLONIZATION RATE (c)[b]	PERSISTENCE RATE (1−e)[c]	ADJUSTED EXTINCTION RATE (e)	OBSERVED INCIDENCE (OVERALL) J(ALL)	COMPUTED INCIDENCE[d]	
SPECIES[a]	0 → 0	0 → 1	1 → 0	1 → 1					J*	J**
Lonicera involucrata	225	12	8	513	0.051	0.985	0.014	0.677	0.773	0.781
Vaccinium ovatum	246	6	3	503	0.023	0.994	0.006	0.671	0.793	0.797
Gaultheria shallon	243	9	5	501	0.036	0.990	0.010	0.656	0.783	0.789
Rubus spectabilis	217	55	43	443	0.202	0.911	0.074	0.649	0.694	0.732
Amalanchier alnifolius	348	23	17	370	0.062	0.956	0.041	0.495	0.585	0.599
Vaccinium parvifolium	351	19	20	368	0.051	0.944	0.053	0.506	0.477	0.489
Rosa nutkana	393	3	3	359	0.008	0.992	0.008	0.475	0.500	0.502
Ribes divaricatum	364	24	29	341	0.062	0.922	0.073	0.489	0.443	0.458
Sambucus racemosa	537	46	49	126	0.079	0.720	0.259	0.231	0.220	0.233
Rubus parviflorus	596	31	25	106	0.049	0.809	0.182	0.173	0.204	0.212
Rubus ursinus	677	13	17	51	0.019	0.750	0.245	0.076	0.071	0.072
Rhamnus purshiana	722	8	8	20	0.011	0.714	0.283	0.041	0.037	0.037
Vaccinium ovalifolium	744	2	3	9	0.023	0.750	0.244	0.019	0.084	0.086

[a] Data are shown for 13 shrub species on the 96 islands for which there are >7 census years (n = 758 transitions).
[b] Calculated as (0 → 1)/(0 → 0 + 0 → 1).
[c] Calculated as (1 → 1)/(1 → 1 + 1 → 0).
[d] Incidence J* is computed as c/(c + e); incidence J** is computed assuming rescue effects, (1 − e + ec) rather than (1 − e), giving a value of e shown in column "Adj e."

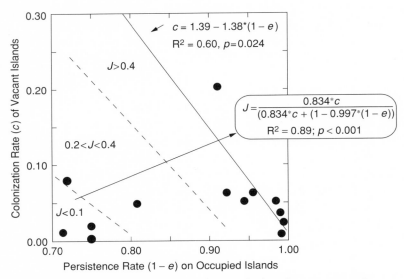

FIGURE 7.11. Relation between colonization of vacant islands (ordinate) versus persistence on occupied islands (abscissa) in 13 berry producing and zoochorous shrub species. The eight species with higher incidence ($J > 0.4$) show a significant negative relationship between colonization (c) and persistence potential ($[1 - e]$; regression statistics shown). Incidence J increases with significant contributions from both c: ordinate, and ($1 - e$): abscissa; see text for discussion.

evergreen habit is an asset to persistence on islands where, especially on smaller and unforested islands, conditions are often drier, less productive, more open (with higher light levels), and warmer. The two evergreen ericads are conspicuously sclerophyllous, with thick, tough leaves (high LSW). And even though *Lonicera involucrata*, another notable persister (Table 7.3), drops its leaves in winter, it too has large and surprisingly robust leaves for a deciduous shrub. This mirrors to some extent the picture of fern persistence in Fig. 7.4, where the incidence gap between persistent and transient populations is generally greater in species with low LSWs. Sclerophylls are at an advantage in warmer environments where, with reduced moisture availability, leaf tissues cannot be cooled by increased transpiration rates (Crawley 1986). But evergreenness and sclerophylly come with a price tag, since, compared to annually renewed deciduous leaves, their photosynthetic efficiency is reduced (Mooney and Gulmon 1982; Wright et al. 2004). This relative inefficiency in turn may commit the more sclerophyllous and evergreen species to fruit production later in the season, relinquishing the early-season fruiting time slots to species such as *Ribes*, jump-started early with new, deciduous leaves, It may also entail a slow maturation of fruit crops into autumn and even winter (as in *Vaccinium ovatum*). Sclerophyllous leaves, while perhaps facilitating island persistence and better accommodating low nutrients and harsh abiotic conditions, must entail high production and maintenance costs as well as low photosynthetic and carbon assimilation efficiency (Mooney 1989; Wright et al. 2004). It might be in this respect that the trade-off between persistence and dispersal abilities is effected, with limited photosynthate resources that are allocated differentially to reproduction (fruits and berries) versus leaf biomass and maintenance. Not surprisingly, reproduction has been shown to be a substantial component of the photosynthate budget, and resources allocated to flowers and fruits will not be available for growth and for the production and maintenance of durable, high-cost leaves (Saulnier and Reekie

1995). The dispersal-persistence trade-off suggests many avenues for additional research and hopefully will become better understood in future years.

HYDROCHORES: DRIFTERS ON THE SEA

Dispersal on sea currents is termed hydrochory by van der Pijl (1982), although more colloquial terms such as "floaters" or "drifters" seem quite appropriate. This form of dispersal is epitomized by coconuts (*Cocos nucifera*), which reach the most remote island atolls by dint of their large, floating drupes. The potential for long-distance dispersal in oceanic floaters is of course enormous, assuming a capacity to resist the effects of immersion in seawater. Robert Lloyd Praeger (1969) talks of the arrival on Ireland's Atlantic coasts of a "tropical jetsam" comprising seeds, mostly legumes, that had ridden the Gulf Stream from the West Indies. Many were fully capable of germination upon their arrival, but presumably they would have slim hopes of extending their geographic ranges in that decidedly untropical region! One of these hopefuls was *Caesalpinia bonduc*, whose large, pebblelike seeds can withstand years of flotation in seawater; this durability undoubtedly contributes to the species' wide, pantropical, distribution. Carlquist (1974) estimated that two-thirds of the plants colonizing equatorial atolls in the Pacific Ocean arrived by means of flotation, while about a quarter of the Hawaiian flora falls into this dispersal category. (N.b., bird endozoochory was, and likely still is, the most prevalent transportation mode to Hawaii, used by approximately 40% of the immigrant taxa.)

While it is a dominant means of long-distance dispersal to remote islands and archipelagoes, oceanic drifting seems not particularly effective for some shoreline plants within the confines of Barkley Sound. Several common species of the mainland sandy beaches and estuaries are absent from the islands, as discussed in chapter 5; recall the suite of four sandy beach species mentioned previously, and common in mainland Keeha and Pachina Bays, that have shown up in no island censuses. It might be argued that a dearth of extensive sandy beach habitat on the islands is the cause of their absence, but lacking any details on their (possibly drift-aided) dispersal modes, it is not possible to resolve the issue of dispersal versus habitat limitation.

Some coastal mustards (fam. Brassicaceae) are unequivocal drifters, and two of them, the alien sea-rockets *Cakile edentula* and *C. maritima*, were introduced in chapter 2. In Barkley Sound they both appear to be annuals, live close to the tide line, and produce inflated waterborne siliques to launch the next generation. While both are very common on the sandy beaches of mainland shores, their island distributions are extremely limited (Cody and Cody 2003). In this case, low island incidence might be a function of time, since these species arrived in the province about a century ago (*C. edentula*) and a half-century ago (*C. maritima*; too recent for inclusion in Hitchcock et al., 1955–1969 [1964] Pacific Northwest flora). Perhaps it takes a while for populations to increase to the point of payoff in the floating diaspore lottery, and that over time their islands distributions will expand. In the early 1980s, I located *Cakile edentula* on six island beaches; in subsequent years, while the cumulative number of islands recording the species increased to eight, only five were occupied post-2000. There were no *C. maritima* on the islands prior to 1998, when two islands (B12: "Haines SE" and C6: "Ross-Sandford Adj") provided the first records. Since then, the cumulative number of islands with this species has increased to four, three of which maintained populations post-2000 (see Fig. 7.12). There is some indication that *C. maritima* will replace *C. edentula* on the islands' beaches, although they coexist at least for the time being on mainland beaches, with the former distributed higher and the latter lower relative to the tide line (Cody and Cody 2003). The replacement scenario seems to be playing out on the island B12: "Haines SE," where a population of about three dozen *C. edentula* individuals was maintained from 1985 to 1998. One individual

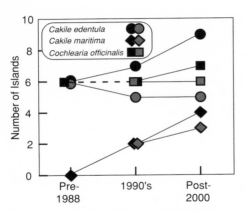

FIGURE 7.12. Island records for three mustard species (Brassicaceae) with sea-drift dispersal. The two *Cakile* species are aliens with invasion during historic times; *C. maritima* appears to be in the process of displacing *C. edentula*. *Cochlearia* is a native of circumboreal distribution, with limited but stable distribution on the islands. Shaded symbols are the number of islands occupied within the time interval, solid symbols denote the cumulative number of islands occupied.

of *C. maritima* was found there in 1998; in 2002 the numbers of *edentula:maritima* were 3:35, and in 2003 they were 0:13. Boyd and Barbour (1993) suggested that earlier maturity, more copious flower production, and possibly its autogamous reproduction system conferred a competitive advantage on *C. maritima* over *C. edentula*.

Another coastal mustard, a native with a circumboreal distribution, is a small and apparently biennial forb, *Cochlearia officinalis*. I have uncovered no details on its dispersal system, but it too would appear to depend on sea currents for dispersal. Like the sea-rockets, it occurs on very few islands (seven to date, spread all across the Sound; Fig. 7.12). But in contrast to the sea-rockets, its populations are generally large and stable, and it persisted apparently unchanged in density on all but one of these seven islands from the early 1980s through to the post-2000 years. Some of the island populations occur on isolated and dry rocks, others on sheltered muddy and cobbly beaches; given its broad habitat ambit, it seems likely that the species' island distribution is dispersal limited.

Several shoreline species belong to the family Caryophyllaceae, one of which, the rocky shore denizen pearlwort (*Sagina crassicaulis*), was discussed extensively in chapter 5. A confamilial with similar high-turnover dynamics is found low down on sandy and cobbly beaches, the sandwort (*Honkenya peploides*). It likely disperses by drifting and has tardily dehiscent capsules, and its brittle stems often break with wave action and often may be found recently washed up on beaches suitable to its establishment. Its incidence is very low on all islands <1 ha in size (Fig. 7.13). On $\mathcal{A}$–$\mathcal{D}$ islands, incidence dramatically increases after the 1 ha size is passed, but on $\mathcal{E}$-group islands, the increase is much more modest. Note also (in Fig. 7.12) that some $\mathcal{A}$–$\mathcal{D}$ islands below the 1 ha size are permanently occupied by *Honkenya*, as also are nearly all islands above this size. In contrast, no $\mathcal{E}$-group islands below 1 ha in size are permanently occupied, and those larger have relatively more transient populations and fewer persisters in comparison to the $\mathcal{A}$–$\mathcal{D}$ islands. It would appear that habitat availability is not a factor in this incidence gap between the more and the less isolated islands, as the species is quite generalized in its substrate preferences, satisfied on practically any larger island where cobble beaches are nearly ubiquitous. There is, therefore, a strong implication of dispersal limitation (though admittedly this was a very basic sort of test). It is possible that the $\mathcal{E}$-group islands in the center of the Sound are not so accessible via transport by seawater currents as those at its periphery.

Note that *Honkenya* incidence jumps, as if surpassing an apparent threshold, after log (AREA) = 4.0, that is, island area > 10,000 m², or >1 ha (see Fig. 7.13). This nonlinearity is not resolved by considering the coastline lengths of islands as the independent variable, rather than island areas. Recall from Fig. 2.5 that coastlines have about the same fractal dimension regardless of island area. Consider a step-length of 30 m, or around 100 ft, to be a reasonable interval to assess the likelihood of recording coastline features such as cobbly or sandy bays. This step-length is at a value of −1.5 on the abscissa of

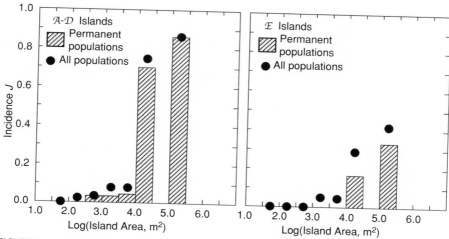

FIGURE 7.13. Incidence of the sandy and cobbly beach caryophyll *Honkenya peploides*, in which dispersal is likely effected by drifing on sea currents. Overall incidence on less-isolated (𝐴–𝒟; left) and more-isolated (𝐸; right) islands is shown (solid dots), and contrasted with the incidence of permanent populations (hatched histograms). The species is scarcer and less permanent on the more isolated islands, despite a uniformity in habitat availability.

Fig. 2.5 and intersects all of the regression lines of the figure such that coastline length can be read with confidence. Then the log(coastline length [in km]) is given by the regression equation $-2.652 + 0.573 * \log(\text{AREA}_{FI})$ with $R^2 = 0.99$, $p < 0.001$. Thus coastline length nearly doubles ($10^{\{0.573*0.5\}} = 1.93$) for every shift of 0.5 on the abscissa of Fig. 7.13, and of course the apparent threshold remains. To pin this point down, *Honkenya* incidence shoots up once island coastlines exceed about 500 m. Further study of coastline topographies might reveal why the chances of *Honkenya* establishment rise so dramatically in shorelines over 500 m in length.

Another caryophyll genus is *Spergularia*, which contains the sand-spurries and is represented in Barkley Sound by three species: *S. canadensis*, *S. macrotheca*, and *S. marina*. There are other species typical of inland and disturbed sites, but these three are considered in the floras to be chiefly species of coastal marshes. All are herbaceous and but a few centimeters tall; *S. canadensis* is a native annual that I have found only in mainland estuaries, *S. macrotheca* is a native perennial, and *S. marina* an alien annual. I have 10 island records for *S. marina* and 5 for *S. macrotheca*, so both are rare on the islands (and neither was reported by Bell and Harcombe 1973). These species well exemplify the ambiguities inherent in assessing dispersal modes. Their seeds are produced in dehiscent capsules and are very small (<1 mm); they are polymorphic in that some are equipped with a circumpheral, narrow wing 0.1–0.3 mm broad, while others are not. Winged versus unwinged seeds is a character under genetic control; winged seeds are reported to be more prevalent (and the wings broader) in denser populations growing in more stable habitats (Sterk 1969; Sterk and Dijkhuisen 1972). The proportions of the different seed morphs vary with habitat, both within and between species, but the utility of the wing in dispersal is not really understood.

Sterk (1969) represents *S. marina* and its relative *S. media* as "polychorous," meaning that they use various and alternative means of dispersal, including hydrochory (the seeds float), anemochory (the winged seeds appear to be aerofoils), and zoochory (the seeds may be eaten and subsequently defecated by herbivorous dabbling ducks). Such seed polymorphisms that collectively cover a number of different dispersal options are rather like bet hedging. A well-known composite example is telegraphweed (*Heterotheca grandiflora*), in which the disk flowers

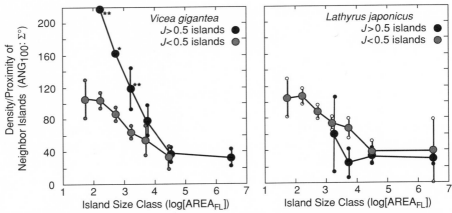

FIGURE 7.14. The two common shore legumes, giant vetch (*Vicea gigantea*) and beach pea (*Lathyrus japonicus*) do not differ in incidence between nearshore and central Sound islands. However, the vetch has significantly higher incidence on islands with many near neighbors, whereas the pea tends to be more common on islands with few neighbors. Bars around means: ±SE. See text for discussion.

produce achenes with a well-developed pappus and disperse well via anemochory, while the achenes of the ray flowers lack a pappus and simply fall to the ground close to the parent plant (Flint and Palmblad 1978).

The two island species of *Spergularia* are quite different in the habitats they occupy there. *S. marina* is found along sheltered, muddy shorelines, in much the same habitats as on the mainland, and it reaches a peak incidence of 0.34 on the largest island size class ($J < 0.03$ elsewhere, on smaller islands); it may be both dispersal and habitat limited. *S. macrotheca*, on the other hand, occurs on just five islands, all of intermediate size, in the $\mathcal{D}$ and $\mathcal{E}$ groups. All are tall, bare rocks, and two are major seabird colonies ($\mathcal{E}83$: Starlight Rf and $\mathcal{E}84$: Baeria Rks). These latter are the only two major bird colonies among the 217 islands in the study; from the binomial, with the probability of "success" = 2/217 = 0.0092, $p < 0.001$ that $k = 2$ successes might occur by chance in a sample grab of $n = 5$ islands. Tall, bare rocks are often topped by perched Glaucous-winged Gulls (*Larus glaucescens*), and while it seems likely that the gull may transport *S. macrotheca* seeds amongst them, many such rocks lack *Spergularia macrotheca*. Dispersal limitation must contribute to the species' extremely limited, though habitat-specific, island distribution.

The two common beach legumes, *Vicea gigantea* and *Lathyrus japonicus*, are almost certainly drifters dependent on seawater transport for dispersal; they belong to the same family as many of the tropical wanderers reaching Ireland's western shores. Both are compound-leaved perennial forbs of similar size, and both are found on the upper reaches of cobbly beaches. The incidences of these two species were described in chapter 3 and Fig. 3.9b, where the data indicated that there were no effects of isolation on the distribution of either species, in that the less isolated $\mathcal{A}$–$\mathcal{D}$ islands appeared to be similar in incidence and dynamics to the more isolated $\mathcal{E}$-group islands.

But distance from mainland, which distinguishes $\mathcal{A}$–$\mathcal{D}$ from $\mathcal{E}$-group islands, is only a crude measure of isolation, and one form of refinement is to consider proximity to neighboring islands. I use ANG_{100} to measure the density and/or proximity of neighbor islands within 100 m of the subject island (units are summed degrees, $\Sigma°$). As most islands with *Vicea gigantea* records have incidence close to or at either 0 or 1, I assess ANG_{100} for incidence $J < 0.5$ and > 0.5, with results shown in Fig. 7.14 (left). Note that (a) ANG_{100} declines with increasing island size, as many small islands have large and influential neighbors, but large islands are minimally affected by small neighbors, and

(b) there are few very small islands with *Vicea* incidence $J > 0.5$, and no islands in the largest island size class with $J < 0.5$. For intermediate island sizes, however, high and low incidence is recorded, and the high-incidence islands are those with significantly more neighbors than the low-incidence islands.

Two possibilities are suggested by the result that incidence is higher on islands with more and closer neighbors. One is that sea currents bearing *Vicea* propagules slow and eddy among such island clusters, and the likelihood of floating seeds being deposited on their beaches is correspondingly higher. The second possibility is that the neighboring islands themselves are the sources of *Vicea* propagules and thus decrease the distance to conspecific populations. I eliminated the second possibility by computing the regression equation for a sample of $n = 48$ islands of $SIZE = 4$, of *Vicia* incidence $J_{Vg} = f(ANG_{100}, D_{CONSP})$, where D_{CONSP} is the distance to the nearest conspecific population. In this test, 11 $SIZE = 4$ islands were discarded, as they lay either closer to the mainland or closer to an uncensused large island rather than to islands on which *Vicea*'s status was known. The contribution on D_{CONSP} to *Vicea* incidence was insignificant, either alone ($p = 0.127$; negative coefficient) or in conjunction with ANG_{100} ($p = 0.669$; negative coefficient). I conclude that *Vicea* populations do not establish with higher probability closer to existing populations, but that drift and currents around higher-density island groups enhance the target effects of island size and significantly boost *Vicea* colonization rates there.

The effect of island neighbors on *Lathyrus japonicus* incidence is shown in Fig. 7.14 (right). In striking contrast to *Vicea*, *Lathyrus* shows higher incidence on islands with fewer, rather than more, neighbors. Though the differences in ANG_{100} between the two incidence classes in *Lathyrus* are not statistically significant, they are consistent over the largest four island size classes (for which both incidence classes are represented). This immediately suggested (to me, at least) that *Lathyrus* incidence is higher on more isolated islands because the less isolated islands are already occupied by *Vicea*, the presence of which inhibits *Lathyrus* establishment. But in fact this seems not to be the case. Taking island $SIZE$ groups 4–6 ($n = 115$), on which in general both species are common with intermediate incidences, *Lathyrus* incidence on the 41 islands where *Vicea* incidence >0.5 is 0.328, and on the 74 islands where *Vicea* incidence <0.5, *Lathyrus* incidence is 0.057. Thus *Lathyrus* tends to be commoner on islands where *Vicea* also is common (Mann-Whitney U-test on J_{Lj} means: $p < 0.001$ that they are the same). A stepwise regression of *Vicea* incidence J_{Vg} and ANG_{100} on *Lathyrus* incidence yields $J_{Lj} = 0.095 + 0.333 * J_{Vg} - 0.0009 * ANG_{100}$ ($R^2 = 0.29$, $p < 0.001$; both coefficients significant). This model supports a positive rather than a negative interaction between the two legume species and reaffirms that *Lathyrus* incidence is lowered on islands that have larger numbers of neighbors. But since both species have higher incidence on larger islands, and island size has not been factored out in this regression, perhaps we have to dig a little deeper.

Finally I attempted to test the hypothesis that dispersal to both more and less isolated islands is similar in both species, but that *Lathyrus* persists better on more isolated islands where *Vicea* incidence is lower. The number of colonization events in *Vicea* ($C = 11$) and *Lathyrus* ($C = 13$) in $SIZE$ 3–5 islands is unrelated to ANG_{100} (Spearman rank correlations; $p \gg 0.05$), so persistence rather than dispersal is implicated in the different species' incidence patterns. *Lathyrus* colonization and extinction events (with $E = 13$) all occurred on 14 islands of $SIZE$ 4–7, on 11 of which *Vicea* was a permanent resident during the study period. Of the three without permanent *Vicea*, D_{35}: "Diplock N" lacked *Vicea* 1985–2002 (CYRS = 5), and *Lathyrus* established between 1990 (absent) and 2002 (rare). On $E68$: "Keith W," *Lathyrus* was present in 1989 and absent the following year when *Vicea* was recorded for the first time, and in 2000 and 2003 both species were present. And on the somewhat larger $C6$: "Ross-Sandford Adj," *Vicea* was present at the west end from

1982 to 1989, after which it became extinct. In the same location, *Lathyrus* colonized in 1985 but was extinct the following year, colonized again in 1987, persisted in 1989 but was extinct in 1996, colonized again in 1998, and was common there in 2002 (in the absence of *Vicea*). On the other 11 islands, on which *Vicea* had permanent resident status, one or another of several scenarios unfolded: *Lathyrus* (a) went extinct during the study period (*E*22, *E*24, *E*49, *E*58), (b) recolonized following an extinction after a period of persistence (*A*8, *B*12), (c) colonized once or twice but was extinct in following censuses (*B*2, *B*6, *E*42), (d) colonized and then persisted until the last census (2002 or 2003; *D*9: persisted for 17 years, *E*4: persisted for 14 years). The data are by no means definitive, but they provide no strong support for the hypothesis that *Lathyrus* persistence is compromised by *Vicea* presence.

ANEMOCHORES: DRIFTERS ON THE WIND

Anemochores are epitomized by the familiar dandelions (*Taraxacum* spp., fam. Asteraceae), whose propagules, a light-weight achene ("seed") supported by a fluffy parachutelike pappus, disperse by drifting in wind currents. The long-distance dispersal success of such anemochores is attested by their reaching some very far-flung oceanic islands, on many of which their evolutionary derivatives show spectacular local adaptations and adaptive radiations. Recall Joseph Hooker's interest in the Galapagos' *Ageratum*, mentioned in chapter 1, and his observations on woody taxa in the family Asteraceae that apparently lack close relatives. Many exemplars of this phenomenon are wind-dispersed taxa whose mainland ancestors we now recognize as weedy, herbaceous composites (Asteraceae) with morphologies and ecologies very different from their island descendants. The 13 species of endemic *Scalesia* shrubs of the Galapagos, with which Hooker was familiar, are now known to have an Andean sunflower (*Helianthus* sp.) as their most likely ancestor (Schilling et al. 1994;

Spring et al. 1999). The spectacular Hawaiian silversword alliance of three endemic genera (*Argyroxiphium, Dubautia, Wilkesia*) and around 20 species is thought to have been derived via long-distance dispersal from Californian montane forbs (perhaps *Raillardella*; tribe Madiinae, Asteraceae; Baldwin et al. 1991; Carlquist et al. 2003). Another Hawaiian endemic genus, *Hesperomannia* (with three species), is a product of really long-distance dispersal, as its closest ancestors are African *Vernonia* (tribe Vernoneae, Asteraceae; Kim et al. 1998). With divergence times between the Hawaiian plants and their putative ancestors estimated from cpDNA to be 17–26 million years, the original colonists (proto-*Hesperomannia*) must have arrived in the late Oligocene to mid-Miocene on islands that are now far to the west-northwest of the present Hawaiian islands and below sea level. When the Emperor seamount and its neighbors rode the Pacific Plate away from the Hawaiian hot spot and eventually sank below the sea, proto-*Hesperomannia* must have maintained position in space and over time by "stepping" (dispersing) onto new islands as they emerged, becoming extinct on older islands as they aged and submerged.

There are two main concepts that island biogeographers have developed from studies of anemochore derivatives on remote islands: the evolution of reduced dispersal capacity, and a growth-form shift from short-lived herbaceous plants to long-lived woody plants, the "weeds-to-trees" theme. Sherwin Carlquist, with an especially broad knowledge of the botany of Pacific Ocean islands, documented both of these themes with acuity and insight. His three books, *Island Life* (1965), *Hawaii: A Natural History* (1970), and *Island Biology* (1974), are replete with examples. A chapter entitled "They Can't Go Home Again" captures the essence and consequences of reduced dispersal (1965), and another on "Insular Woodiness" (1974) details the evolution and anatomy of secondary woodiness in diverse taxa whose closest relatives are herbaceous. He found relatively low representation of plants derived from anemochore ancestors in

Hawaii (1.4%), the Juan Fernandez Islands (2%), and the Galapagos (4.3%), but in Samoa, closer to continental sources upwind in the prevailing westerlies, the proportion is much higher (15.1%). Clearly, windborne propagules reach such isolated islands only rarely, allowing long time intervals for local adaptive shifts, often to dramatically different morphologies, before any repeat colonization happens.

Among many other examples, Carlquist (1974) discusses adaptive radiations of plants in the Juan Fernandez Islands, where Skottsberg (1953 and earlier) did much of the pioneering botanical work. These islands are located over a hot spot at the subduction zone of the oceanic Nazca Plate under the South American Plate (Kopp et al. 2002), about 700 km off the Chilean coast opposite Valparaiso. This small archipelago, chiefly three small islands (Robinson Crusoe or Mas-a-Tierra, Alejandro Selkirk or Mas-a-Fuera, and Santa Clara) with total area around 10,000 ha, supports about 200 native species, 60% of them endemic, and three-quarters of them extinct, endangered, threatened, or rare (Stuessy et al. 1992, 1998). The archipelago seems rather limited in both size and number of islands to generate adaptive radiations, but two distinct clades of Asteraceae provide spectacular examples of the phenomenon. One is derived from *Senecio*-type ancestors (tribe Senecioneae) and is now represented there by one genus and three subgenera (*Robinsonia, Rhetinodendron, Symphyochaeta*) and seven species (Sang et al. 1995). Collectively the group displays growth form variation from subshrubs to trees and occupies habitats from wet forest to montane heath; one species, *R. evenia*, is even an obligate epiphyte on tree ferns (on *Blechnum* and *Dicksonia* spp.). The second clade is derived from *Hieracium-* or *Sonchus*-like ancestors (tribe Lactuceae) and comprises 11 species in the genus *Dendroseris* (encompassing three related taxa of near-genus status, *Phenicoseris, Hesperoseris,* and *Rea* in the latest revision by Esselman et al. [2000]). These species encompass an even wider range of growth forms and habitats, from beaches and dry cliffs to scrub vegetation and rainforest. A third, minor radiation has produced *Centaurodendron* (two woody species, up to 4 m tall), derived from thistlelike ancestors (presumably weedy and herbaceous anemochores) of the tribe Cynareae.

Besides their radiation into species of diverse growth forms and habitats and their evolution from herbs to trees, the Juan Fernandez composites also show changes in the morphology of the achene-pappus dispersal unit. In *Centaurodendron* and in contrast to its mainland relatives, the achenes are large, and most of the relatively short pappus is "caducous," that is, falls off rather than persists to aid in dispersal. In *Robinsonia* and its island relatives, some species have much enlarged achenes, but all have an ineffectual pappus that is very short, caducous, bristly rather than fluffy, or some combination of these modifications. In *Dendroseris* and relatives, while some achenes are small relative to typical mainland relatives, others are grossly enlarged and carunculate, but all show an extreme reduction in the pappus, which is short, sparse, caducous, and clearly no longer a functional parachute (Carlquist 1974; Figs. 11.7, 11.8). Reduced dispersal ability is clearly an evolved trait in these island plants derived from erstwhile anemochores.

I chose these particular examples of island adaptation and radiation because there are close relatives of the Juan Fernandez endemics in the weedy component of the Barkley Sound flora. These are species that are probably close, ecologically and morphoplogically, to the ancestral colonists of the Juan Fernandez endemics: groundsels, *Senecio* species (*S. vulgaris, S. sylvaticus*), sow-thistles (*Sonchus* spp.), wall-lettuce (*Lactuca muralis*), hawkweed (*Hieracium*), and a thistle (*Cirsium vulgare*), among other island composites. Asteraceae, with 21 species, is the second largest family (after Poaceae) in this study, and about half of them are anemochorous. Besides these composites, anemochory is the chief dispersal mode of the local fireweeds (*Epilobium* spp., fam. Onagraceae). Beyond this limited roster, the willows (*Salix*, 3 spp.) have silky, wind-dispersed seeds, and the forest and edge trees are also anemochorous, but probably weakly so (all of our conifers except *Taxus* have winged seeds, as do the maples [*Acer*] and the alders [*Alnus*]). Apart from the trees, anemochory is not a prevalent dispersal strategy in and around the coastal coniferous forest vegetation, and those species that pursue it seriously (i.e., excluding the trees) are species of open, often disturbed sites and are mainly native or alien weeds (in the classical, nonpejorative, sense). It is undoubtedly the case that areas such as ours, with 3 m of annual rainfall and tall forest as the climax vegetation, will support rather few plants dependent on warm, dry breezes for effective aerofoil operation and successful dispersal.

I list all the island and nearby mainland Asteraceae, regardless of dispersal mode, in Table 7.4, which includes also the two *Epilobium* species, in which the regional, local, and island status for the species is given. The *Epilobium* species are native perennial forbs with fruiting capsules that split to release small seeds attached to a coma of fine hairs to aid anemochory. *Epilobium angustifolium* is holarctic in distribution, *E. watsonii* nearctic, and both are typical of open, disturbed sites, with the former twice as common on the islands as the latter (Table 7.4). Their incidence functions are shown in Fig. 7.15, from which it appears that *E. angustifolium* has a wider island distribution than *E. watsonii*. There are only minor differences in incidence between the $\mathcal{A}$–$\mathcal{D}$ and the $\mathcal{E}$ islands. I note that, for example, the $J = 0.667$ of *E. angustifolium* on the three $\mathcal{E}$-group islands of $SIZE = 7$ is not statistically distinct from the $J = 1$ on the four $\mathcal{A}$–$\mathcal{D}$ islands of $SIZE = 7$ (by the binomial, $p = 0.132$ that this could occur by chance). Greater incidence in *E. angustifolium* might be a product of either a greater colonization rate, a greater persistence rate, or both. Certainly the observed residence times, while increasing for both species with increasing island size (Fig. 7.16), tend to be somewhat longer in *E. angustifolium*, but this may misleading as *E. watsonii* reaches fewer islands on which its persistence is tested. I revert to calculations of colonization rate (c) and persistence rate ($1 - e$) from observed transitions (0→0), (0→1), (1→0) and (1→1) on the 96 islands with $CYRS > 7$ (see, e.g., Table 7.3). The numbers of each sort of transition are 571, 46, 39, and 103 in *E. angustifolium*, and 684, 13, 12, and 49 in *E. watsonii*. With adjustments for rescue effects, colonization and extinction rates c and e are computed to be 0.075 and 0.297 in *E. angustifolium*, 0.019 and 0.200 in *E. watsonii*. The higher colonization rates in *E. angustifolium* correspond to its much longer seed-bearing coma (0.8–1.2 cm long, vs. 0.2–0.3 cm on *E. watsonii*), but no detailed measurements are available. Overall, colonization rates are low in both species and are likely the larger limiting factor in incidence. Here again there is an indication of a trade-off, with better persistence but poorer colonization in *E. watsonii*, the reverse in *E. angustifolium*. The predicted incidences from c and e values for the two species on these islands and given by $J = c/(c + e) = 0.201$ and 0.086, respectively. The numbers are very close to the observed incidence values of 0.193 and 0.088 (slightly different in this island subset from the all-island values shown in Table 7.4).

Table 7.4 lists the 27 species of Asteraceae I have found on the Barkley Sound islands and on the immediately adjacent mainland. Bell and Harcombe (1973) list an additional three species (*Artemisia suksdorfii*, *Cirsium arvense*, *Leontodon nudicaulis*), all rare and found on a single or on several large $\mathcal{E}$-group islands that I did not census. The 27 species tabulated include 4 with beach, estuary, or bog habitat preferences with no island records; a fifth species (*Senecio vulgaris*) is very common in disturbed sites on the mainland but, again, has no island records. Of the remaining 22 species, 7 (*Arctium*, *Chrysanthemum*, *Eriophyllum*, *Gnaphalium*, *Lapsana*, *Solidago*, *Sonchus arvensis*) are represented by single records on large islands (e.g., 3 on $C36$: Sandford, 1 each on Dodd, on $\mathcal{A}14$: Dixon, and on $\mathcal{E}49$: Clarke). Two other species listed as rare have two or three island records, one of these (*Baeria maritima*) being restricted to the two large seabird colonies of $\mathcal{E}83$ and $\mathcal{E}84$, Baeria

TABLE 7.4
Species in the Family Asteraceae (Compositae) and Onagraceae in the Barkley Sound Area

	COMMON NAME	LONGEVITY	LOCAL STATUS	DISPERSAL MODE	ISLAND STATUS	INCIDENCE A–D	J E
Asteraceae							
Achillea millefolium	Yarrow	P	Native	Epizoochory	Abundant	0.672	0.632
Ambrosia chamissonis	Burweed	P	Native	Epizoochory	Mnld beaches only	0.000	0.000
Anaphalis margaritacea	Pearly everlasting	P	Native	Anemochory	Uncommon	0.061	0.039
Arctium minimus	Burdock	P	Native	Epizoochory	Rare	0.009	0.000
Aster subspicatus	Douglas aster	P	Native	Epizoochory	Common	0.150	0.059
Baeria (Lasthenia) maritima	Goldfields	A	Native	Epizoochory	Rare	0.000	0.022
Bellis perenne	English daisy	P	Alien	Epizoochory	Rare	0.024	0.000
Chrysanthemum leucanthemum	Oxeye daisy	P	Native	Epizoochory	Rare	0.000	0.011
Cirsium vulgare	Bull thistle	B	Alien	Anemochory	Uncommon	0.048	0.060
Eriophyllum lanatum	Wooly eriophyllum	P	Native	Epizoochory	Rare	0.008	0.000
Gnaphalium purpureum	Purple cudweed	P	Native	Anemochory	Rare	0.008	0.000
Grindelia integrifolia	Gumweed	P	Native	Floater	Rare	0.041	0.021
Hieracium albiflorum	Hawkweed	B	Native	Anemochory	Rare	0.072	0.011
Hypochaeris radicata	Hairy cat's-ear	P	Alien	Anemochory	Common	0.188	0.023
Jaumea carnosa	Fleshy jaumea	P	Native	Floater	Mainland estuaries	0.000	0.000
Lactuca (Myocelis) muralis	Wall lettuce	B	Alien	Anemochory	Common	0.251	0.085
Lapsana communis	Nipplewort	A	Alien	Epizoochory	Rare	0.000	0.005
Matricaria matricaroides	Pineappleweed	A	Alien	Epizoochory	Rare	0.017	0.030
Microseris (Apargidium) borealis	Apargidium	P	Native	Anemochory	Mainland bogs only	0.000	0.000
Senecio sylvaticus	Wood groundsel	A	Alien	Anemochory	Uncommon	0.072	0.096
Senecio vulgaris	Common groundsel	A/B	Alien	Anemochory	Mainland only	0.000	0.000
Solidago canadensis	Goldenrod	P	Native	Anemochory	Rare	0.008	0.000
Sonchus arvensis	Perennial sow-thistle	P	Alien	Anemochory	Rare	0.024	0.011
Sonchus asper	Prickly sow-thistle	P	Alien	Anemochory	Rare	0.070	0.013
Sonchus oleraceus	Common sow-thistle	A	Alien	Anemochory	Rare	0.055	0.011
Tanacetum bipinnatum	Dune tansy	P	Native	Floater?	Mnld beaches only	0.000	0.000
Taraxacum officinale	Dandelion	P	Alien	Anemochory	Rare	0.017	0.027
Onagraceae							
Epilobium angustifolium	Fireweed	P	Native	Anemochory	Common	0.211	0.163
Epilobium watsonii	Purple-leaved willowherb	P	Native	Anemochory	Uncommon	0.101	0.068

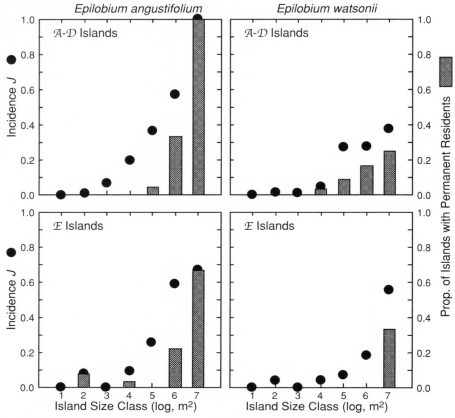

FIGURE 7.15. Incidence J in two *Epilobium* species (ordinate, left; solid dots) on less isolated (*A–D*) and more remote (*E*) islands (upper vs. lower panels), as a function of island size (abscissa). The shaded histograms give proportions of the island size classes (ordinate, right) occupied by permanent resident populations.

Rks and Starlight Rf. Of the 12 commonest species (with an overall $J > 0.02$), 8 are anemochores, but none is nearly as common as the ubiquitous yarrow (*Achillea millefolium*), a species of shoreline rocks and present on almost two-thirds of the islands (see also Fig. 3.9a). Its achenes, devoid of a pappus, mature late in the season, falling onto the parental rocks by mid-August in time for the hordes of migrating shorebirds that begin moving south through the Barkley Sound islands at that time. The species seems most likely epizoochorous, and the shorebirds seem the most likely agents of its dispersal. Other probable epizoochores are the Douglas aster (*Aster subspicatus*) and pineappleweed (*Matricaria matricaroides*), though dispersal of the former's hairy achenes might be also wind aided. Incidence over island size and isolation of these species, plus the floater *Grindelia integrifolia*, is shown in Fig. 7.17.

There are only minor differences in these species' incidences with island isolation; with the exception of *Grindelia*, which is absent from the largest *E*–group islands, none shows a significant relation between incidence and distance from mainland. Figure 7.17 includes a bar on which, for the sample of 95 oft-censused islands with $CYRS > 7$, the proportion of known populations that persisted for the duration of the study is shown (P: solid bar), and the number of colonization events per island (C: close-hatched, upper bar) and extinction events (E: open-hatched, lower bar) are shown to scale. All of the

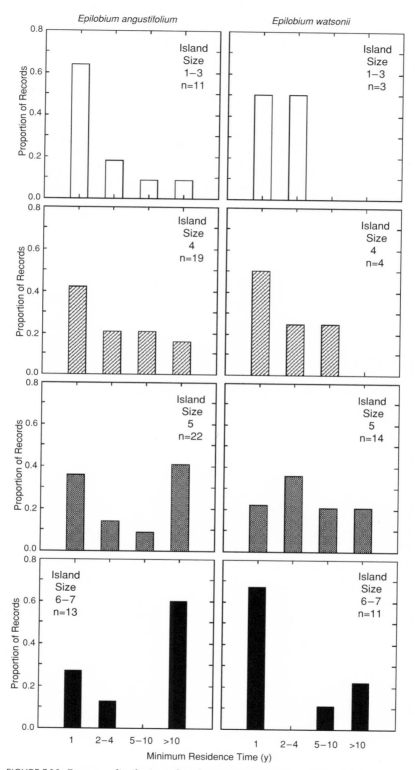

FIGURE 7.16. Frequency distributions of residence times in four classes (abscissa) for two *Epilobium* species on islands of increasing size (top to bottom). Residence times shift to larger values (i.e., left to right) as island size increases. *E. angustifolium* (left column) has somewhat longer residence times, and greater persistence on smaller islands, than does *E. watsonii* (right column). Sample sizes per island size group are shown.

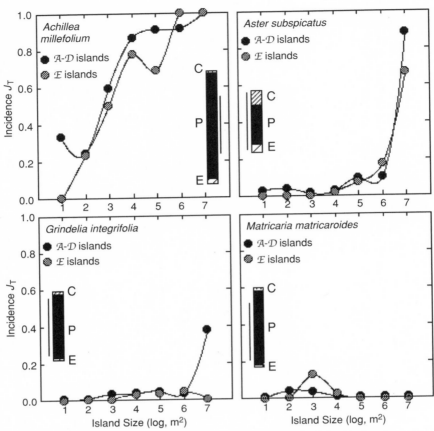

FIGURE 7.17. Distribution of four composite forbs (Asteraceae) over island size (abscissa; log scale) and isolation (A–D vs. E islands). Dispersal is by drifting (*Grindelia*) or epizoochory (others). The vertical bar shows the proportion of occupied islands with persistent populations (P: solid shading), and the numbers of colonization events (C: close hatching) and extinction events (E: open hatching) per island on 95 oft-censused islands. The vertical line beside the C/P/E bar denotes a value of 0.5. These species show little difference in incidence with island isolation and are relatively persistent with slow dynamics.

censused populations show good persistence, but *Achillea* is outstanding, with 96% of the known populations, no matter how small, lasting the full length of the study. Collectively, there are relatively few colonization and extinction events in these species (although *Aster* populations are rather more dynamic), and incidence is clearly dominated by persistence. Note the peculiar distribution of *Matricaria* over island size. It is quite absent from the large islands and reaches peak incidence on *SIZE* 2, 3 islands, that is, on islands <1/10 ha. In this respect, it is a classical "tramp" species, likely gull dispersed as are other shoreline plants of small, bare rocks.

The true anemochores are shown in Fig. 7.18. All have a well-developed pappus, although in pearly everlasting (*Anaphalis margaritacea*), it is rather sparse and lax. This does not hamper its long-distance dispersal, however, as it was one of the early and common colonists, along with *Epilobium angustifolium* and *Senecio sylvaticus*, of the lava flows on Mount St. Helens following the 1980 eruption, establishing on sites as far as 20 km from likely source populations (Wood and del Moral 2000). The seed rain on Mount St. Helens also included conspicuous contributions from *Hypochaeris radicata*, *Hieracium albiflorum*, and *Epilobium watsonii*, all

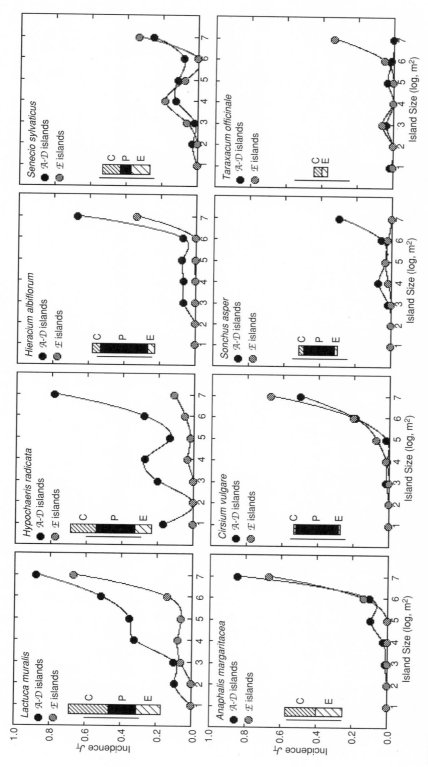

FIGURE 7.18. Incidence (ordinate) of eight anemochorous composite forbs (Asteraceae) over island size (abscissa) and isolation (𝒜–𝒟 vs. 𝓔 islands). The three commoner species (*Lactuca*, *Hypochaeris*, *Hieracium*; top row) have higher incidence on the less-isolated-𝒜–𝒟 islands. Note that there are extensive colonization and extinction dynamics in most species (see vertical bars and caption to Fig. 7.17), and persistent populations are relatively infrequent, and absent entirely in *Anaphalis* and *Taraxacum*.

of which are also prominent among the Barkley Sound anemochores. The commoner anemochores are shown in the top row of Fig. 7.18, and the three species with widest incidence, *Lactuca*, *Hypochaeris*, and *Hieracium*, all show a considerably reduced incidence on the more isolated *E*–group islands. Four of the eight species are perennial, three typically biennial, and one (*Senecio*) is annual; individual longevity differences do not seem to contribute to the variation among species shown in the figure. The two species with no persistent populations, *Anaphalis* and *Taraxacum*, are both perennial; the three with the longer persisting populations, *Hypochaeris*, *Hieracium*, and *Cirsium*, are perennial (first mentioned) and biennial (last two).

Note that all of these anemochores are weedy, edge species, and although they differ somewhat one to another in the substrates on which they are usually found, it seems improbable that they are habitat limited across the Sound. With the overall low incidences in several species and the reduced incidence on the more isolated islands of the commoner species, it would appear that they may be dispersal limited. Anemochore dispersal is achieved by the aerial drift of pappus/achene propagules that vary considerably over species, in size and structure of the pappus and in size (mass) and shape of the achene. In some species (e.g., *Hypochaeris*, *Taraxacum*), the achene has a beak or shaft that attaches it to the pappus, in others the pappus is a fluffy ring directly attached to the top of the achene. While the measurement of mass in the tiny achenes was beyond our technological capacity, we used measurements of achene dimensions and pappus radii to evaluate dispersal capacity in several of the Barkley Sound anemochores.

Pappus and achene morphology vary among mature propagules within individuals, vary between individuals within populations, and vary in average measurements between populations at different locations. Variation is also apparent within populations in different (adjacent or subadjacent) years; as this variation shows up in parallel across species, it is likely attributable to yearly variations in summer weather conditions. And finally, within some populations there is unidirectional change in propagule morphology over the longer term, of several generations. This longer-term temporal variation is more interesting, as such phenotypic shifts signal changes in dispersal ability. In some long-established island populations in the Sound, changes in propagule morphology are consistent with the hypothesis of a rapid evolution to reduced dispersal potential in island populations.

A synopsis of the morphological variation in the local anemochore propagules is given in Table 7.5, where the 10 species on which we made measurements are listed (see Cody and Overton 1996 for a previous report on some of the following results). Achenes were measured in length (l) and width (w; two orthogonal widths w_1, w_2 in achenes not circular in cross section) to the nearest 0.01 mm, and pappus radius (r) to the nearest 0.1 mm. To characterize a population, 20 propagules were measured, from 20 different individuals if available. Achene volume (A_V) was assumed ellipsoidal and derived as $(4/3)\pi l w_1 w_2/6$, and pappus volume (P_V) assumed spherical: $(4/3)\pi r^3$. We reasoned that dispersal potential, *DISPER*, is enhanced by larger pappi and by smaller achenes, and the ratio P_V/A_V is used as its estimator.

The distributions of achene and pappus volumes for six of the commoner measured populations are shown as confidence ellipses in Fig. 7.19, to which *Cirsium vulgare*, an outlier with very large pappi and achenes but few data, is added for reference only. Across species, P_V increases faster than does A_V: a 1 mm³ achene in *Lactuca* apparently requires a 1 cm³ pappus, but a 10 mm³ achene in *Cirsium* is equipped with a 100 cm³ pappus. However, within species the two variables P_V and A_V are statistically unrelated in all except *Sonchus asper* (where the Pearson correlation coefficient is 0.88; $p < 0.01$). Mean *DISPER* values vary over populations by a factor of about 20, but these alone are not good indicators of island incidence; the extreme

TABLE 7.5
Propagule Morphology and Its Variation Within and Among Species of Anemochorous Asteraceae in Barkley Sound

SPECIES	SITES	NO. POPS. YRS.	ACHVOL (A_V)[a] MEAN±SD	PAPVOL (P_V)[a] MEAN±SD	DISPER P_V/A_V[a] MEAN±SD	DROP TIME T_D (s/2m)[b]	CORREL. T_D-DISPER[c]
Cirsium vulgare	Islands	2	7.24 ± 3.97	92750 ± 24590	14706	6.47±1.09	0.27 (n=9)
Hieracium albiflorum	Islands	8	0.44 ± 0.11	772 ± 282	1867±773	5.25±1.03	0.27 (n=14)
Hypochaeris radicata	Mainland	4	1.17 ± 0.19	4331 ± 542	3971±962	3.91±1.35	0.95[d]
	Islands	29	1.46 ± 0.38	2847 ± 1257	2202±1179	4.45±0.71	0.87[d] (n=12, 20)
Lactuca muralis	Mainland	3	0.79 ± 0.07	1245 ± 209	1497±217		
	Islands	49	0.74 ± 0.11	939 ± 207	1314±421	5.51±0.88	0.95[d] (n=8)
Senecio sylvaticus	Mainland	1	0.35 ± 0.13	906 ± 387	3274		
	Islands	30	0.39 ± 0.09	939 ± 207	2640±746	8.01±0.84	0.23 (n=20)
Senecio vulgaris	Mainland	2	0.39 ± 0.10	947 ± 275	2461	9.61±3.20	0.86[d] (n=20)
Sonchus arvensis	Islands	3	1.54 ± 0.41	5880±642	4463±2244	10.00±1.16	0.53 (n=10)
Sonchus asper	Mainland	1	0.39 ± 0.03	1168 ± 279	2961		
	Islands	7	0.73 ± 0.31	1968 ± 386	3293±1367	8.03±2.44	0.92 (n=4)
Sonchus oleraceus	Mainland	2	0.66 ± 0.09	1481 ± 392	2636		
	Islands	4	0.67 ± 0.12	1678 ± 242	2936±966		
Taraxacum officinale	Mainland	3	1.90 ± 1.39	1161 ± 416	737±256	6.16±1.95	0.85[d] (n=18)
	Islands	2	0.95 ± 0.08	1031 ± 313	1463		

[a] Achene and pappus volume (ACHVOL, PAPVOL) are averaged across all measured populations, as is the measure of dispersal potential, DISPER, the quotient of PAPVOL/ACHVOL.
[b] Mean drop times T_D of propagules are measured for single populations as seconds (s) per 2 m drop in still air.
[c] Propagule morphology related to variation in drop time.
[d] Denotes significance at the $p < 0.001$ level.

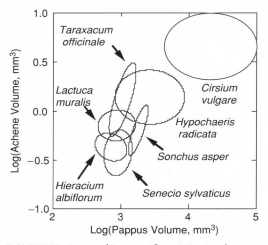

FIGURE 7.19. Inter- and intraspecific variation in achene and pappus volumes among anemochorous species of Asteraceae. The data used in the figure are mean values of achene and pappus volumes, derived from 20 mature propagules from 12–20 individual plants and assessed over 150 island and mainland populations between 1987 and 1990. In species with >30 measured populations (*Lactuca, Senecio, Hypochaeris*), 90% confidence ellipses describe the distributions; in the remaining species with fewer measured populations, 50% confidence ellipses are used. Note that an increase of 1.5 orders of magnitude in achene volume is countered by proportionately a somewhat larger increase in pappus volume.

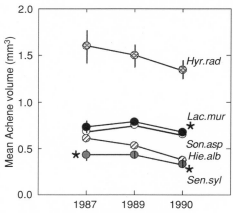

FIGURE 7.20. Mean achene volume (all measured populations) varies among years in a similar fashion over species, with smaller achenes in 1990. Species codes and sample sizes for the three years of measurements are *Hypochaeris radicata* (Hyr.rad: 6, 9, 17); *Lactuca muralis* (Lac.mur: 7, 22, 24); *Sonchus asper* (Son.asp: 3, 2, 3); *Hieracium albiflorum* (Hie.alb: 1, 2, 5); *Senecio sylvaticus* (Sen.syl: 4, 14, 15). Asterisks at 1990 data denote a significant difference between 1990 and 1989, at 1987 datum between 1987 and 1990.

DISPER values (high in *Cirsium*, low in *Taraxacum*) occur in two species with the most limited island distributions.

We measured most achene/pappus morphologies in 1987, 1989, and 1990. Figure 7.20 is redrawn from Cody and Overton (1996) to show parallel variation in achene volumes, among species between years, and in particular a synchronous decline in A_V in 1990; the reduction in achene size that year is statistically significant in the two species (*Lactuca, Senecio*) with large sample sizes. Variation in pappus volume was much less pronounced among years, and the smaller achenes in 1990 drove *DISPER* values higher in that year in all but one species (*Hypochaeris*). Since 1990 was a relatively wet year and 1987 and 1989 relatively dry years (see Fig. 2.2), it may be that in dry years achene mass is larger (with more sunny days for growth), perhaps enhancing seedling establishment. On the other hand, dispersal of these larger achenes, which produce a greater pappus loading, might be reduced in dry years—yet another trade-off situation.

If the ratio of pappus to achene volume, *DISPER*, measures dispersal potential, then propagules with higher *DISPER* values should stay aloft longer and descend more slowly in still air. We measured the descent times of propagules from specific populations over a 2 m drop, giving them an initial half meter to reach their terminal velocities. Each propagule was timed two to five times, and its mean drop time T_D (in seconds) was recorded. Table 7.5 gives overall mean drop times for nine species, which vary among species by a factor of two. This is much less than the interspecific variation in *DISPER*, and clearly there is a good deal more to pappus efficacy than is captured by a measure of its volume. Indeed, pappus structure varies considerably among species, in the arrangement and density of the filaments and their secondary branching.

In six of the nine species with measured drop times (Table 7.5), drop time was significantly correlated with *DISPER*, but in the

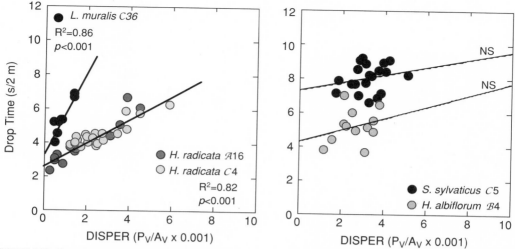

FIGURE 7.21. Drop time in anemochore propagules (ordinate; seconds per 2 m drop) is closely related to *DISPER* (abscissa), the ratio of pappus to achene volume, in *Lactuca muralis* and *Hypochaeris radicata*; drop time increases as pappus volume increases and achene volume decreases (left). In *Senecio sylvaticus* and *Hieracium albiflorum* (right), no significant relation was found between drop time and *DISPER*.

remaining three, *DISPER* was not a good predictor of drop time. In Fig. 7.21, drop time T_D is plotted as a function of *DISPER*; regressions are significant for *Lactuca* and *Hypochaeris*, but not for *Senecio sylvaticus* and *Hieracium*. We are finally in a position to examine the possibility that dispersal potential (*DISPER*) is reduced on islands. This occurs, hypothetically, via selection in established island populations that eliminates the better-dispersing propagules, with higher pappus volume and smaller achene volume. On islands with a narrow annulus of suitable edge habitat, dispersal for any distance except along the annulus would mean either dying in the forest, or getting blown off the island and likely perishing at sea. We require species (a) that have a suitably high island incidence (for good sample sizes) and (b) in which there is evidence that *DISPER* reflects dispersal potential (i.e., is correlated with propagule drop times). Further, for selection against *DISPER* to become effective, (c) island populations must persist over sufficient time that the results of such selection can be measured, ideally as a function of island residence time. Three species meet the first criterion: *Senecio sylvaticus*, *Hypochaeris radicata*, and *Lactuca muralis*. But the *Senecio* fails the second criterion, as T_D and *DISPER* appear to be uncorrelated in this species (Table 7.5; Fig. 7.21). My colleague J. Overton noticed, while measuring these *Senecio* propagules, that the pappus was fragile and seemed readily caducous in some populations, falling easily from the achene (this property was noted above in *Centaurodendron* and other endemic composites of the Juan Fernandez Islands). It is possible that the degree to which the pappus is caducous is under selective control, and that variation in dispersal potential is effected via this phenotypic property, but we collected no systematic data to test that hypothesis. It is also the case that *Senecio sylvaticus*, while reaching islands frequently, is a very poor persister (90% of new colonists failed to survive until the next census); poor persisters clearly cannot benefit from selection for reduced dispersal.

The second species passing the first criterion is *Hypochaeris radicata*, and it also passes the second, with a tight association between T_D and *DISPER* (Table 7.5, Fig. 7.21). It is a particularly good persister, but it colonizes relatively infrequently. Thus we have many island populations of unknown age that have

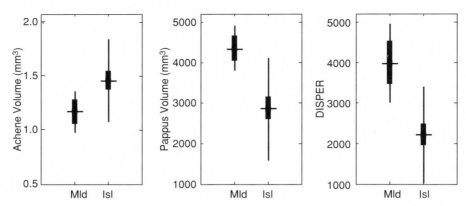

FIGURE 7.22. Variation in achene and pappus morphology (ordinate) between mainland and island populations (abscissa) of *Hypochaeris radicata*. Means, standard errors, and standard deviations are indicated for n = 4 mainland populations and n = 29 long-established island populations. Each populational mean is derived from measurements of 20 mature propagules. Island achenes tend to be larger, island pappi are significantly smaller, and their ratio $DISPER$ (= P_V/A_V) is significantly reduced on islands populations, where lower dispersal potential is selectively advantageous.

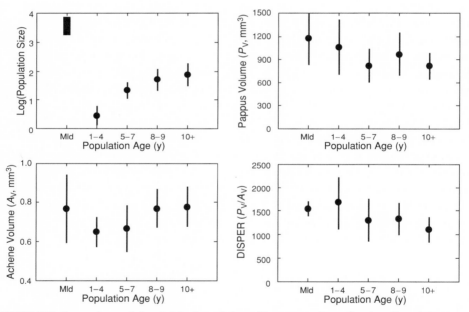

FIGURE 7.23. Propagule morphology in island populations of *Lactuca muralis* as a function of population age. Mainland data are shown to the left on the abscissa, and island population age increases to the right. Island populations increase in size after colonization (upper left). New colonists tend to have smaller achenes (lower left) and higher $DISPER$ values (lower right) than mainland populations. With increasing residence time on the islands, achene volume increases, pappus volume decreases (upper right), and $DISPER$ steadily declines. Populations aged 10+ y are significantly lower in $DISPER$ than mainland populations, indicating the results of selection for reduced dispersal on the islands.

persisted for the duration of the study, perhaps even since well before the start of the study, but few newly established populations. The data permit a comparison of achene and pappus morphology between mainland and islands populations. Island achenes tend to be larger than mainland achenes ($t = 1.47$, df $= 31$, $0.2 > p > 0.1$), pappus volumes are smaller on the islands ($t = 2.26$, df $= 31$, $p = 0.03$), and DISPER is on average 80% higher on the mainland ($t = 2.78$, df $= 31$, $p < 0.01$). Table 7.5 reports the data, and Fig. 7.22 depicts the differences graphically. While the majority of the island populations, having existed since the beginning of the study, are of unknown age, it is clear that they have persisted long enough for selection to shift the balance of phenotypes to those with a combination of larger achenes, smaller pappi, and a reduced dispersal potential.

It was the third species, *Lactuca muralis*, which precipitated our studies on achene/pappus morphology, as I first noticed conspicuously large achenes relative to pappus size on the long-term resident population on AII: Nanat. The species is widely distributed on the islands and shows a high correlation between DISPER and drop times. It has higher colonization rates than *Hypochaeris*, and higher persistence rates than *Senecio*. This translates to an intermediate turnover rate through time and allows us the capacity to identify island populations in which the year of their establishment on an island is known or accurately estimated. Thus we segregated the various island populations into subgroups of various ages, evaluated in 1990: recent colonists (populations established for 1–4 years: $n = 8$), established 5–7 years ($n = 12$), 8–9 years ($n = 9$), and long established (10+ years: $n = 20$). A priori, we expect recently established populations to originate from propagules with high DISPER, but with selection operating to reduce dispersal potential on the islands, we expect DISPER to decrease over time. This is in fact what is observed. In populations known to have been established for just 1–4 years, achene volumes are lower and DISPER is higher than in mainland populations (Fig. 7.23, adapted from Cody and Overton [1996]). Over the course of increased island residence time, pappus volumes decrease, achene volumes increase, and their ratio DISPER decreases to a value (1109 ± 253 SD) about two-thirds that of mainland populations (1537 ± 148) by the time the island has been occupied for 10 years or more. Thus, in just a decade or so, selection had reduced dispersal potential to significantly lower levels in the island residents ($t = 2.73$, df $= 21$, $p = 0.015$). While this does not compare to the dramatic adaptive shifts in reproductive characters of plants on oceanic islands, isolated by thousands of kilometers and millions of years, it does attest to the strength of selection for reduced dispersal. The *Lactuca* story emphasizes the value of continental islands for evolutionary as well as ecological studies and provides an example of adaptive change from field observations in just a single decade.

8

Ecological and Evolutionary Shifts on Continental Islands

The topic of this chapter really began at the end of the previous one (chapter 7) where I described, for some of the wind-dispersed forbs in the family Asteraceae, the changes in the morphology of the achene/pappus on islands relative to mainland populations. The changes were statistically detectable over time intervals of a decade or so after initial establishment and were interpreted in the light of strong selection for a reduced dispersal capacity on islands. This, then, is just the same phenomenon, the same process, and indeed same result that we had discussed earlier in plants on much more isolated oceanic islands such as the Juan Fernandez. There also plants arrived by dint of good anemochorous dispersal but, after long establishment and time for the evolution of morphological modifications, the achene/pappus unit showed evidence of a severely reduced dispersal potential. The differences between these changes in endemic plants on remote oceanic islands with the Barkley Sound results are simply those of scale: here we have shorter dispersal distances, and more subtle morphological changes over shorter time intervals.

There are several reasons why finding support for short-term evolution on continental islands of Barkley Sound is somewhat surprising. First, consider that the island residence times of weedy, edge species are rather limited, and the majority of newly established island populations last a few years at most. Second, both the spatial and temporal scales of isolation of plants on these Barkley Sound islands are narrow: a few kilometers, a dozen or fewer generations. These lead to a third point of obvious contrast: unlike oceanic islands, where strong isolation produces very different species composition and dramatic species disharmonies, and the maritime climate is likely quite different from that of the colonist's continent of origin, the shores of Barkley Sound islands look and feel much like the nearby mainland. This environmental similarity must ameliorate the selection pressures on continental island populations; why change if abiotic and biotic environments island to mainland are very much alike? And also, fourth, oceanic islands with extreme isolation mean long time intervals between successful colonization events, within which there are opportunities for evolutionary change unconstrained by genetic swamping via repeated invasions. On the minimally isolated islands of Barkley Sound, one would expect repeated conspecific invasions to be routine, and that in consequence most indicators of island selection would be suppressed because of gene flow. So it is quite reasonable, then, to expect little if anything in the way of evolutionary, or even ecological, shifts on these continental islands.

Considering that the forest floras (and also faunas, in large part) are mainly relictual, that the edge floras have high species turnover, and that the distances between colonist sources and target islands are very modest, it was perhaps unexpected to encounter this classical island-mainland contrast of reduced dispersal in the island anemochores. However, there are indications here and there in other local taxa of such fundamental island biogeographic processes as niche shifts, density compensation, and morphological change with a genetic basis. In this chapter I collate and discuss some of the evidence for them, bolstering the plant cases with references to some animal data. Some of my examples are somewhat qualitative; on others we collected data during the Island Biogeography courses that I taught at BMSC over the years. Note that, while there may be evidence for behavioral, or physiological, or ecological, or even morphological changes in island populations, there is only inferential evidence that the phenotypic changes are correlated with genetic modification. Bearing this limitation in mind, at the very least I hope to provide a stimulus for the next generation of Barkley Sound biogeographers and students to follow up on some of the interesting questions that remain unresolved.

ALIEN INVADERS

Over the last couple of centuries, with the enormous increase in commercial and other long-distance traffic, thousands of plant species have expanded their geographic ranges beyond the historical confines where they are thought to be native (Mooney and Hobbs 2000). Thus the makeup of many local floras has been altered by the addition of nonnative species, some by deliberate introductions and others by accident. This "species pollution" (in Rapoport's evocative terminology), is a global phenomenon, but some regions are particularly susceptible to invasions of alien plants. The five regions of the world, all on different continents, that share Mediterranean-type climates (the "MTE[cosystem]s") are prime examples, each reciprocally supportive of aliens from other MTE regions. Thus California's flora (5862 species: Hickman 1993) currently includes about 1023 (17.5%) alien species; while many aliens there and elsewhere are roadside weeds with limited representation in natural and undisturbed plant communities, others are aggressively invasive, posing threats to native plant diversity and also constituting an economic liability (Cox 1999; Mooney and Cleland 2001).

Undoubtedly, habitat modification and disturbance have been factors in the persistence of many plant newcomers in the host or "importing" region; thence a greater abundance of disturbed sites can better serve as an effective launching pad for the potential "naturalization" of aliens into less disturbed native vegetation. Regions that have a very long history of anthropogenic disturbance and utilization of native vegetation, such as the Old World Mediterranean Basin, are replete with annual plants with viable life histories under intense disturbance conditions. This Old World region is both the cradle of civilization and not coincidentally the source of most of the alien plants in California and other MTEs such as Chile and South Africa. In the native shrub habitats of California (chaparral) and Israel (maquis or phrygana), annuals plus biennials (therophytes) plus hemicryptophytes (perennials that die back to ground level in the off season) make up 66.7% and 77.6% of the floras, respectively. But the number of plant taxa involved in these growth-form categories in Israel (1205 species) dwarfs that in California (271 species), and further, 60% of California's annual plants are nonnatives (Shmida 1981). Naveh and Dan (1973) made a strong case for the anthropogenic degradation of near-eastern landscapes, where the present-day life-form spectrum of the vegetation is skewed heavily toward short-lived and nonwoody species and is thought to be the product of intensive human use over many millennia. Ironically, continued heavy grazing in these landscapes may be required in order to maintain the high diversity of the ruderal (weedy) plants, whose high species diversity likely originated in the first place via intense grazing (Seligman and Perevolotsky 1994).

Islands, with impoverished floras by mainland standards, are particularly susceptible to alien plant invasions, especially if the aliens are given a foothold in disturbed, grazed, or cleared sites. Carlquist (1974) analyzed the adventive (naturalized, alien) components of the floras of the Juan Fernandez (146 native species), Canary (826 natives), and Hawaiian Islands (1472 natives), showing high percentages of aliens (37% and 42% respectively of the total floras; the 10% figure for Hawaii cited by Carlquist has since been revised up to 42% [Eldridge 1999]). Moreover, the successful aliens were predominantly therophytes or hemicryptophytes, plants typically dominant in open and disturbed habitats; these two growth-form categories comprised around 90% of the naturalized aliens in these three archipelagoes.

While Barkley Sound islands appear superficially pristine, and only minor modifications of the landscape due to anthropogenic influences are apparent, there are several ongoing activities that foster alien plant establishment and undoubtedly will continue to do so. Logging practices, largely forest clear-cuts, impact both mainland sites and a few unprotected and larger islands. In addition, expanding coastal settlements, fisheries, and aquaculture activities along shorelines, and tourist traffic amongst the islands, constitute factors favoring further potential increases in the incidence of alien plants. Until recently, the native eelgrass (*Zostera marina*) was the only representative of its genus in shallow and sheltered coastal waters on the British Columbian coast. But in 1957 *Zostera japonica* was found in Willapa Bay, Washington, and it has since spread south into Oregon and north into southern British Columbia and is almost certainly now present in Barkley Sound (Bigley and Barreca 1982; Harrison and Bigley 1982; Bigley pers. comm.). This invader is thought to have been introduced originally (and by accident) with Japanese oyster stock imported for cultivation. Like the sea-rockets *Cakile* in the upper intertidal (see above), it seems at least possible that the two *Zostera* species might eventually coexist by partitioning the subtidal waters relative to depth below the sea surface. Whereas the annual *Z. japonica* predominates in the upper subtidal, the native perennial *Z. marina* persists in deeper water.

Weedy adventive aliens will find and exploit opportunities to become established wherever there is a high turnover of individual plants in the vegetation, due to either natural or anthropogenic disturbances. In the case of the islands in Barkley Sound, high levels of natural disturbance characterize the edge habitats, thus affording colonization opportunities to alien plants with good dispersal. In the closed forest island interiors, one finds no alien plant species, and very few along the shorelines (the *Zostera* and *Cakile* already mentioned, along with *Spergularia marina*, fill out this limited roster). Virtually all of the aliens on the islands, which constitute 22% of the island flora of 310 species (appendix A), are strictly edge species, and all except a few shrubs that are garden escapes (*Ilex, Sorbus aucuparia, Cotoneaster*) are forbs and grasses. In the weedy edge habitats, in disturbed sites, and along roadsides on the mainland (typified here by my casual surveys around Bamfield village), there are many additional alien species that do not reach the islands. Some of these are abundant on the mainland, and in some cases close relatives that are less common on the mainland are widely distributed on the islands. For example, *Senecio vulgaris* handily outnumbers *S. sylvaticus* on the mainland but has never been recorded on the islands where *S. sylvaticus* (see chapter 7) is quite common. Thus, weediness and a high mainland visibility do not automatically convey access to island edge habitats. Grime (1979) classified plants in terms of

TABLE 8.1
Numbers of Alien and Native Grasses (Poaceae) on Five E-group Islands

		NUMBER OF GRASS SPECIES		
ISLAND	LOG AREA (m^2)	TOTAL	NATIVE	ALIEN
E49: Clarke	5.373	33	17 (0.515)[a]	16 (0.485)[a]
E76: Turtle	6.059	17	12 (0.706)	5 (0.294)
E76: Willis	5.676	17	12 (0.706)	5 (0.294)
E44: Walsh	4.929	17	10 (0.588)	7 (0.412)
E54: Owen	4.827	15	12 (0.800)	3 (0.200)
Total for four islands other than Clarke:[b]			0.697	0.303

[a] The expected number of natives is 23; of aliens, 10. $\chi^2 = 5.17$, df = 1; $p < 0.05$.
[b] The last four islands lack extensive camping facilities. Clarke has twice as many grass species and a higher proportion of alien grasses (49% vs 30%), relative to other islands.

their strategies for persistence in vegetation: as good competitors, or good stress tolerators, or good at coping with disturbance (weeds or ruderals). In his scheme, the weedy ruderals might have low edge incidence either because of competition with natives or other ruderals, or because of abiotic stress factors (such as harsh rocky substrates or saltwater spray).

Of course, there are additional native species in natural mainland habitats that also lack island records; the overall percentage of alien species in the "regional" species list of appendix A is 19.7%, not significantly different from that in the island subset ($\chi^2 = 1.00$, df = 1, $p \gg 0.05$). Many of the aliens are composites (Asteraceae: 12/29 species) or grasses (Poaceae: 23/58 species); the proportions of aliens in these two families in the island flora are about double that in overall regional flora.

When anthropogenic disturbance is added to natural disturbance in island edge habitats, the incidence of alien weeds increases. Camping beaches in the E-group islands are specific areas officially designated for this purpose by Parks Canada. They are also the most productive places to look for weedy plants, and aliens are especially numerous in such places. For example, the only island record of the alien composite *Lapsana communis* (nipplewort) is from the camping beach on the northeast corner of Dodd Island; the only record of the alien grass *Festuca arundinacea* is on the popular sandy beach on D24: "Big Stud." A mecca for campers and kayakers in the Broken Group is beachside at the northwestern corner of E49: Clarke, the old homestead site. In Table 8.1, I compare the grasses and composites of Clarke with those of four other large E-group islands, two larger in area and the other two smaller, all four of which are very little, if at all, affected by campers. We have found 33 grass species on Clarke, 15–17 species on the other four comparison islands. Clarke's grasses are split about 50:50 natives: aliens, whereas the split is 70:30 on the other islands. There are significantly more grass species on Clarke, and significantly more of them are aliens. I should add that virtually all of the supernumerary grasses on Clarke, natives and aliens alike, were recorded in the camping area at its northwest corner, where shrub clearance and human trampling have created some inviting grassy swards.

Some of the alien plants in Barkley Sound have undoubtedly been there for a very long time, perhaps a century or more. Many are truly naturalized and widespread in native and naturally disturbed habitats. Others are more recent arrivals, and the numbers of new arrivals, or at least new island records of them, continue to expand yearly. In Table 8.2, I consider species

TABLE 8.2

Changes in Incidence over Time of Native and Alien Composites and Grasses on the A and B Islands Groups, Barkley Sound

		1981–1982 (n = 275)	1989–1991 (n = 435)	2002–2003 (n = 405)
Composites, Native	No. Island Records	55	65	63
	Group Incidence J	0.200	0.149	0.156
	Chi square	> 89–91, > 02–03	< 81–82, = 02–03	< 81–82, = 89–91
Composites, Alien	No. Island Records	22	30	40
	Group Incidence J	0.080	0.069	0.099
	Chi square	= 89–91, = 02–03	= 81–82, < 02–03	= 81–82, > 89–91
Grasses, Native	No. Island Records	124	120	126
	Group Incidence J	0.451	0.276	0.311
	Chi square	> 89–91, > 02–03	< 81–82, = 02–03	< 81–82, = 89–91
Grasses, Alien	No. Island Records	38	70	82
	Group Incidence J	0.138	0.161	0.203
	Chi square	= 89–91, < 02–03	= 81–82, < 02–03	< 81–82, < 89–91

Note: Number of island records is summed over each target group of five species, with denominator being number of censuses during the time period. Numbers of censuses in 1981–82: 55; 1989–91: 87; 2002–03: 81. Chi-square tests for differences in incidence among years show results > significantly greater, < significantly less, = no significant difference.
Native composites (Asteraceae) surveyed: Achillea millefolium, Anaphalis margaritacea, Aster subspicatus, Grindelia integrifolia, Hieracium albiflorum. Alien composites: Hypochaeris radicata, Lactuca muralis, Senecio sylvaticus, Sonchus spp., Taraxacum officinale.
Native grasses (Poaceae) surveyed: Agrostis exarata, Calamagrostis nutkaensis, Festuca rubra, Hordeum brachyantherum, Puccinellia pumila. Alien grasses: Aira praecox, Holcus lanatum, Poa annua, Poa pratensis, Poa trivialis,

records from the 𝒜- and ℬ-group islands that collectively provide a good sample of censuses in early, middle, and late periods of this study. Censuses in 1981–82 are used as the early date, 1989–1991 as the midpoint, and 2003–04 as the latest period (n.b., there are some minor differences between numbers of islands censused and in island identities among time periods). Within composites Asteraceae and grasses Poaceae, each with numerous widespread native and alien species, I selected five of the commoner native species and five alien species in each family. The table (Table 8.2) gives the number of island records of each species group in each of the three time periods, and the overall group incidence J for the time period. I contrast the numbers of island records among pairs of time periods, using chi-squared tests (using $p < 0.05$ to indicate statistical significance), with results shown in the table and summarized here. There were significantly more island records for native composites in the early period relative to middle and late periods, but no significant difference between middle and late period. For native grasses the results were the same, namely, more island records early in the study. In contrast, alien composites were not significantly different between the early and either of the two later periods, but in the middle period relatively fewer records were scored than in the latest (2003–04) period. Thus they appear to have increased incidence of late. For the alien grasses, the latest census period scored more island records than either middle or early periods, although these last two were not significantly different from each other. My conclusion is that alien plants in these two families have expanded their distributions on the islands in Barkley Sound over the last 20 or more years, while their native confamilials show a declining incidence over the same time interval. It is hard (but not impossible) to avoid the conclusion that the expansion of the aliens is responsible for the contraction of the natives. After all, space at the island edges is a limited resource, island species counts show a degree of regulation via island areas and isolation distances, and competition between confamilials of similar growth forms is a strong possibility.

Note that this test, based as it is on presence-absence data, is of the coarsest possible type. Presumably there would be a stage in the replacement process intermediate between no effects and local extinctions, at which the numbers of individuals of resident species would be reduced by the incoming alien colonists destined to replace them. However, my data are inadequate to detect such numerical influences, and thus only the broadest inference of impending ecological change can be drawn here.

ECOLOGICAL SHIFTS IN IMPOVERISHED BIOTAS

DENSITY CHANGES AND HABITAT SHIFTS

The biotas of remote islands are invariably species poor relative to comparably sized mainland areas of the same habitat types, and in these circumstances, species common to both island and mainland may display different ecologies. Ecologists have documented numerous examples of niche shifts between mainland and island, such as changes in morphology, ecology, or behavior, in species that encounter different selective regimes on impoverished islands relative to their mainland counterparts in richer biotas. Island phenotypes may differ because of changes in the abiotic environment, because of differences in the resource base, or because of fewer competitors or fewer predators or herbivores. The weeds-to-trees paradigm described in the previous chapter relates to herbaceous colonists shifting morphology and associated life history traits via adaptation to new growth-form opportunities that result from the impoverishment of the woody floras of islands. Basic ecology texts are replete with many other examples of character shifts in insular populations, and most of the dramatic examples surely require the evolution of new genotypes over many generations of strong selection.

While some adaptive changes in island phenotypes, especially those involving adaptive morphology or physiology, may require substantial genetic change, perhaps new alleles or different gene frequencies, others may not. Some niche shifts might require no more than an opportunistic switch in behavior, or may simply reflect a capacity for survival in island sites not usually available to our opportunist; where such sites are occupied by a broader range of resident species, for example, the survival of the generalized opportunist might be precluded. In species-poor sites, the realized niche may be closer to a potential maximum, unconstrained by competitors or by other factors that confine it elsewhere. A plant species might occur in a different habitat on an island simply because many of the usual inhabitants of the habitat are rare or absent; no specific new properties, characters, or genetic reorganization may be necessary for this to occur, other than a general predisposition toward or tolerance of the new habitat's conditions.

Competitive release in species-poor situations may allow broader niches in plants through the exploitation of new habitats, and in mobile vertebrates the use of new foraging sites and opportunities. In many sorts of vegetation, bird species partition resources such as foliage insects by species-specific foraging heights or sites (MacArthur 1958; Cody 1999). But on an island, with fewer avian insectivores or perhaps just a single species, a foliage-gleaning bird might forage over a much broader range of vegetation heights in the absence of its usual competitors (Cody 1983b; Cody and Velarde 2002). Such behavioral shifts may require no or at most minimal genetic changes.

SONG SPARROWS

To amplify this last notion, Song Sparrows (*Melospiza melodia*) are common residents on Barkley Sound islands (see appendix E for a synopsis of island birds). On larger islands they live in scrubby edge habitats, but on small islands, down to 1/10 ha or so in size, they utilize the whole island. Indeed, very small islands with but a single bird species are inevitably occupied by Song Sparrows. In mainland habitats, Song Sparrows coexist with a dozen or more additional bird species, but as more and more of the mainland species drop out of the picture on islands of decreasing size, Song Sparrow density increases. This is the classical picture of density compensation, as the Song Sparrow, a consummate generalist species that occurs coast to coast across North America and from Alaska to Mexico, can avail itself of food resources that would otherwise go to the absent species. On islands in Puget Sound, some 125 km east of Barkley Sound, Yeaton and Cody (1974) showed that Song Sparrow territory size declines more or less linearly as the number of coexisting species falls off on smaller and smaller islands. Song Sparrows can indeed reach rather alarming densities on small islands, and as the species is resident (rather than migratory), this makes for some interesting population dynamics, as Smith et al. (1996) have described for Mandarte Island off the southeast tip of Vancouver Island. On the mainland, Song Sparrow foraging activity is strictly delimited; it feeds on and within a meter or so of the ground, generally in thick, edge vegetation. On islands such as A8: "San Jose NW," Song Sparrows feed in the canopies of the Douglas firs 25 m above the ground and also spend a good deal of time foraging within the intertidal zone, especially at low tide (as they do elsewhere on the islands). With the wide range of foraging opportunities opened up to them on monopolized small islands, on which they can capitalize because of their flexible behavior and catholic tastes, small wonder that an island Song Sparrow can subsist on a territory a tiny fraction of the size typically defended by mainland birds.

CHANGES IN PLANT DENSITIES

To return to plants, and to the question of whether island plants might occupy atypical habitats on islands that are species impoverished by dint of small size and isolation, there are some relevant data on this subject, though they are mostly qualitative. Island A8: "San Jose

NW," just mentioned, had until recently Douglas fir as its only tree species, with the notable absence of Sitka spruce, western red cedar, and western hemlock. One small (2.5 m) spruce was found in 1981, and a second became established in 1986. Though this is a low, well-drained, and cobbly island, it is patently obvious that no comparable piece of mainland would be forested essentially without representation by the three dominant tree species (listed above), as indeed are no other islands of similar topography. Similar quirks in plant relative abundances have been recorded on various other islands, which have unusually high densities of plant species that are normally far less prevalent. Thus some islands have a striking predominance of one or another of such plants as bitter cherry (*Prunus emarginatus*), red cedar, wall-lettuce (*Lactuca muralis*), columbine (*Aquilegia formosa*), harebell (*Campanula rotundifolia*), *Spergularia macrotheca*, *Tolmiea menziesii*, to name a few standouts. Most of these species are patchily distributed over the islands and usually occur as scattered individuals, but on some few islands, perhaps because of chance, or early precedence, or for some other reason, they reach high abundance and become major components of the vegetation.

Parenthetically, we have also noted unusually high densities on specific islands in several nonplants. On some small Barkley Sound islands, spiders are impressively abundant, and their webs festoon the vegetation. Spiders reach high densities on Bahamian islands, also, and are especially common in the absence of their usual predators, *Anolis* lizards (Schoener and Spiller 1995; Spiller and Schoener 2001). On small islands in the Sea of Cortés, very high spider densities have been attributed to high levels of spider prey, such as detritivorous and scavenging insects and other arthropods, their abundance boosted by marine inputs from seabirds and decomposing seaweeds (Polis et al. 2002). Both reduced predation (from fewer birds) and more food resources (from marine inputs) may play a role in elevating Barkley Sound spider densities.

On several midsized islands, salamanders (Northwestern Salamander [*Ambystoma gracile*], Clouded Salamander [*Aneides ferreus*], Western Red-backed Salamander [*Plethodon vehiculum*]) reach high densities that they never seem to approach on the mainland. Garter snakes (*Thamnophis* spp.) are reasonably common on the mainland, but nowhere as abundant as along the sunny edges of the islands. We have even met with garter snakes swimming between islands. There may be an element of competitive release and density compensation here, but it seems more likely that it is a release from higher mainland predator pressures that allows island reptile and amphibian populations to reach high densities.

SPHAGNUM BOGS

A good example of a rare and extremely patchy habitat is a sphagnum bog, so defined for the peat-forming *Sphagnum* mosses (of many species) that are their indicator plants (Larsen 1982; Strong 2002). Bogs develop on poorly drained, acidic soils associated with peat formation and are "common features of the glaciated landscapes of the entire Northern Hemisphere [with] a remarkably uniform structure and composition throughout the circumboreal regions" (Curtis 1959, p. 378). Bogs are scattered through the coastal coniferous forests of Vancouver Island and the British Columbia mainland (Gignac et al. 2000; Beilman 2001) and are in fact of worldwide distribution at high latitudes and high elevations. From the northern forests of the Holarctic region, sphagnum bogs extend onto tropical mountaintops such as 2900 m Mount Oko in Cameroon, West Africa (Maisels et al. 2000), and the high paramo in Central America (McQueen 1992). They recur in suitable antipodean sites in New Zealand, southeastern Australia, and southern South America. Bog habitats occur on the five larger Hawaiian islands of Kauai, Molokai, Oahu, Maui, and Hawaii. They are found mostly at high elevations, where they develop in flat or depressed areas of high rainfall, in sites associated with extinct or quiescent volcanoes where claylike silts have accumulated and drainage is poor.

Bogs support a distinct flora of specialized plant species that are able to cope with the unusual conditions presented by bog habitats: high insolation, low nutrients, and waterlogged soils. Many studies of a physiological nature have investigated adaptation in bog plants in relation to these challenges, in particular addressing the apparent anomaly of xeromorphic (basically thick or sclerophyllous) leaves (Marchand 1975; Moore 1980), photosynthetic efficiency (Reader 1978; Small 1972), and nutrient uptake and transport (Damman 1978; Sprent et al. 1978). Many of these specialized plants, as well as *Sphagnum* species themselves, have very broad geographic distributions, such that a list of bog plants from central Russia (P'Yavchenko 1964) looks very much like a list from the bogs of British Columbia. The Hawaiian bogs share *Sphagnum* (but only on Hawaii) and the sundew *Drosera anglica* (conspecific with North American taxa), and numerous congenerics of Bamfield bog plants listed in appendix D, including species of *Lycopodium, Deschampsia, Panicum, Rhynchospora, Sisyrinchium, Viola,* and *Vaccinium*. On the one hand, these ties are perhaps surprising, given that the Hawaiian flora is regarded as overwhelmingly of Malesian (southwest Pacific) affinity (>80% according to Wagner et al. 1990). But on the other hand, probable dispersal agents make the associations more credible, since birds breeding in Alaska routinely overwinter in Hawaii (e.g., Pacific Golden Plover [*Pluvialis dominica*], Bristle-thighed Curlew [*Numenius tahitiensis*]) and could serve as seed vectors for various North American plant taxa.

Although several of these Hawaiian bog genera have species representatives in the Barkley Sound region bog flora, the origins of the Hawaiian species are uncertain, and not necessarily all North American. *Rhynchospora*, for example, may have reached Hawaii from China or the western Pacific, and *Sisyrinchium* perhaps from Central America, rather than the northeast Pacific. Plants now found in or associated with Hawaiian bogs show various and diverse habitat associations. Apart from presumptive bog conspecifics or relatives of North American bog specialists, there are several Hawaiian bog occupants that have their origins in wet forest habitat (e.g., dwarf *Metrosideros polymorpha* and *Cheirodendron*, shrubby *Pelea, Myrsine,* and *Styphelia*). The last-mentioned genus, in the family Epacridaceae, shows strong southwest Pacific ties, yet this family is related to, and sometimes included in, Ericaceae of northern temperate bog dominance. Other Hawaiian bog species are fairly generally distributed in high-elevation grasslands or alpine habitats (e.g., the *Argyroxiphium* and *Dubautia* spp. of Californian tarweed affiliation, and the grass *Deschampsia*). Thus a variety of less-specialized plant taxa has moved into the Hawaiian bogs, where they may be predisposed to find suitable conditions, since alpine habitats share several physical features with bogs (such as low temperature, high radiation, wet conditions, slow nutrient recycling; Billings and Mooney 1968).

Bogs are generally considered to be a seral habitat, a transitional stage between open lakes in topographical depressions and their eventual occlusion by forest as the lake fills in by deposition of sediments aided by the encroaching plants at the lake margin. Various alternative sequences have been documented, in all of which bogs are a temporary, seral stage (e.g., restioid mire to *Sphagnum* bog to heathland in Tasmania; Whinan 1995). However, in some circumstances of more substantial freshwater flows through the lake, it has been suggested that bogs might be rather permanent fixtures, retaining their indicator plants over very long time spans (Klinger 1996). Regardless of their projected life spans, most open lakes have a few bog-specialist plants at their margins, and in those at later successional stages as the open water areas shrink, the expanse of low and open bog vegetation advances from the margins on floating vegetation mats and eventually replaces the lake. The bogs show up in aerial photographs as conspicuous pale patches within the dark forest matrix, and we readily identified the bogs around Bamfield and on the islands in this manner.

Together with my colleague Dr. Adolf Ceska, an avid bog-walker and aquatic plant expert, we

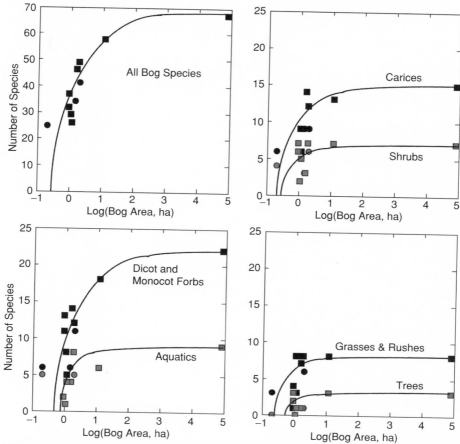

FIGURE 8.1. Numbers of bog plant species are plotted as functions of bog area. All species are grouped in the upper left panel, and six components of the bog floras segregated by growth form and habit are shown in the remaining three panels. Note that bogs down to 1 ha in area retain about the same number of species as in bogs four or five orders of magnitude larger, except for the category labeled "dicot and monocot forbs," in which species richness falls sooner with decreasing area. Island bogs are represented by the circular symbols, mainland bogs by the square symbols.

censused 12 mainland bogs in the Bamfield area, and all 3 of the island bogs, on Nettle, Effingham, and E76: Turtle. Bogs are essentially habitat islands, and the island bogs are really islands within islands. The complete list of species we found in the bogs is included in appendix D, and a summary of the bog floras is given in Table 8.3; all listed plants are regarded as native species. Bog areas were measured from aerial photographs, and are included in the table. These range from the huge "Bamfield Meadows" bog, about 82,500 ha, down to the tiny 0.2 ha bog on Turtle Island. Species are assorted into categories based on growth form, taxonomy, or subhabitat affiliation in Table 8.3, but the group labeled "Primitives" in the table, intended to include ferns and grape-ferns (Ophioglossaceae), horsetails (Equisetaceae), Lycopodiaceae and mosses, is incomplete. In all, we recorded 67 species in the largest bog, 25 species in the smallest, and a total of 92 species are listed in appendix D.

Bog floras can be considered relictual, as all of the bogs were presumably larger in the past and become reduced in area over time. The bog plants show excellent persistence with decreasing bog area, as Fig. 8.1 illustrates. There is a reduction in species numbers, overall and within each of the growth-form categories of Table 8.3, of two- to threefold over a reduction in bog area by a factor of five orders of magnitude. Species

TABLE 8.3

Numbers of Plant Species in 12 Mainland and 3 Island Bogs

Bog/Plant Species Group	Bamfield Meadows	Bamfield #1 Bog	Tepaltos Bog	Bamfield Lake*	Grappler Bog	Kichha Lake*	Nettle Island Bog	Bamfield Huu-ay-aht Bog	Effingham Island Bog	Lighthouse Bog	Calamity Lake*	Ninemile Lake*	Eight-Mile Bog	Cape Beale Bog	Turtle Island Bog
Primitives	3	3	1	3	3	3	3	2	3	2	2	1	1	2	1
Carices	15	13	12	11	14	7	9	7	9	7	9	6	9	6	6
Rushes	4	5	5	5	4	3	2	2	5	0	3	2	5	1	2
Grasses	4	3	3	3	3	4	4	2	3	1	3	2	3	2	1
Aquatics	9	6	8	6	4	8	5	13	4	2	4	10	1	4	5
Dicot/monot forbs	22	18	12	12	14	9	11	6	6	11	5	6	8	5	6
Shrubs	7	7	7	5	3	7	6	3	3	7	3	3	2	5	4
Trees	3	3	1	3	1	3	1	3	1	2	1	0	0	1	0
Total bog spp:	67	58	49	48	46	44	41	37	34	32	30	30	29	26	25
Area (ha):	82500	12	1.93	13	1.6	3	2	0.95	1.5	0.94	2	5	0.11	1.17	0.2

Note: Island bogs appear in gray columns. Four of the mainland bogs are lake edge bogs (indicated by *) and are not included in Fig. 8.1. See appendix D for complete species lists.

richness is retained in most of the plant categories (Fig. 8.1) right down to 1 ha bogs, below which species are lost rapidly. Note that the largest category, that of dicot and monocot forbs, contains species apparently most sensitive to decreasing habitat area, such that bogs of around 1 ha have only half the species found in the largest bog. There is no compelling reason to either predict a priori or adduce from the data (Fig. 8.1) that island bogs are more impoverished in plant species than mainland bogs. The two larger island bogs, Nettle and Effingham, support (with 41 and 34 species, respectively) species numbers quite consistent with their areas (2 and 1.5 ha) relative to the mainland bogs. Turtle Bog, with a 25 species total, appears not particularly species poor; it has one fewer species than the 5 times larger Cape Beale Bog, and four fewer than the smaller (0.11 ha) Eightmile Bog, but this latter is just downstream from Ninemile Lake and likely gains species richness from this adjacency.

What does distinguish the island bogs is the species that occupy them. While some species of *Carex*, *Scirpus*, *Juncus*, and *Vaccinium* are true bog specialists, and are not found outside bog habitats, others are more generalized wetland plants and can be found in damp ditches along mainland roads, in freshwater seeps at island edges, and even in damp pockets on rotting intertidal logs. Just as in the Hawaiian bogs, the smallest bogs in the Bamfield data set tend to have a higher proportion of more generalized species and a lower proportion of the true bog-specialist species. I have assembled a list of 29 species of sedges and rushes (Cyperaceae, Juncaceae) found in seven mainland and the three island bogs in Table 8.4, where the bogs are ranked left to right in terms of the total number of species in these two taxa. The species in the table are ranked top to bottom in terms of their moments around the right-hand edge of the table (high moments: species extends into larger bogs; low moments: species restricted to small bogs). The specificity of plants to bog habitats can be evaluated using the criterion of whether or not they occur on the islands outside of the three island bogs. Thus *Carex lyngbyei* is the most widely distributed island sedge and occurs on some 24 islands; it is found in just two bogs, namely, the two smaller of the three island bogs. *C. obnupta* is slightly less broadly distributed (on 22 islands); it occurs in all three island bogs but is absent from two of the three smallest mainland bogs. A third sedge with a wide island distribution is *Carex pansa* (18 islands), but it occurs in much drier sites and does not show up in the bogs. A fourth is the very widespread and generalized spike-rush *Eleocharis palustris*, with just two bog records, both of them island bogs. The rush *Juncus arcticus* has a broad island distribution (with occurrences on 27 islands), and it was recorded in the two smaller island bogs and in one small mainland bog. *J. effusus* is another rather generalized wetland rush, rare on the islands but common on the mainland, that was found in two island and two mainland bogs.

Nettle Bog appears from Table 8.4 to be rather species poor. It lacks all three *Scirpus* species, many of the more specialized *Carex* species of smaller mainland bogs, and has just two specialized *Juncus* species (*J. oregonus*, *J. ensifolius*). Moreover, the incidence of the more generalized carices and rushes is very low in the Nettle Bog, perhaps because of its interior location on a very large island and thus isolated from the potential sources for generalist plants along the coast. Both Turtle and Effingham Bogs are much closer to their islands' shorelines (on which the more generalized sedges and rushes might be found). Note that Turtle Bog and similarly sized Lighthouse Bog on the mainland share just two species in common. Contributing conspicuously to the nonnestedness of the distributions in Table 8.4 are generalized species with incidence only in the smallest bogs, and Turtle Bog is the best example of this phenomenon. For each bog, I evaluated its contribution to nestedness by computing the proportion of nongaps to gaps in its component species. A bog with n species and m bogs to its left in the ranking has a maximum of $n * m$ 1's or 0's on its species' rows. Thus Nettle

TABLE 8.4

Incidence of Carices and Rushes in Mainland and Island Bogs

SPECIES (29)	BAMFIELD MEADOWS	BAMFIELD #1	NETTLE ISLAND BOG	TEPALTOS BOG	GRAPPLER BOG	EFFINGHAM ISLAND BOG	CAPE BEALE BOG	HUY-AY-AHT BOG	LIGHTHOUSE BOG	TURTLE ISLAND BOG	TOT. # BOGS	Σ MOMENTS
Carex phyllomanica	1	1	1	1	1	1	1	1	1	1	10	55
Carex livida	1	1	1	1	1	1	1	1	1		9	54
Rhynchospora alba	1	1	1	1	1	1	1	1		1	9	54
Carex obnupta	1	1	1	1	1	1		1	1		8	48
Carex viridula	1	1	1	1	1	1		1			7	48
Juncus oregonus/supiniformis	1	1	1	1	1	1	1				7	48
Carex rostrata/utricularia	1	1	1	1	1	1		1			7	47
Dulichium arundinaceum	1	1	1	1	1	1					6	45
Juncus ensifolius	1	1	1	1	1	1					6	45
Scirpus caespitosus	1	1		1	1		1	1	1		6	37
Scirpus subterminalis	1	1	1	1	1						5	36
Carex leptalea	1		1	1	1	1				1	5	32
Juncus falcatus	1	1			1	1					4	30
Carex sitchensis	1			1	1	1	1				5	30
Carex physocarpa/saxatalis	1	1			1	1					4	30
Juncus acuminatus	1	1		1		1	1				4	23
Juncus effusus	1	1		1				1		1	4	22
Carex pluriflora	1			1					1		3	19
Scirpus vallidus	1	1									2	19
Juncus brevicaudatus		1	1								2	16
Carex angustior/echinata		1			1						2	15
Juncus arcticus				1		1				1	3	13
Carex pauciflora	1										1	10
Eleocharis palustris			1							1	2	9
Carex buxbaumii		1									1	9
Carex lyngbyei						1		1			2	6
Carex stylosa					1						1	6
Eriophorum angustifolium									1		1	2
Carex canescens										1	1	1
Total # Species:	19	18	11	17	18	13	7	9	7	8		

Note: Bogs are ranked left to right in decreasing area. Species are ranked by their moments about the right edge of the matrix. Island bogs (3: Nettle, Effingham, Turtle) are outlined. Note the low species number in

Bog is well nested, with its 11 species with 19 records and just 3 gaps to the left; $19/2 * 11 = 0.864$. The seven bogs to the right of Bamfield Meadows Bog (first column) average 0.773 ± 0.046 SD in this nestedness measure. The second smallest mainland bog, Lighthouse Bog, returns a value of 0.661, significantly less nested on account of its unique record of the cotton-grass *Eriophorum angustifolium*. Turtle Bog, however, is especially distinct. Its nestedness value of 0.375 is about half that of the other bogs, because it is populated by numbers of generalized species that are patchy or absent on larger bogs. Thus the composition of small and/or isolated bogs shifts from a predominance of bog specialists to a predominance of their generalist congenerics. Such habitat shifts in these sedge and rush generalists are paralleled in many island biotas where species routinely shift their habitat type or use to avail themselves of opportunities provided by absent competitors. Such competitive release is precluded in richer floras, just as Song Sparrow foraging behavior (above) is much more narrowly constrained in richer mainland avifaunas.

Besides the sedges and rushes, there are both generalist and bog-specific shrub species in the genus *Vaccinium* (fam. Ericaceae). Appendix D lists the bog-specific species, namely, *V. alaskense*, *V. uliginosum*, and *V. [Oxycoccus] oxycoccus*. The former, tall and erect, is found only in Nettle Bog on the islands, the latter, a prostrate shrub, occurs in all three island bogs, and the third (*V. uliginosum*) is found on no islands. With this impoverishment of the island bogs, there would seem to be opportunities for the more generalized *Vaccinium* species, namely, *V. ovatum*, *V. parvifolium*, and *V. ovalifolium* to move into bog habitats there. The appendix also includes two additional ericad shrubs that are bog specialists, Labrador tea (*Ledum groenlandicum*) and Western bog-laurel (*Kalmia polifolia*). Two additional species of generalized ericads, *Menziesia ferruginea* (mostly forest understory) and salal (*Gaultheria shallon*) (mostly edge habitats) are also widespread on the islands and are potential bog colonists in suitably conducive situations. The more widespread salal exhibits a wide degree of variation in leaf morphology, having larger and thinner leaves on larger islands, and on the northern aspects of islands relative to southern aspects (Nguyen-Phuc 1996). Burns (2004b) collected distributional and leaf morphology data on ericad shrubs from two species-rich mainland bogs and one species-poor island bog (that on Effingham Island). From transects from bog center across the bog-forest ecotone and into the forest, he determined that the bog specialists *Ledum* and *Kalmia* were commonest in the bog center and rarest in the adjacent forest in the two mainland bogs. The four generalist species *Vaccinium ovatum*, *V. parvifolium*, *G. shallon*, and *M. ferruginea* were all commonest in the forest and least abundant in the bog center. *V. ovatum* and *G. shallon* were common also at the bog-forest ecotone at mainland bogs, and in the island bog (which lacked *Ledum* and *Kalmia*) these two species were far commoner within the bog itself, a significant habitat shift. Leaf specific weight (g/m^2) in these ericads varied both among and within species; within species, all showed relatively thicker, heavier leaves per area in sites of high insolation, that is, in bogs or at least outside of the forest shade. The two bog specialists *Ledum* and *Kalmia* are both relatively thick leaved; of the nonspecialist ericads, the thicker-leaved is *V. ovatum* followed by *G. shallon*. Thin-leaved *Menziesia* and *V. parvifolium* showed no capacity to shift into the bog habitat where shrub diversity was reduced on the island, but both thicker-leaved species were able to do so, in proportion to leaf thickness and the extent to which this feature predisposed them to do so. Note that intraspecific variation in leaf specific weight, an index of the potential to occupy a variety of habitats, is not genetically fixed but a product of the growth site of the individual shrub. Burns (2004b) conducted seedling transplant experiments that showed a capacity in both *Menziesia* and *V. parvifolium* to develop thicker leaves in bog habitats, and in *Ledum*, *Kalmia*, and *V. ovatum* to develop thinner leaves in forest habitats. Such developmental plasticity must be an asset to species presented with opportunities to utilize a wider a range of habitat types, and in filling vacant habitat niches in species-poor sites such as islands.

EVOLUTIONARY SHIFTS IN ISOLATED POPULATIONS

DEER MICE

Small animals live at higher densities than large animals, and with higher numbers per area they survive well in relictual island populations. My last examples of evolutionary shifts on the continental islands of Barkley Sound involve two common forest animals, mice and slugs. While the latter are discussed here in much more detail than the former, there is an extensive literature on the evolution of body size shift of island rodents, and the Barkley Sound data are presented in this larger perspective.

The coastal coniferous forests are populated with deer mice (*Peromyscus maniculatus*; fam. Cricetidae, recently relegated to the fam. Muridae) belonging to the most diverse genus of rodents in North America. Hall (1981) lists 50 species of *Peromyscus*, many of which co-occur in various habitats across the continent. I trapped three species in the chaparral of the Santa Monica Mountains of southern California (*P. maniculatus, P. californicus, P. boylii*) and another three in the Mojave Desert not far away (*P. crinitus, P. eremicus, P. truei*). These mice are omnivores, eating both plant and animal material, and their generalist food habits undoubtedly serve them well in their colonization of islands. Populations derived from seven of the eight *Peromyscus* species of the Sonoran mainland (to the east) and Baja California peninsula (to the west) of the Sea of Cortés are found on various islands within it (Lawlor et al. 2002). They are now classified into nine endemic species and many endemic subspecies and are recorded on islands with and without recent land-bridge connections. The mice are certainly capable of dispersal across seawater gaps of a dozen kilometers or more, but the extent of endemism argues that such events are very infrequent.

The deer mouse of the coastal forests is a truly cosmopolitan species and lives in a wide variety of other habitats, grasslands to forests, from Alaska to Mexico. Deer mice are divided into around 70 subspecies based on size, tail and foot lengths, and coloration. Over one-half of these subspecies are insular endemics, many on islands off the Pacific coast of North America, from the Sea of Cortés and the Pacific islands off Baja California to the California Channel Islands, Vancouver Island and islands in the Georgia Strait and Puget Sound, on north to the Queen Charlotte Islands and thence to Alaskan islands. Vancouver Island supports two endemic subspecies, one interior and montane and the other, *P. m. angustiae*, in coastal regions. This latter occurs on the larger islands in Barkley Sound such as Æ49: Clarke and Æ77: Willis. The deer mice are known to do particularly well in fragmented forest landscapes, and their densities are high even in clear-cuts, where species that are more closely tied to forests such as *P. oreas* are scarcer (Sanger et al. 1997).

We trapped mice on Willis Island in July 1991, as a class exercise for the student biogeographers. Our trapping success (with 17 traps over two nights) was 71%, indicative of an extremely dense (or at least extremely gullible) population. Eleven females averaged 28.52 ± 2.29 g, and 13 males averaged significantly smaller: 25.91 ± 0.99 g ($t = 3.55$, df = 22, $p <$ 0.01). The island mice collectively were almost 20% heavier than the average of mainland mice caught and weighed at odd intervals over the years; this body size difference is similar to that of the *Peromyscus* on Sea of Cortés islands, where the island mice are recognized endemics and known to be genetically distinct. Thus there are "giant" island mice in Barkley Sound, although they are gigantic to a rather modest degree. But in this respect they mirror the island gigantism found in insular populations of many small rodents, especially cricetids (now murids) and microtines, as well as in other vertebrates such as lizards and birds. A survey by Foster (1964) showed that nearly 90% of insular rodents were larger than their mainland counterparts, whereas 85% of the carnivores and artiodactyls were smaller. He was particularly concerned with mice on the Queen Charlotte Islands, where two species are found, *P. m. cancrivorus* on the larger islands and *P. sitkensis* on smaller and more isolated islands. The latter is a large species, twice the weight of the local *P. maniculatus*, and besides the Queen Charlottes it occupies islands of the Alexander Archipelago in Alaska.

Ecologists have written extensively on the reasons for size shifts in island mammals. Foster's (1964) observations were formalized into an "island rule" that small mammals become larger, large mammals become smaller (Lomolino 1985), but Lawlor (1982) suggested that only small generalist species, and not the specialists, become larger on islands as they benefit more from reduced competition. Angerbjörn (1985) showed that the body size of island mice (*Apodemus*) in the Baltic was enhanced with increased isolation from the mainland. Lomolino (in Brown and Lomolino 1998, Fig. 14.24) depicted body size shifts of *P. maniculatus* as a function of island isolation in both the Queen Charlottes and on islands in the Gulf of British Columbia. In this latter region, body weights (which are [again] overall 20% higher on the islands) increased from a range of 14–22 g at mainland sites to around 27–28 g on islands with ≥ 5 km isolation from mainland sources. Our results are entirely consistent with Lomolino's synthesis: Willis is some 6 km from the nearest point on the mainland, and mean mouse mass in 27.1 g.

But although the island body size patterns show a high degree of consistency, the reasons for larger body sizes in island mice are still not understood with any precision. Explanations potentially include components of competition (reduced interspecifically, increased intraspecifically), predation (reduced), and changes in island resource levels (potentially higher). There are also differences in life histories and social structure in island populations, and differences in the selection regime associated with high densities of individuals in strictly confined populations. A comprehensive study of body size/density/isolation/food resources in these island mice would surely prove rewarding, but it has not yet been conducted.

BANANA SLUGS

A last example of evolution on the continental shelf islands of Barkley Sound involves another nonplant, the banana slug *Ariolimax columbianum*; see Gordon (1994) for an introduction to its natural history. This magnificent mollusk, beloved mascot of the University of California's Santa Cruz campus, shares the islands with two less obvious and shelled mollusks, the yellow-shelled *Haplotrema sportella* and the somewhat smaller, maroon-shelled *Vespericola columbiana*. Both are rather inconspicuous detritivores found within the forests and restricted to larger islands (generally >1 ha), although *Haplotrema* incidence falls off less rapidly than *Vespericola* incidence with decreasing island size (Heubner 1996). In contrast to the snails, the banana slug is a very prominent island resident; it occurs on all larger islands and on smaller islands down to 1/30 ha in area, it is found over all island habitats from shorelines through edges to forest, and it is a voracious herbivore of broad dietary scope. Some alien slugs (*Arion, Limax* spp.) occur on the mainland, but except for one record of a single individual alien (on an island with a camping beach), they are absent from the islands.

Banana slugs are western North American in distribution and are divided into five clades, now considered distinct species. Our species, *A. columbianum*, ranges from Alaska to central California, whence it is replaced by *A. stramineus* from there down into Mexico; three additional and narrow-ranged species (*A. californicus, A. brachyphallus, A. dolichophallus*) are restricted to the San Francisco peninsula. The various species have different (but always bizarre) sexual habits; they are hermaphrodites, either reciprocally and simultaneously or serially, and they frequently chew off each other's penises after copulation (again reciprocally), a procedure known as "apophallation" (Leonard 2001; Leonard et al. 2002). This may allow them to concentrate better on motherhood, although postapophallation they are still able to copulate as females. One might have thought, a priori, that preventing further female function made more sense. These slugs are true omnivores, ascending bushes to eat soft fruits and berries in season, feeding on a wide range of low, leafy forbs at the forest edge (Apiaceae and Leguminosae are popular) and on the leaves of deciduous shrubs (such as *Vaccinium parvifolium*) but not the evergreens. They also feed avidly at piles of otter and mink feces, and on carrion (dead fish, birds, and mustelid orts). Cates and Orians (1975) tested the palatability of plants to slugs by presenting them with a wide variety of leaf disks; they (authors and slugs) ranked early successional annuals highest in palatability, early successional perennials second, and late successional perennials as the least palatable. Preferences notwithstanding, the list of acceptable food plants was very long. In late summer, when many of the forbs have dried up on the island edges, the slugs commonly feed on various fungi within the forest. One forest mushroom, *Clitocybe flaccida* (fam. Tricholomataceae), belongs to a genus with many species that are very toxic to humans. It produces a compound, clitolactone, that effectively protects it from slug herbivory (Wood et al. 2004); it might be that many of the toxins for which mushrooms are widely respected have evolved as antislug protection.

The slugs themselves seem well protected against their own predators, by a mucus covering over the body and a large, caudal mucus gland that operates when the slug is agitated. Naïve robins will attack banana slugs and then spend some minutes attempting to wipe the mucus from their beaks. More successful as slug predators are the larger salamanders *Ambystoma gracile*, which, over the course of an hour or two, will slowly ingest banana slugs of nearly their own size.

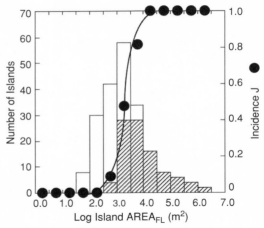

FIGURE 8.2. Distribution of banana slugs *Ariolimax columbianum* on islands in Barkley Sound. With essentially no dispersal, the slugs are relictual and persist on islands as a function only of island size (n.b., the smooth logistic incidence function showing the close control of incidence by island area alone). Critical island area for the slugs is about 0.1 ha, below which they are extremely rare, and above which they rapidly become ubiquitous. The histogram shows the overall frequency distribution of islands by area, and the shaded portion indicates those islands occupied by banana slugs.

Most of the islands censused for plants were also searched for banana slugs, chiefly in 1987, 1989, and 1990. The slugs are relatively easy to find and are broadly distributed on the larger, cooler, and damper islands. On smaller and drier islands, the slugs are relatively inactive in midsummer, when they congregate at the one or two sites that seem to us the coolest and wettest, such as under a shaded rock or beneath a rotting log. In this way, with a "think like a slug" approach, we were able to locate the slugs and determine their incidence over a wide range of islands with some confidence. Island size, it turns out, very precisely determines incidence (Fig. 8.2). All islands >1 ha (log $AREA_{FL} > 4$) have slugs, only a few islands with $3.5 < \log AREA_{FL} < 4.0$ lack slugs, half those of size $3.0 < \log AREA_{FL} < 3.5$ have slugs, and on islands < 0.1 ha fewer than 1 in 10 are occupied. The smallest occupied island is Æ20: "Tiny SW," $AREA_{FL} = 345$ m².

In several large populations (as well as in many smaller ones) we measured slug body mass with Pesola™ scales, and the frequency distribution of mass for three sites combined is

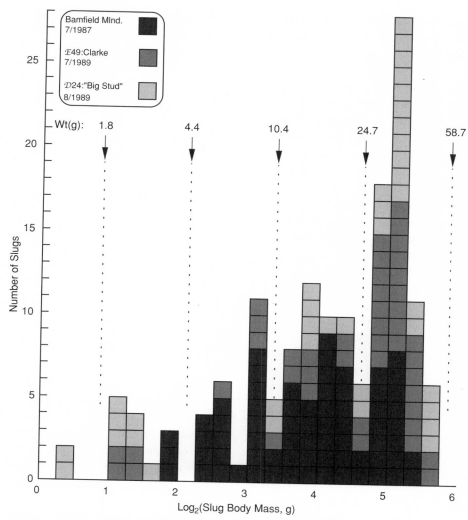

FIGURE 8.3. Frequency distribution of slug body mass at three sites, Bamfield mainland, $E49$: Clarke, and $D24$: "Big Stud." Discontinuities in the distribution likely indicate age classes, suggesting that the slugs live for five years. The division into size classes is at intervals of $2^{1.25} = 2.38$, the factor by which mass increases between adjacent hypothetical year classes. The representation of different "age" or size classes is not uniform across sites, indicating that recruitment varies over years and sites.

shown in Fig. 8.3. Slugs fall into some five fairly well defined size groups, which we interpret as age or year classes, with an increase in mass between putative year classes by a factor of about 2.38 (an interval of 1.25 on the $\log_2$-scaled abscissa of the figure). It was also apparent that representation of slugs in the different size classes, although naturally biased in favor of the larger and more easily found individuals, was not at all uniform, and some islands, especially smaller ones, lacked individuals of particular size classes. It seemed to be that reproduction in the slugs was variable among sites and years.

The slug distributions, with such a close correspondence with island area, have to be considered strictly relictual. Further, as slugs essentially dissolve into a ball of froth when placed in seawater, there is surely very little, if any, colonization taking place within the Sound. If there were, we presumably would find new island populations (which has not happened), the incidence function of Fig. 8.2 would be less regular, and it would be less likely that island slugs could adapt morphologically to their particular island's environment and diverge in appearance between islands (see below). Without colonization, the extant slugs are most likely those that have managed to persist on their islands for the last 10,000 years or so; obviously this persistence is strongly island and thence population size dependent. It must be added, however, that to maintain that slug colonization events simply never occur may be somewhat foolhardy, and there are counterindications of a sort. We found slugs on some five islands with elevations above the *Fucus* line of under 4 m, the degree to which tide lines were estimated to be higher than present on the outer coast (Friele and Hutchinson 1993) 5000 years BP. It remains possible that islands within Barkley Sound experienced different rates of uplift than those on the outer coast, in which case 4 m high islands may have remained emergent throughout the period of rising sea levels in the Holocene. We found slugs on $B12$: "Haines SE," $C4$: "Mid Ross," and $E20$: "Tiny SE," all with ELEV of 3 m, and also on $B1$: Wizard and $A8$: "San Jose NW," with ELEV between 3 and 4 m. It is uncertain, then, whether slugs survived on these low islands in situ when sea levels were higher than at present, or whether in fact they colonized during the last one or two thousand years, conceivably by riding on floating logs or gull's legs.

During the two decades of the study, we believe that one island population, that on the 993 m^2 $B26$: "Diana Ddgr bN," became extinct. Slugs were noted on this island in 1985, and 22 individuals were found in a central rock crack, a midsummer refuge, in 1987; they were still there in 1990, but searches for them in 1996, 1998, and 2002 were unsuccessful. We feared that the same fate had befallen the slug population on $D23$: "M Studs Big," which, although it is a larger island (AREA$_{FL}$ = 3820 m^2), is very dry and open with little shrub or tree cover. We found six slugs in 1987 (August 12) at their summer retreat, a central, damp crevice, but could find no slugs there in 1989. However, in 1990, several tiny slugs were discovered in the usual midsummer spot, and it appears that, although the larger slugs had all died off, they had left eggs (which hatch into sluglings the following spring) that had extended the longevity of this precarious population. On $D28$: "W Studs nC" (AREA$_{FL}$ = 1340 m^2) a single slug was located in 1987 after a 20 minute search, but in 1989 we were able to locate 12 slugs, 10 of which appeared to be in their second or third year posthatching (3–8 g). This population was bouncing back after going through a rather severe crash. Note that the extinction event and the severe population bottlenecks occurred during and after a series of extremely dry summers. Rainfall in 1985 was a record low (Fig. 2.1), and 1986–1989 were also below average years for precipitation. While the slugs on $D23$ and $D28$ recovered, those of B26 apparently did not. It would appear that the slug incidence is still undergoing relaxation, and the function shown in Fig. 8.2 therefore is not strictly equilibrial. A series of years with dry summers will likely trim incidence from the smaller islands with tiny population sizes (e.g., 20 or fewer individuals), from islands that are dry, lacking taller vegetation, and more susceptible to desiccation during low rainfall periods. Clearly, the slugs thrive better in cooler and damper conditions that in warmer and drier times. This is apparent also in slug activity patterns, as they forage throughout the day on wet or overcast days, but activity is greatly reduced on sunny days.

What is especially interesting about the banana slugs is the variation in their color patterns, both within and among populations. The "body" or foot is generally of some lighter color, on which are scattered various spots and blotches that are usually black, sometimes dark brown. The "background" body color varies from a creamy white through a yellowish green to a darker olive green. The mantle is usually unspotted, and in color similar to that of the body, but some individuals have a single black spot, small to large, in the (dorsal) center of the mantle. It is in the degree of spotting that slug individuals differ most conspicuously. In some, the spots are

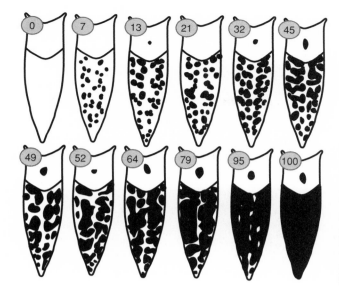

FIGURE 8.4. Variation in banana slug (*Ariolimax columbianum*) spotting patterns on islands in Barkley Sound. The figures are scanned from field-drawn "slug-o-grams" and digitized to obtain the percent of black spotting on the foot or "body." The circled figures are percentage of spot coverage and range from 0 to 100%.

large enough so that they run together and nearly obscure the lighter body color, the slug appearing virtually solid black; in others the spots are distinct and well separated, in others they are merely small dots, and yet others slugs are completely unspotted. Some components of the variation are clearly interdependent, such that heavily spotted slugs have darker background coloration, lightly spotted slugs have paler background colors, and mantle spots occur most frequently in heavily spotted slugs, rarely in lightly spotted or unspotted slugs. All of the evidence, from the distribution of the variations within and among islands and over a range of island sizes and conditions, suggests that slug coloration and spotting patterns are genetically controlled, but their heritability has not been confirmed experimentally.

We recorded slug spot variations in the field by drawing on blank charts, called "slug-o-grams," with marker pens. Nearly one thousand individuals were so recorded, from the mainland and some 54 islands. With the slug-o-grams scanned and digitized, spot density on the foot (or "body") could be quantified as a percentage of spot cover; the breadth of variation in this metric, 0–100% with various intermediates, is shown in Fig. 8.4. Much of this variation can been seen in a single large population, such as on the Bamfield mainland or on very large islands such as $E49$: Clarke, $E76$: Turtle, $E77$: Willis, or $D11$: Friend. Histograms of percent spotting are shown in Fig. 8.5 for 30 islands plus the Bamfield mainland. Variation decreases from left to right across the figure, and generally with decreasing island area. However, average spot coverage (SP_{AV}) as well as the maximum spot coverage recorded within the population are both best predicted by vegetation height (*VGHT*): $SP_{AV} = 22.88 + 0.55 * VGHT$ ($R^2 = 0.20$, $n = 43$; $p = 0.003$) and $SP_{MAX} = 32.01 + 1.25 * VGHT$ ($R^2 = 0.42$, $n = 42$; $p < 0.001$). Mean spot density is shown as a function of both (log) island area and vegetation height in Fig. 8.6; the distribution of the two island variables is shown in the upper part of the figure, and in the lower part the SP_{AV} values are superimposed with confidence ellipses. Smaller and more open islands with less tall vegetation tend to house slugs with ever decreasing spot coverage, and on some very small islands slugs are quite unspotted. On smallish but damp and well-vegetated islands (a rare combination, best illustrated by $A11$: Nanat), slugs become very heavily spotted, nearly all black. The smaller the island, the more homogeneous its slugs'

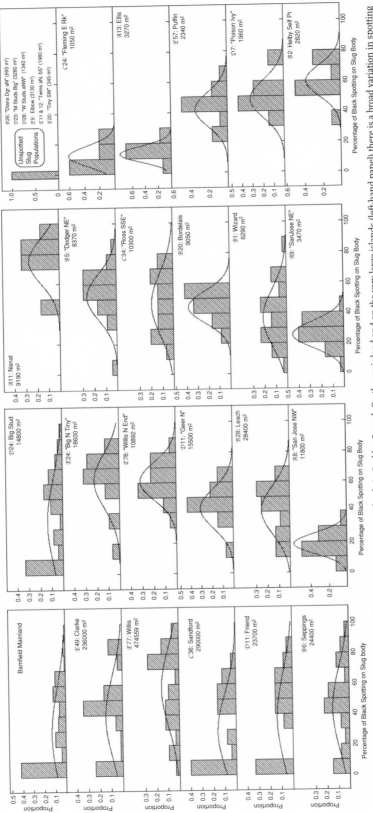

FIGURE 8.5. Frequency distributions of banana slug spotting patterns on islands in Barkley Sound. On the mainland and on the very large islands (left-hand panel) there is a broad variation in spotting patterns, from unspotted slugs to nearly or completely black slugs. With decreasing island size (center panels), variation in spotting pattern within islands begins to decrease, with taller and well-vegetated islands supporting heavier spotting patterns and lower, drier islands with more sparsely spotted slugs. On the smallest islands with slugs (right panel, top), slugs are completely spotless, and on other very small islands slugs are rather uniformly densely or sparsely spotted, depending on whether the island is dry and open, or damp and forested. See text for further discussion.

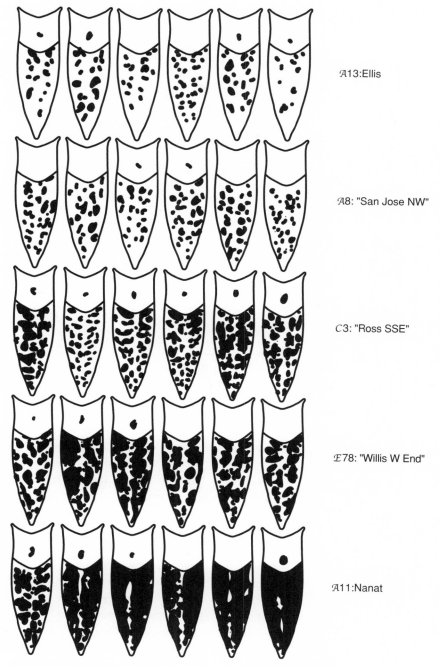

FIGURE 8.6. Representative banana slugs from five Barkley Sound islands. See Fig. 8.5 for frequency distributions of slug spotting patterns on these islands. As on many mid- to small-sized islands, these slug populations exhibit patterns that are relatively homogeneous and island specific.

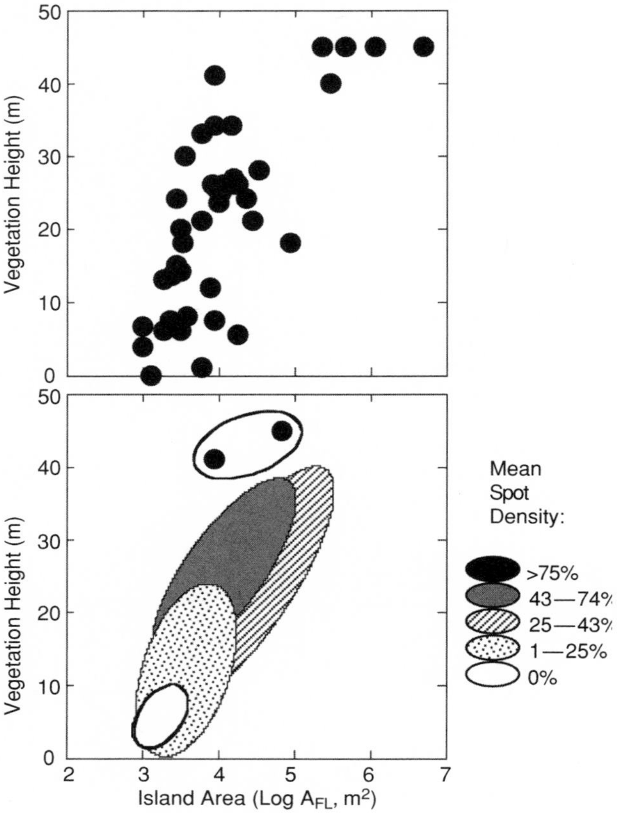

FIGURE 8.7. Vegetation height increases with island area (upper graph), and slug spotting patterns vary by island area and especially by the vegetation height on the islands (lower graph). Note that unspotted slugs occur only on small islands with low vegetation, and that mean spot density increases with vegetation height, and varies also with island size. Only small, damp, and well-forested islands have uniformly black or nearly black slugs (2 islands). 50% confidence ellipses are used for three spot density categories with $n \geq 8$, 35% for $n < 8$.

patterns (Fig. 8.5), and on very small islands all slugs look nearly identical. The islands with unspotted slugs (Fig. 8.5, upper right) are either very small or not quite so small but with very low vegetation cover. Representative slugs from five of the smaller island populations are shown in Fig. 8.6, with populational spotting averages ranging from very low (13.0% on 𝒜13: Ellis) to very high (75.3% on 𝒜11: Nanat). Standard deviations in spot density (SP_{SD}) derived from the histograms of Fig. 8.5 increase significantly with both island area and vegetation height: $SP_{SD} = -8.16 + 0.12 * VGHT + 5.08 * \text{Log}(AREA_{FL})$, with $R^2 = 0.38$, $n = 42$, $p < 0.001$. A plot of spotting variation as a function of these two variables is given in Fig. 8.7.

In mid-July 2003, we placed HOBO™ remote sensors 1 m above ground level in the forest interiors of nine islands of similar sizes, around 1 ha, across the Sound. The sensors recorded daily maximum and minimum temperature and relative humidity for two weeks. Slug-o-grams from these islands were then analyzed; mean spot density in their slug populations was unrelated to island position, distance from mainland, and to the small variations in island area. Mean spot density was also unrelated to mean daily maximum and minimum relative humidity, and to mean minimum temperature, but it was strongly associated with mean maximum temperature. The coolest and the warmest islands differed by 3 °C, and mean spot density differed on these islands by about 75%, with the least spotted slugs on the warmest islands and the most heavily spotted slugs on the coolest islands (Fig. 8.8).

The covariance of slug spotting with mean maximum temperatures fits in with observations of which habitats we generally associate with the different spot morphs. On the mainland or in large-island populations, the extreme color variants, the unspotted and the black or

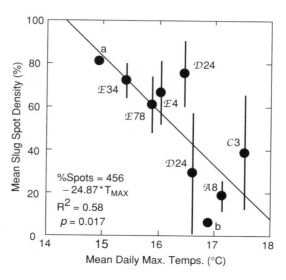

FIGURE 8.8. Mean spot density in island populations of banana slugs (ordinate) decreases as the mean maximum temperature on the island increases (abscissa). On island $E54$: "Big Pig," only a single slug was measured (point 'a'). On $B12$: "Haines SE" (point 'b') only tiny hatchling slugs were measured, and as spot density is not reliably recorded in very young slugs, this datum is tentative. Islands are named; means ± SD are shown for the measured populations, with $5 \leq n \leq 18$.

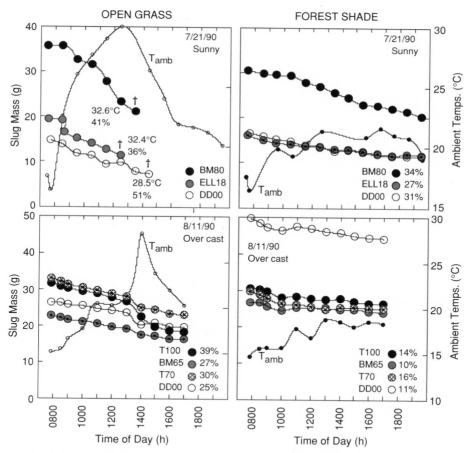

FIGURE 8.9. Survival and weight loss of banana slugs in open, grassy sites (left) and shaded, forest sites (right) on two days (upper, lower) in 1990. Three slugs were tested on sunny July 21, four on overcast August 11, with origins (BM: Bamfield Mainland, ELL: $A13$ Ellis, DD: $B26$ Diana Ddgr aN, T: $E76$ Turtle) and spot percentage (e.g., 100, 80, 70, 65, 18, 00) as indicated in the key. Percent weight loss at life's or day's end is included in the figure. Ambient temperature recorded 10 cm above the ground is given in the small-dotted lines (ordinate scale at right). See text for discussion.

nearly black slugs, are clearly not randomly distributed but segregate according to the openness of the habitat: least spotted slugs in more open sites, heavily spotted slugs in the shaded and damp forest interior. A transect across the northeast peninsula of $\mathcal{D}_{11}$: Friend in 1990, from open coast to forest interior, uncovered 19 slugs with spot density (SP) ranging from 0 to 80%. Measurement of the soil moisture (SM) content 5 cm below ground surface showed that spotting patterns and soil moisture were significantly related (SP = 103 −6.89 * SM; R^2 = 0.23, n = 19; p = 0.04); the least spotted slugs were in the drier, more open, and least vegetated sites.

Slug color patterns appear to be primarily associated with, and an evolutionary response to, temperature regulation and ultimately water conservation. We tested the ability of slugs of different spot patterns to survive and conserve water in two 1990 experiments. Slugs of different spot patterns were confined in two habitat types, open/grass and forest/shade, the former reaching midday temperatures at 10 cm above ground level some 10 °C higher than the latter. All slugs lost weight faster on the sunny day compared to the overcast day, and lost weight faster in the open than the forest habitat (Fig. 8.9). In the open, unshaded habitat, the unspotted slug kept cooler and survived longest in sunny weather, and lost the least water (25% body weight) on the overcast day, when the all-black Turtle slug lost 39% of its body weight in the same 9.5 h period. In the forest habitat, slugs lost more weight on the sunny day when peak temperatures were 4 °C higher relative to the overcast day, but slugs of all spot patterns lost mass at fairly similar rates and reached comparable body temperature highs in the afternoon, 1200–1600 h. It might be anticipated that dark slugs have an advantage over unspotted slugs in the forest, but it was possible for us to distinguish consistent performance differences among the spot morphs only in the open habitat.

Banana slugs do not appear to select sexual partners with any regard to spot patterns, and we have seen all combinations in copulo (as even unspotted slugs occasionally meet and mate with darkly spotted partners). The genetics and heritability of the spot patterns is not known, but presumably spot patterns in the offspring reflect those of the parents, and individuals seek and settle in microhabitats in which they are most comfortable, ultimately those in which they survive best. On large islands with a wide variety of microhabitats, a wide range of morphs is retained; small islands are mostly warm and dry and clearly select for the unspotted slugs most tolerant of higher temperatures, whereas small islands that are cool and damp favor dark morphs presumably because their thermoregulatory performance is advantageous there. With the strong correlation between spotting patterns and habitats, selection apparently dominates any effects of genetic drift, and on each island slugs evolve to and maintain the mean and variance in spotting patterns reflecting the habitat mean and range.

The morphological variations within and amongst the island slug populations remind one of the classical ecological genetics studies of Arthur Cain and Paul Sheppard a half century ago, on the shell colors and patterns of Cepaea nemoralis snails of English woodlands (e.g., Cain and Sheppard 1950). That work, and efforts to distinguish amongst the effects of selection, drift, and glacial refuges on snail morphology, is still ongoing (e.g., Davidson and Clarke 2000), and it might be expected that genetics studies on the island slugs would be similarly productive and multifaceted. The banana slugs dramatically display the consequences of island-specific selective regimes on small but persistent populations, strong morphological divergence related to small-island temperature regimes, and correlation between morphology and habitat in heterogeneous large-island populations. These engaging animals are eminently suitable subjects for further research in genetics, physiology, and ecology.

9

Synopsis

LESSONS FROM A CONTINENTAL ARCHIPELAGO

What seas what shores what gray rocks and what islands
What water lapping at the bow
And scent of pine and the woodthrush singing through the fog
What images return . . .

T. S. ELIOT, "MARINA," 1930

THE BARKLEY SOUND SCENE

Much of the material I have presented on island plants was strongly numerical and necessarily analytical; after the facts are bared, the data analyzed, the patterns and processes sketched out and at least partially understood, it seems useful to sit back and breathe deeply, metaphorically at least, and attempt to pick out the main lessons derived from the study. In this last chapter, I attempt an overview of the island biogeographic picture in Barkley Sound plants and point out where the more interesting population biology and ecology was found. Some of the results of the study were surprising and others routine, but it seems that there is a great deal of interesting biology taking place on these islands, and that abundant opportunities remain for future research to bring it into a sharper focus.

The foremost characteristic of islands in Barkley Sound is they are land-bridge islands, part of the continental shelf, and therefore they have had limited time to develop insular characteristics from the downsized forest patches and drowned hilltops of their origins. Recall in chapter 7 that some plant colonists reached early Hawaiian Islands well over 20 million years BP, and therefore they have experienced a history of isolation thousands of times longer than our islands. Moreover, the Hawaiian Islands flora began from scratch and was built up by plant species that were able to overcome the impressive dispersal barriers of around 3500 km from either coastal California or the Aleutian Islands, 5500 km from Japan. This isolation is a formidable dispersal filter to all but the most effective (and fortuitous) colonists and results in an eclectic mix of species within which there is strong impetus, and substantial opportunity, for evolutionary and ecological transitions. The continental islands of Barkley Sound are again about a thousand times less isolated from colonist sources, and they began their isolation history not de novo but already clothed in a "harmonious" forest vegetation, on which has been superimposed over time new shoreline and edge elements.

An island biogeography course conducted in Hawaii would have quite different emphases from one in Barkley Sound, as Fig. 1.1 predicted. In Hawaii and on other well-isolated oceanic islands, the attractions are the older endemics, the adaptive radiations, examples of phyletic speciation (i.e., accumulated change within a single lineage), startling new growth forms and other novelties, and dramatic niche shifts. Much of this high drama is precluded in the relatively subdued confines of Barkley Sound, where islands differ (from Hawaii) by three orders of magnitude in their spatial and temporal scales of isolation. But not all—there is plenty to discern and discuss here, and the action simply moves down from the upper boxes to the lower in Fig. 1.1. On the continental islands, colonization and extinction events are very much evident (e.g., over 1000 such events were recorded on the *A*-group islands alone, averaging >40 per island in 22 years), and the dynamics of species turnover are rapid. The influences of colonization and extinction on the statics of species numbers and distributions are found to depend on a wide range of variables, some subset of which was most relevant to a particular subset of the flora. Island size and elevation, isolation and position, bleak exposure versus sheltering neighbors, all seemed to play a role somewhere in the system. History was not entirely discounted either, short though it has been, and the legacies of past island connections, of shared positions with the Sound, of plant arrivals, dispersals, expansions, all were shown to have contributed in some way to the present-day picture. Given the relatively easy logistics, a plethora of islands, and time and affinity for the glorious seascapes and landscapes of Barkley Sound, field studies within this system can be certainly productive and rewarding; they can also reveal a few puzzles and surprises.

Having many islands makes for convenient replication and large sample sizes (Figs. 2.3, 2.4) and opens up the possibility of subdividing the islands into subsets within which area, or isolation, or position within the Sound is narrowly restricted. Consideration of a taxon such as plants, with some hundreds of local species of many quite different lifestyles, growth forms, habitat preferences, and dispersal syndromes, predisposes the subdivision of this flora. Thus we identify floral components that comprise different vegetation types, in different abiotic habitats in which the dynamics of component species also differ. There are advantages, then, in our continental island system, that derive from assessing different parts of the island and plant rosters independently, and in a long-term study of making comparisons among different time intervals within it.

Few if any archipelagoes are exempt from outside influences, which, while they appear modest within the Barkley Sound island system, are not entirely absent. There are natural disturbances of a sort resulting from the retreat of the ice sheets and the rising sea levels, but distant in time. There is a measurable degree of global warming detectable in increasing mean annual temperatures (Fig. 2.2), there are climate changes of intermediate periodicity apparently associated with sunspot cycles, and there are yearly, seasonal challenges to plants, such as more or less dry summers and the impacts of winter storms. Anthropogenic disturbances are also present, as discussed in chapter 8, and many of these are much more subtle than a clear-cut but just as apparent in their consequences. Human traffic amongst the islands certainly seems to move the propagules of weedy plants around, and logging on the adjacent mainland can reduce sources of propagules for forest plants and increase sources for weedy, edge, or early successional plants. The overall incidence of alien plants seems to be increasing too, in terms of both species numbers and those of individuals, and there is every reason to believe it will continue to do so.

IS THERE AN EQUILIBRIUM OUT THERE?

Clearly abiotic factors change in Barkley Sound in the shorter and the longer term, with some recurrent regularities and others far more sporadic or episodic as in e.g., sunspots versus tsunamis; such events can exert a very considerable influence on the island environments and the processes that determine which plant species occupy them. For example, weather affects dispersal potentials in some species, as with anemochores being favored by drier and warmer summers; winter storms must have a large role in maintaining plentiful edge habitat on exposed islands. Anthropogenic factors also can influence dispersal, and human activities within and around the Sound further impact the composition of the regional flora and add a component of alien plants to it. So we might ask to what extent, against a continually shifting background potential to dampen or quicken the dynamics of plant colonization and persistence, do measures of plant diversity and distribution provide evidence for constancy over time. At the outset, I outlined the elegant M/W theory of island biogeography, along with some of its embellishments, as a standard by which the data, and both the static and dynamic patterns they describe, could be assessed. One such standard, central to the theory, is that of the equilibrium: are species numbers on islands at equilibrium, and are island numbers per species likewise constant over time?

The numbers of plant species on islands were shown as functions of time in Figs. 4.2–4.5, for several different island sets reflecting a range of sizes and positions within the Sound. About three-quarters of these islands, selected for their relatively high census frequency and temporal expanse, showed no evidence of change in species numbers over time. To the extent that equilibria in species counts can be established over just two decades, these islands support the position of no time trend in species numbers. This conclusion holds despite considerable variation in species numbers, and despite high colonization and extinction rates and therefore extensive species turnover within the island lists. But on 20 of 72 islands in these figures, there was statistical support for changing counts over time, and most of the changes were to increasing species numbers. Islands with increasing counts were predominantly within the A–C groups, thus in islands less isolated with respect to the mainland. They ranged widely in size, but many are small in this sample, some larger but sparsely vegetated. We noted that smaller islands with more open vegetation have higher proportions of weedy and edge species, and on these islands species counts are more variable (Fig. 4.6) with higher turnover rates.

Fewer islands showed a species decline over time, and a preponderance of these are small- to midsized islands in the E group (e.g., Pigot Islands, Fig. 4.4 upper). A broader picture of trends in species numbers was given by Fig. 4.7, where on 40 of 154 islands censused in ≥ 5 years, regression coefficients b (slope) differ significantly from zero. The highest coefficients (largest increases over time) were found on large islands of minimal isolation from the mainland, whereas negative coefficients (decreasing species over time) are associated with islands of greater isolation, less strongly related to decreased island size.

These results make sense in the light of the different components of the island floras, each with different history and behavior. The forest component dates from or predates the onset of insularity of the island habitats, a result of rising Holocene sea levels, which produced nearly all of the present-day islands after 11,000 years BP and after forests had recolonized the postglacial landscape. This component therefore can be considered relictual; its species are strongly nested relative to island area (Fig. 5.2), and more species have been retained on the larger and taller islands (Fig. 4.12, top) that maintain cooler and more humid conditions in the summer (Fig. 2.13). There is little evidence that forest species have been lost from the islands over the time span of this study, although some species that require cool and humid conditions, such as the orchid *Listera caurina*, the fern *Athyrium felix-femina*, and the yew *Taxus brevifolius* (Fig. 5.5) have contributed to species loss on the Pigot Islands (see above). Population declines in these and other species similarly dependent on cool temperate forest habitats may signal further species losses in the years to come.

It is possible, even likely, that many forest species have already been lost from the islands, since the forest florulas do appear to be depauperate with respect to certain taxonomic groups. The lilies and orchids (Liliaceae, Orchidaceae) of forest-floor habitats seem especially sparse in the island forests. Pojar and MacKinnon's (1994) guide book can be consulted for plants that might be expected in the island forests by dint of geographic range, low elevation, and affiliation with the coastal forests. There are at least eight lily and four orchid species that are apparently absent from the Sound, found neither by us nor by Bell and Harcombe (1973). The list includes species of *Streptopus, Smilacina, Disporum, Trillium, Clintonia, Stenanthium, Veratrum, Calypso, Corallorhiza,* and *Platanthera*. These monocots seem particularly prone to early disappearance as the island forests relax toward lower and equilibrial species counts. The other conspicuous group of absentees, unrecorded from the islands, is bog specialists, including four or five species of *Gaultheria, Vaccinium, Rubus,* and *Andromeda*. These are plants that, by the same criteria as above, might be expected to occur in the Barkley Sound area, but presumably they were lost when past bogs were inundated by rising sea levels, and the tiny, extant bogs proved incapable of maintaining viable populations.

Two other components of the island flora, shoreline and edge species, are added to the forest component with island formation; they must have gained in species numbers postinsularity, reaching or at least striving for an equilibrium from below (Fig. 3.11). Shoreline species are mainly hydrochorous, disperse well with few or no distance effects on incidence (Figs. 3.7, 5.11–5.13, 7.12–7.14), and most are likely at equilibrium with island areas and the shorelines that are their dispersal targets. Recent novelties in the shoreline flora, such as the alien sea-rockets, *Cakile* spp. (Fig. 7.13), will presumably require more time to reach all potential sites among the islands, as may the alien grass *Poa annua* (Fig. 5.13), but these recent additions seem to be making rapid strides in that direction.

It is the increased incidence of edge species that accounts for trends to higher species numbers on some Barkley Sound islands. Edge species are particularly favored on islands with extensive areas of open habitat, and islands heavily forested down to the shoreline provide little opportunity for the establishment of weedy and herbaceous forbs. The edge flora is nonnested, indicating that most species can occur on most islands regardless of size (or other island feature). Amongst the edge species, turnover in time and space is extensive; longevity in most species (and individuals) is relatively short, and many populations are either transient or maintained by colonization and recolonization. Dispersal success, in turn, depends on both island area and isolation. Larger islands constitute easier targets for passively dispersing propagules, and colonization rates are in general constrained by isolation distance. Increased distances between islands and mainland sources of colonists reduces incidence and contributes significantly to reduced species numbers on more distant islands (Figs. 4.12, 7.1). Thus islands in the middle of the Sound are both less suitable for and less reachable by edge species in general, and equilibrial levels of edge species there will be lower than on the more accessible and the generally drier and more open islands closer to the mainland. It remains possible, however, that the incidence of edge species will yet increase on mid-Sound islands, especially if colonization rates are boosted and habitat availability enhanced by continued human activities there.

While subdivision of the flora into habitat-related components, the forest, edge, and shoreline species, respectively, segregates relictual and possibly retracting species from those of more recent and expanding taxa, another sort of subdivision was introduced in chapter 6 (Fig. 6.1). Species can be characterized as long-, mid- or short-term island residents (signified by **L, M, S**) on a given island, depending on what proportion of the study period they were represented in censuses there. While long-term residents constitute the largest component of the island floras throughout the Sound, they are fewer in absolute numbers in two islands groups, the most isolated $\mathcal{E}$-group islands and the most exposed $\mathcal{B}$-group islands. On the latter islands, **L** species are bolstered by many **S** and especially **M** species, but on the former ($\mathcal{E}$-group) islands these two categories of shorter term residents are quite poorly represented. There are clearly many more candidates for **S** species classification on the $\mathcal{A}$- and $\mathcal{B}$-group islands; by assessing plants on islands collectively over expanding areas, additional S species are accumulated rapidly on nearby islands in the $\mathcal{A}$ and $\mathcal{B}$ groups, but much more slowly on the neighbors of $\mathcal{E}$-group islands (Fig. 6.9). Clearly the incidence, distribution, and dynamics of transient species differ considerably with more isolated and less isolated island groups.

The characterization of species by island residence times (**L**, **M**, or **S** species) is not one that translates across islands. Species that are long-term residents on one island may be mid- or short-term residents on another, the change in status due to factors such as island size or position. The changing classification of species across gradients was depicted diagrammatically in Fig. 6.10, where position on an "environmental gradient" determines island longevity. In the most obvious instances, across gradients of island size, many forest trees and ferns show up as transients on small islands, but only on much larger islands can permanent resident populations of these species be maintained.

In the same way that species numbers on some islands are demonstrably changing over time, the roster of islands occupied by certain species is also changing. Recall that, with fixed colonization and extinction rates that are properties of an island's size, isolation distance, and other at least temporary constants (elevation, number and proximity of neighbors, microclimates, and habitat types), incidence J is constant and determined by $c/(c + e)$ (Fig. 3.3). Smaller and midsized forested islands are likely to continue losing forest species as permanent residents, and the distributions of species more heavily dependent on the deep-forest microhabitats will suffer increasingly curtailed distributions over the longer term. On the other hand, edge species are likely to become established on more islands over time, as forest habitats shrink owing to some combination of disturbance, relaxation, and possibly global warming. An assessment of the expanding distributional ranges of alien plants within the Sound (Tables 8.1, 8.2) illustrates the temporal dynamics of incidence patterns. Some species doubtless will expand their ranges in the short- and midterm, and several factors will contribute to it. Cyclical patterns in precipitation and temperature, demonstrable at several temporal scales (chapter 2), will likely oversee temporary waxing and waning of species' ranges, whose dispersal and persistence is affected by these abiotic variables. Over the longer term, the regional flora will change though time, and the representation on nonnatives in the Sound will likely increase, with expanding ranges of the aliens and concomitant range reductions in the natives most similar in growth form.

COLONIZATION AND EXTINCTION DYNAMICS

Another key feature of M/W theory, beyond the equilibrium notion, is that of the dynamical nature of the equilibrium: *ex hypothesi*, it is maintained by a balance between colonization and extinction rates. It is chiefly the lack of observable colonization and extinction that has caused students of oceanic islands to regard M/W theory as inapplicable, since neither sort of event is recorded on very isolated islands except extremely infrequently. On the continental islands of Barkley Sound, colonization and extinction events are extremely common and were recorded on all islands and in nearly all taxa (though at very different frequencies in different species). Thus another criterion is satisfied, and further, colonization events are more or less balanced by extinction events (see, for examples, Figs. 3.4,

3.9a,b, 4.1, 5.10, 5.25, Table 5.3). But in striking contrast to the classical model, both colonization and extinction rates are highest on large, close (nonisolated) islands, and both lowest on small, distant (most isolated) islands. Island area is the predominant influence on both rates, and although the effect of isolation is weak on extinction rates, it is still a statistically significant contributor to them (see Fig. 3.5). Thus the data support some aspects of the classical M/W model, but they differ in others.

The lively dynamics generated by high colonization and extinction rates produce a substantial turnover of species through time, such that islands change in species composition as some erstwhile residents go extinct and others successfully colonize. It is important to distinguish amongst different components of the island flora in their contributions to species turnover. On tiny islets, like the three "Hosie Rocks" of Fig. 6.5, with mean species numbers of 1.7 to 6.8, all or most of the species censused may become replaced ("turn over") in two decades, during which time the cumulative species numbers steadily rise. On the very smallest islet of the group, species actually begin to recycle through the island roster after this length of time, as the number of candidate species for residence on really small rocks is very limited (see Fig. 6.8, lower right). Species turnover occurs in space as well as in time and is measured by the accumulation of new species gained from adjacent islands as these are added successively with increasing distance. The florulas of neighboring islands are much more different from each other in the A–D groups than in the E group (Fig. 6.8; see also the Ross Islets of Fig. 5.16), and thus new species are accumulated as a significant function of distance only in the latter. This lack of systematic turnover with distance is approximately the obverse of "distance decay," where cumulative species lists are continuously enhanced by increasing survey distances. The lack of extended "distance enhancement" by neighbor contributions to the A–D islands is owed to several characteristics that distinguish them from the E-group islands. There are many more edge species on these less-isolated islands, and their spatial distributions and the composition of coexisting subsets are largely stochastic. Thus species' identities are not well predicted by island separation distances, and most additional edge species are gained from a few close neighbors, with little or no further accumulation of species from more distant neighbors.

One important qualification of species turnover is that most island species, the permanently resident **L** species, are immune to it. The long-lived forest species on all but the smallest islands, the perennial shoreline species such as *Potentilla villosa* and *Plantago maritima* (e.g., Fig. 5.11) and the large well-established shrubs of the edge vegetation, all contribute little if anything to species turnover. Turnover occurs mostly in the marginal edge species, those with small populations, some of them recently established, and many of limited longevity. There are substantial numbers of such species on the edges of large islands, but they are still outnumbered by permanent residents; on small islands with few if any permanent residents, the whole island list can contribute to turnover. By subdividing the flora into components of high- and low-turnover species, each with its own specific dynamics of colonization and persistence, one can mimic the extent to which cumulative species counts increase over time, on different sized islands, rather closely (Fig. 6.3, 6.4).

Because large islands make large targets, and small islands are hit-or-miss "long shots," there is a higher predictability of species on large relative to small islands. This was illustrated in species turnover differences between the individual members of island twins, matched for size and elevation and in their very close proximity to each other (Fig. 6.11). Florulas may diverge considerably between the smallest matched twins, which share only about one-third of their species, but turnover between large twins is much lower, with around four-fifths shared species (Fig. 6.12). An evaluation of the extent to which close neighbor islands are more or less similar than are comparably sized islands selected at random without regard to proximity led to discovering "neighborhood" effects. Island adjacency can beget higher than expected species overlaps, but it does not always do so. Adjacency effects were found among shoreline species in general, perhaps not surprising since shoreline species numbers are enhanced where there are many neighboring islands (Fig. 4.12, lower panels), and where island groups influence the currents that bear their propagules. Thus groups of adjacent islands are uniformly very rich on the North Hosie islands, all richer than average on the San Jose islands, and all of strictly average species richness on the Brady Beach islets depicted in Fig. 5.15. One sees adjacency effects also among weedy edge species, which from an initial foothold on a single island may be recorded later on adjacent islands reached successively over time (Fig. 6.7). These effects notwithstanding, there is still impressive unpredictability in lists of edge species, even those from adjacent islands a few tens of meters apart. Note that even in some shoreline species, such as the ubiquitous but ephemeral pearlwort (*Sagina crassicaulis*) (Fig. 5.12 and accompanying text), with extremely rapid turnover and vigorous recolonization, there were no identifiable adjacency effects. Incidence in the giant vetch (*Vicea gigantea*) increases on islands with many close neighbors relative to more isolated islands, as discussed in chapter 6. But this was not an adjacency effect attributable to the proximity of conspecific populations that might constitute propagule sources, but rather to a purely topographic influence on dispersal and establishment.

In my analysis of the plant data on these continental islands, I have referred consistently, and perhaps with a degree of deliberate disengenuity, to the M/W theory. This was done not to antagonize readers with less than complete faith in M/W tenets—indeed, I am one of you. I did not expect to verify its wholesale applicability and have not done so, but still I find M/W to be a most useful, comparative tool with which to assess the plant dynamics. In some respects, many in fact, there were concordances between observation and M/W theory. We have quasi-equilibrial species counts on islands, plenty of colonization and extinction, and a rough balance between rates in these latter two processes. However, points of departure soon developed, chief amongst which was the dependence of both colonization and extinction chiefly on island area, a positive relationship in both variables. It appears that, on these islands, the target effect dominates the distance effect in colonization by these passively dispersing plants, and extinction rates track colonization (and trail it in time, see above) in securing a balance between the two and thereby regulating species numbers as functions chiefly of island area. A further, and major, point of departure is that the continental islands support relictual floras of forest species. These species are unlikely to display other than a gradual attrition in species numbers over the longer term, and on larger islands the forest species constitute a bloc of persistent species essentially exempt from colonization/extinction dynamics. The simulation model of Figs. 6.3 and 6.4 attempted to reflect this distinction. The lesson here was that we cannot treat all species as equals in the numbers game, there are critically important biological differences amongst them. Ample turnover certainly characterizes many shoreline species, especially those of short individual longevity, but it most accurately pertains to the edge floras. Some shoreline and edge species come very close to fitting the metapopulation metaphor developed for refuging species over patchy habitats. In this respect they are furthest from the classical picture of oceanic islands, on which external inputs are very minor components of the system, and nearly all of the interesting dynamics is internal or autochthonous. Figure 9.1 is very simplistic, but may help to focus the largely size-dependent abundances of species in different ecological groups, and in distinguishing the

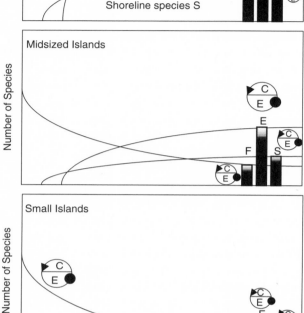

FIGURE 9.1. The composition of island floras is strongly dominated by island areas, with the persistence of relictual forest species (F) high on large islands and much lower on small islands. New floral components, since island emergence, include both shoreline (S) and edge species (E). The former are minimally affected by island isolation, but the latter are attenuated with increasing island isolation. The magnitude of species turnover is represented by emblems with C for colonization and E for exinction. Turnover is low in forest species on large islands, large on small islands; in edge species it is high on large islands, especially those of low isolation from the mainland, on which colonization rates are highest.

very different dynamics of relictual versus expansive components of continental island floras.

COEXISTENCE AND A POTENTIAL ROLE FOR COMPETITION

The classical M/W theory had little or nothing to say about biological interactions among species, as it regarded species essentially as interchangeable entities whose biological attributes were of little consequence to the system either in colonization or persistence potentials. Here I have documented profound differences among species in their abilities to tolerate, on the one hand, or exploit, on the other, the Barkley Sound environment. This is perhaps not very surprising since plants, though a single taxon, encompass the largest and longest lived of terrestrial organisms and span much of the range down to the most minute and ephemeral. Differences among species have been one of the principal themes of this book, which I hope has been enlivened by the specifics of natural history as well as by numerology. We have found important differences even amongst closely related species and those of similar growth forms; recall the two species of *Epilobium*, and the morphologically similar species of *Prunus* and *Malus*, that differed in their trade-off strategies of better colonization versus better persistence, revisited in more detail below.

Certainly differences amongst species and at least some degree of interaction between them were implicit in M/W theory. Why else would extinction rates climb as species numbers on an island increased, if not because island resource "pies" become subdivided into ever smaller slices as more consumers take smaller shares and their populations become more size limited and extinction prone? On Barkley Sound islands, the general correspondence between colonization and extinction rates on islands invites speculation on causality, and some evidence exists to support the connection, namely, elevated extinction rates a few years after colonization events have been more frequent (above, chapter 6). Another form of support for interspecific influences at the scale of the vegetation is in the bog plants, where generalist species rather than specialist congeners are often recorded as replacements in the smallest bogs (e.g., in *Juncus, Carex,* and *Vaccinium*; see above, chapter 8). But these effects, to the extent that they exist, are not very explicit; they are system- or island- or community- (or vegetation-) wide, and beg the more interesting questions of how they operate at the level of individual species.

I made a number of attempts in the foregoing chapters to document interspecific effects and identify a role for competition. Since this theme has been a major topic of vertebrate communities on islands and mainlands both, from the documentation of checkerboard patterns of distribution to niche shifts and density compensation, I was determined to uncover whatever evidence might exist for its operation in the island plant communities. If there are interspecific interactions such that the persistence or chances of colonization in one species are impaired by the presence on an island of another species, they are not likely to be discovered among the forest or shoreline species. These components of the vegetation are surely comprised of species with long histories of coexistence and mutually coadaptation. In chapter 5 the distributions of the three common forest trees, cedar, spruce, and hemlock, were examined on small islands where their joint incidences were generally low. The conclusion was that the species tended to be coincident in the islands they occupied rather than disjunct; no one species excluded another, and they readily shared islands that all three found suitable. In chapter 7 (also Fig. 7.14), the two shoreline legumes *Vicea* and *Lathyrus*, similar in habitat, habit, dispersal mode, and growth form, were assessed as good candidates for mutual influences on each other's island distribution. A rather exhaustive analysis of the question reached the conclusion that differences in islands occupied by the two taxa were not attributable to preemption by the other, and that the persistence of one species on a co-occupied island was not jeopardized by the other's presence.

I expected to be more successful in finding examples of interspecific influences, such as mutual competition and exclusion, among the species of the edge flora. This expectation was generated by considering that edge species are almost by definition generalized species that share catholic habitat preferences as well as, in many cases, similarities in growth form and life history characteristics. Broad similarities among edge species notwithstanding, some species groups supply good examples of niche partitioning, indicating (weakly, for sure) that over time the member species have been factors in the evolution of each other's ecologies. Amongst forest species, the segregation of coexisting fern species by frond morphology and leaf specific weight (chapter 7) is a good example of this. In shoreline species, the partitioning of beach habitats by congeneric species with higher versus lower distributions relative to the tide line is quite common. Examples of such displacement patterns occur in *Potentilla* (*P. villosa* vs. *P. pacifica,* respectively), *Cakile* (*C. maritima* vs. *C. edentula*), *Plantago* (*P. macrocarpa* vs. *P. maritima*), and the three species of *Puccinellia* (*P. nuttalliana, P. nutkaensis, P. pumila*), amongst others.

In the edge vegetation, the distinct microsite differences among the four commoner umbels

(*Oenanthe, Angelica, Heracleum,* and *Conioselinum*) might be loosely construed as an example of niche partitioning. However, a far more compelling example was presented by species of fruit-bearing zoochorous shrubs, illustrated in Fig. 7.9 by ericad and *Rubus* species. There is a convincing displacement among species in the timing of fruit maturation, such that collectively the shrubs encompass a long spring-to-winter fruit production span, but each individual species occupies a particular narrow time slot within this span. Given the similarities, as described in chapter 5, amongst the shrubs in overall size and growth form, similar patterns of incidence over islands (Fig. 5.17), and similar responses to island area, isolation, and elevation (Fig. 5.18), one might have expected a more direct interaction among species in their respective occupancy of specific islands. A test for such interactions was devised using "incidence shifts" (Fig. 5.19). Basically, island subsets were identified, for each species sequentially, on which a certain species is persistent ($J = 1$), and then we assessed whether the incidence of the other shrub species changed on these island subsets. They usually did; in almost every case, shrub species showed higher incidence on island subsets where one or another of their number was persistent. A shift to lower incidence would be expected if there are negative interactions among species, but like the forest trees on small islands (above), the shrubs all agreed on which islands were best (Fig. 5.20), and they coexisted on these islands amicably.

Composites (fam. Asteraceae) are prominently represented in the edge vegetation. There are many species of similar habits and sizes, a majority of which share anemochory as their dispersal mode. However, there are conspicuous differences in distribution among species (Figs. 3.9, 7.17, 7.18), some of which seem to be attributable to differences in dispersal mode and efficacy. To test for possible interspecific effects on incidence, I identified particular island subsets within "hysteresis zones," where the different species maintained intermediate levels of incidence, $1 > J > 0$ (Fig. 5.23). I then examined whether incidence within the hysteresis zones was related to presence or absence of other composite species on those islands. Though most species' pairwise combinations showed no interspecific effects, in three species (*Lactuca, Anaphalis, Senecio*) incidence was negatively affected by *Achillea* presence, and in the last two of these three, incidence was positively affected (i.e., related to) by the presence of *Lactuca* and *Aster* (Table 5.5, upper part). In a somewhat more sensitive test, I related the residuals from the fitted logistic incidence curves of Fig. 5.23 to the presence or absence of other composite forbs on those islands. The results, reported in the lower half of Table 5.5, showed that most of these relationships (18/24), and all six of the statistically significant relationships, were negative. These edge forbs, then, do seem to influence each other's incidences (finally!), and competition for or preemption of space would appear to be the likely means by which they do so. I might add that, while edge habitats might appear sparsely occupied and thus space might not be obviously limiting, there are relatively few growing sites along the edges of rocky shores, and soil-filled cracks suitable for persistence must be at a premium.

There is one last point about potential interspecific interactions I shall make in this section, and it concerns the form of niche partitioning that is constituted by tradeoffs. We normally think of niche partitioning as a result of both bygone and ongoing selection for species to specialize in a certain way, to become better at doing a more limited range of tasks. These specialists thereby become unassailable in a particular part of niche space (habitat, microsite, season, regeneration tactics, growth form and its attendant constraints), and relegate other species to dominate other parts, if they can (Grubb 1977; Cody 1986b). The case of the sequential fruiting of edge shrubs described above is an excellent example of niche partitioning. With a slight shift of perspective, we can think of tradeoffs as a form of niche partitioning, and in this study the potential tradeoff between dispersal and persistence abilities has surfaced repeatedly. This is a form of ecotypic (rather than archetypic) selection, with related species maintaining their viability within the system in quite different ways. An example was provided by the fireweeds *Epilobium*, with two common species in the Barkley Sound region and both well represented on the islands. One species, *E. angustifolium*, displayed superior dispersal abilities, and the other, *E. watsonii*, had poorer dispersal but better persistence once it reached an island. Another involved two similar and confamilial small trees that are both quite common at the forest edge, crabapple *Malus* and bitter-cherry *Prunus* (fam. Rosaceae). Neither species seems especially disadvantaged or otherwise by the different but obviously zoochorous fruits, nor does either seem better preadapted for island persistence by stature or leaf size or shape, which are similar. Yet there is an apparent tradeoff, since *Malus* is the superior colonist but *Prunus* is superior in persisting once it reaches an island (chapter 5 and Fig. 5.25). The tradeoff seems particularly well balanced, as the two species maintain island incidences that are quite similar, but exactly how it works, at the level of phenotypic characteristics, is not as yet known.

The last example of this sort of dispersal-persistence tradeoff was suggested by differences in colonization rates and persistence abilities among the fruiting edge shrubs. A number of species have low incidences on the islands as a consequence of both poor colonization and low persistence (Fig. 7.11), but there are eight species that are quite widespread, with incidences $J > 0.4$. Amongst these latter, the two components of island incidence are negatively related (Fig. 7.11, regression line). Positions along this tradeoff gradient are clearly associated with leaf morphology, and with having evergreen or deciduous leaves in particular. The late-season fruiters are evergreens, while early fruit production apparently requires a kick-start from high-efficiency deciduous leaves. The seasonality of fruit production has implications for dispersal, since the availability of bird dispersers changes seasonally, and survival on islands is higher in evergreens that are better able to cope with drier and/or nutrient-poor conditions.

ADAPTATION, EVOLUTION, CONSERVATION

The enormous attraction of oceanic islands to most biologists is surely the endemics, the wonderful oddballs and peculiarities found nowhere else. They have provided an enormous impetus to island biogeography studies, as well as helping fuel the ecotourism industry and fothering the popular science writers. These endemics may be the only extant examples of evolutionary lines that are extinct elsewhere, they may be trees whose closest relatives are weedy mainland annuals, or they may be speciation products of local adaptive radiations or repeated invasions; various examples of these phenomena have already been cited. Unfortunately, the continental islands of Barkley Sound have nothing quite this dramatic; but while there are no endemics, the evolutionary play is nevertheless underway. If the ecotourist (or better, the inquisitive island biology student) seeks ready examples of niche shifts, ecotypic selection, divergent phylogenies over a few thousand years of isolation, or divergent dispersal morphologies after a decade or so of intense island selection, they can be found on these islands.

The island biotas have been separated from each other and from the mainland by at most a little over 10,000 years on tall and deepwater islands, by at least a few thousand years in the lowest islands. The longer times translate to perhaps 20 generations in forest trees, but several thousand generations for mice and slugs and short-lived plants. The spatial isolation among islands reflects genetic isolation among their populations only to the extent that there is no exchange amongst the islands, no transfer of pollen, and no dispersing propagules. This may be accurate for a few taxa that seem particularly poor dispersers, such as slugs and some plant rarities such as forest-floor monocots or other narrowly confined habitat specialists, but surely not for the vast majority of the island plants. In the shoreline species and the most common and persistent edge and forest plants, the island populations are very likely cohesive with a good deal of genetic exchange.

Even in species for which we expect little if any genetic isolation amongst island populations, such as the ferns of chapter 7, there may be striking differences among populations, such as the morphological variability in leatherleaf fern (*Polypodium scouleri*) (Figs. 7.6, 7.7). If the striking inter- and intra-island differences in frond morphology are achieved without variation in the genetic constitution of the fern individuals, as we suspect they are, then there exists a degree of developmental plasticity (itself heritable) that is quite remarkable. The external controls of alternative developmental pathways in these ferns would be well worth further investigation.

The morphological divergences recorded in slugs on different islands (Fig. 8.5) indicate that there is little if any gene transfer among their populations, even between neighboring islands. The size shifts in the island mice suggest that they are genetically isolated at least from mainland mice, if not from adjacent island populations. But the divergence in morphological characteristics in island composites is more surprising, in that it signals genetic isolation and an absence of incoming new and repeat colonists over time periods of some decades at least. Thus small populations of edge species are freed to diverge under local selective regimes for as long as their longevity permits, buffered from genetic swamping. Given this degree of isolation in island populations of small size, and perhaps in small populations on even the larger islands, there is a potential for selection to change phenotypes in various ways. Local adaptations in physiology, behavior, morphology, and ecology may well be evidenced in these island populations, if they persist long enough for genetic change; I can imagine, in retrospect, putting together a team of congenial fellow investigators, physiologists and geneticists included, that might have shared this project over the last two sunspot cycles and whose contributions would have greatly strengthened and broadened the output of the study. There are surely many potential applications here for the plethora of new genetics techniques that have already contributed much to modern island biogeography, and for the elaboration of physiological and anatomical fine-tuning that plant ecologists can measure. Such laboratory-based studies could certainly provide many new insights to Barkley Sound biogeography; surely their practitioners will construct their own form of discovery curves and realize too, like Kenneth Graeme's Rat, that "there is nothing—absolutely nothing—half so much worth doing as simply messing about in boats."

APPENDIX A

Plant List for Barkley Sound Islands and Adjacent Bamfield Mainland

FAMILY	COMMON NAME	OLD/CURRENT BINOMIAL	SYNONYM	NATIVE/ALIEN	MAINLAND/ISLAND
Lycopodeaceae (Clubmosses)	Elk-moss	*Lycopodium clavatum*		N	I
	Bog clubmoss	*Lycopodium inundatum*		N	I
	Fir clubmoss	*Lycopodium selago*		N	I
Selaginellaceae	Selaginella	*Selaginella wallacei*		N	I
Equisetaceae (Horsetails)	Common horsetail	*Equisetum arvense*		N	I
	Swamp horsetail	*Equisetum fluviatile*		N	M
	Giant horsetail	*Equisetum telmateia*		N	M
Ophioglossaceae	Grape-fern	*Botrychium multifidum*		N	M
	Rattlesnake fern	*Botrychium virginianum*[a]		N	I
Polypodiaceae/Blechnaceae (Ferns)	Maidenhair fern	*Adiantum pedatum*	*A. aleuticum*	N	I
	Spleenwort	*Asplenium trichomanes*		N	I
	Lady-fern	*Athyrium felix-femina*		N	I
	Deer-fern	*Blechnum spicant*		N	I
	Bladder-fern	*Cystopteris fragilis*		N	I
	Wood-fern	*Dryopteris austriaca*	*D. expansa*	N	I
	Male fern	*Dryopteris felix-mas*		N	I

APPENDIX A (continued)

FAMILY	COMMON NAME	OLD/CURRENT BINOMIAL	SYNONYM	NATIVE/ ALIEN	MAINLAND/ ISLAND
	Licorice-fern	Polypodium glycyrrhiza	P. vulgare	N	I
	Leather-leaf polypody	Polypodium scouleri		N	I
	Sword fern	Polystichum munitum		N	I
	Bracken	Pteridium aquilinum		N	I
Taxaceae (Yews)	Pacific yew	Taxus brevifolius		N	I
Cupressaceae (Cedars)	Yellow cedar	Chamaecyparis nootkatensis		N	M
	Western red cedar	Thuja plicata		N	I
	Common juniper	Juniperus communis		N	M
Pinaceae (Pine, Fir, Hemlock, Spruce)	Silver fir	Abies amabilis		N	M
	Grand fir	Abies grandis		N	M
	Sitka spruce	Picea sitchensis		N	I
	Shore pine	Pinus contorta		N	I
	Western white pine	Pinus monticola		N	M
	Douglas fir	Pseudotsuga menziesii		N	I
	Western hemlock	Tsuga heterophylla		N	I
Aceraceae (Maples)	Big-leaf maple	Acer macrophyllum		N	I
	Douglas maple	Acer douglasii		N	I
Apiaceae (Parsleys, Carrots)	Kneeling angelica	Angelica genuflexa		N	I
	Sea-watch	Angelica lucida		N	I
	Water hemlock	Cicuta douglasii[a]		N	M
	Pacific hemlock-parsley	Conioselinum pacificum	C. gmelinii	N	I
	Cow-parsnip	Heracleum lanatum		N	I
	Western lilaeopsis	Lilaeopsis occidentalis		N	I
	Pac. water-parsley	Oenanthe sarmentosa		N	I
	Sweet cicely	Osmorhiza chilensis[a]		N	I
	Sanicle	Sanicula crassicaulis[a]		N	I

Family	Common name	Scientific name	Abbreviation	Origin	Status
Aquifoliaceae	English holly	Ilex aquifolium		A	I
Asteraceae/Compositae (Asters, Daisies)	Yarrow	Achillea millefolium		N	I
	Silver beachweed	Franseria chamissonis[a]	Ambrosia c.	N	I
	Pearly everlasting	Anaphalis margaritacea		N	I
	Apargidium	Apargidium boreale	Microseris borealis	N	M
	Burdock	Arctium minus		A	I
	Mugwort	Artemisia suksdorfii[a]		N	I
	Douglas aster	Aster subspicatus		N	I
	Hairy goldfields	Baeria maritima	Lasthenia m.	N	I
	Oxeye daisy	Chrysanthemum leucanthemum	Leucanthemum vulgare	A	I
	Canada thistle	Cirsium arvense		A	I
	Bull thistle	Cirsium vulgare		A	I
	Wooly eriophyllum	Eriophyllum lanatum		N	I
	Cudweed	Gnaphalium purpureum		N	I
	Gumweed	Grindelia integrifolia		N	I
	Hawkweed	Hieracium albiflorum		N	I
	Hairy cat's-ear	Hypochaeris radicata		A	I
	Fleshy jaumea	Jaumea carnosa		N	I
	Wall-lettuce	Lactuca muralis	Mycelis m.	A	M
	Nipplewort	Lapsana communis		A	I
	Fall dandelion	Leontodon nudicaulis[a]		A	I
	Pineappleweed	Matricaria matricaroides	M. discoidea	N	I
	Wood groundsel	Senecio sylvaticus		A	I
	Common groundsel	Senecio vulgaris		A	M
	Goldenrod	Solidago canadensis		A	I
	Prickly sow-thistle	Sonchus asper		A	I
	Perennial sow-thistle	Sonchus arvensis		A	I
	Common sow-thistle	Sonchus oleraceus		A	I
	Dune tansy	Tanacetum douglasii	T. bipinnatum	N	M
	Common dandelion	Taraxacum officinale		A	I
Betulaceae (Alders, Birches)	Red alder	Alnus rubra		N	I
	Sitka alder	Alnus sinuata	A. crispa	N	I
Boraginaceae	Forget-me-not	Myosotis laxa		N	M

APPENDIX A (continued)

FAMILY	COMMON NAME	OLD/CURRENT BINOMIAL	SYNONYM	NATIVE/ALIEN	MAINLAND/ISLAND
Brassicaceae/Cruciferae (Mustards)	Tower mustard	Arabis glabra		A	I
	Hairy rock-cress	Arabis hirsuta		N	I
	Wintercress	Barbarea orthoceras		N	I
	Bitter wintercress	Barbarea vulgaris		A	I
	Field mustard	Brassica campestris	Brassica rapa	A	I
	American searocket	Cakile edentula		A	I
	European searocket	Cakile maritima		A	I
	Shepherd's purse	Capsella bursa-pastoris		A	I
	Bitter-cress	Cardamine oligosperma		N	I
	Scurvy-grass	Cochlearia officinalis	C. groenlandica	N	I
	Wild radish	Raphanus sativus		A	I
Callitricaceae	Water starwort	Callitriche sp.[a]		N	M
Campanulaceae	Harebell	Campanula rotundifolia		N	I
Caprifoliaceae (Honeysuckles, Elderberries)	Twinflower	Linnaea borealis		N	I
	Hairy honeysuckle	Lonicera hispidula		N	I
	Black twinberry	Lonicera involucrata		N	I
	Red elderberry	Sambucus racemosus		N	I
	Snowberry	Symphoricarpos albus		N	I
Caryophyllaceae (Pearlwort, Pinks, Chickweeds)	Field chickweed	Cerastium arvense		N	I
	Nodding chickweed	Cerastium nutans		N	I
	Sticky chickweed	Cerastium viscosum	C. glomeratum	A	I
	Common chickweed	Cerastium vulgatum	C. fontanum	N	I
	Honkenya	Honkenya peploides	Honckenya p.	N	I
	Pearlwort	Sagina crassicaulis	S. maxima	N	I
	Sandwort	Sagina saginoides		N	I
	Canada sand-spurry	Spergularia canadensis		N	M
	Beach sand-spurry	Spergularia macrotheca		N	I

Common name	Scientific name			
Saltmarsh sand-spurry	Spergularia marina		A	I
Northern starwort	Stellaria calycantha		N	I
Wavy-leaf starwort	Stellaria crispa		N	I
Saltmarsh starwort	Stellaria humifusa		N	I
Long-leaf starwort	Stellaria longifolia		N	M
Chickweed	Stellaria media		A	I
Chenopodiaceae (Saltbush, Glasswort)				
Orache	Atriplex patula		N	I
Lamb's-quarters	Chenopodium album		A	I
American glasswort	Salicornia virginica		N	I
Convovulaceae (Bindweeds)				
Bindweed	Convovulus sepium	Calystegia s.	A	M
Beach morning-glory	Convovulus soldanella		N	M
Cornaceae (Dogwoods)				
Bunchberry	Cornus canadensis	C. unalaschensis	N	I
Red-osier dogwood	Cornus stolonifera		N	I
Crassulaceae (Sedums)				
Oregon stonecrop	Sedum oreganum		N	I
Broad-leaf stonecrop	Sedum spathulifolium		N	I
Droseraceae (Sundews)				
Great sundew	Drosera anglica		N	M
Round-leaf sundew	Drosera rotundifolia		N	I
Empetraceae				
Crowberry	Empetrum nigrum		N	I
Ericaceae (Cranberries, Heaths, Salal, Manzanita)				
Hairy manzanita	Arctostaphylos columbianum		N	I
Bearberry	Arctostaphylos uva-ursi		N	I
Hybrid manzanita	Arctostaphyllos "x media"		N	I
Salal	Gaultheria shallon		N	I
Gnome plant	Hemitomes congestum		N	I
Bog laurel	Kalmia polifoila		N	I
Labrador tea	Ledum groenlandicum		N	I
False azalea	Menziesia ferruginea		N	I
Alaska blueberry	Vaccinium alaskense		N	I
Oval-leaf blueberry	Vaccinium ovalifolium		N	I
Evergreen huckleberry	Vaccinium ovatum		N	I
Bog cranberry	Vaccinium oxycoccus	Oxycoccus o.	N	I

APPENDIX A (continued)

FAMILY	COMMON NAME	OLD/CURRENT BINOMIAL	SYNONYM	NATIVE/ALIEN	MAINLAND/ISLAND
	Red huckleberry	*Vaccinium parvifolium*		N	I
	Bog blueberry	*Vaccinium uliginosum*		N	M
Fabaceae (Peas, Beans, Clovers)	Beach pea	*Lathyrus japonicus*		N	I
	Black medic	*Medicago lupulina*		A	I
	Low hop-clover	*Trifolium campestre*		A	I
	Small hop-clover	*Trifolium dubium*[a]		A	I
	Red clover	*Trifolium pratense*		A	I
	White clover	*Trifolium repens*		A	I
	Springbank clover	*Trifolium wormskjoldii*		N	I
	American vetch	*Vicea americana*		N	I
	Giant vetch	*Vicea gigantea*	*V. nigricans*	N	I
	Hairy vetch	*Vicea hirsuta*		A	I
Gentianaceae (Gentians)	Swamp gentian	*Gentiana douglasii*		N	M
	King gentian	*Gentiana sceptrum*		N	I
Geraniaceae (Geraniums)	Dovefoot geranium	*Geranium molle*		A	I
	Small-flowered cranes-bill	*Geranium pusillum*		A	I
Grossulariaceae (Gooseberries)	Stink currant	*Ribes bracteosum*		N	I
	Wild gooseberry	*Ribes divaricata*		N	I
Haloragaceae	Eurasian milfoil	*Myriophyllum sibiricum*		A	M
Hydrophyllaceae	Mistmaiden	*Romanzoffia tracyi*		N	I
Hypericaceae	Bog St. John's wort	*Hypericum anagalloides*		N	I
Lamiaceae/Labiatae (Mints)	Bugleweed	*Lycopus uniflorus*		N	I
	Peppermint	*Mentha piperita*		A	I
	Self-heal	*Prunella vulgaris*		N	I
	Hedge nettle	*Stachys mexicana*		N	I

Family	Common name	Scientific name		Status	Origin
Lentibulariaceae (Bladderworts)	Flat-leaf bladderwort	Utricularia intermedia		N	M
	Lesser bladderwort	Utricularia minor		N	I
	Greater bladderwort	Utricularia vulgaris		N	M
Menyanthaceae (Buckbeans)	Buckbean	Menyanthes trifoliata		N	I
	Deer-cabbage	Nephrophyllidium crista-galli		N	M
Myricaceae	Sweet gale	Myrica gale		N	I
Nymphaceae (Water-lilies)	Watershield	Brassenia schreberi		N	M
	Yellow pond-lily	Nuphar polycephalum		N	I
Onagraceae (Willowherbs)	Fireweed	Epilobium angustifolium	Chamerion a.	N	I
	Purple-leaf willowherb	Epilobium watsoni	E. cilliatum	N	I
Orobanchaceae	Ground-cone	Boschniakia hookeri		N	I
Plantaginaceae (Plantains)	Narrowleaf plantain	Plantago lanceolata		A	I
	Alaska plantain	Plantago macrocarpa		N	I
	Broadleaf plantain	Plantago major		A	I
	Sea plantain	Plantago maritima		N	I
	Hybrid plantain	Plantago "X macrocarpa"			
Polygonaceae (Dock, Bistort)	Knotweed	Polygonum aviculare		A	I
	Sheep sorrel	Rumex acetosella		A	I
	Curly dock	Rumex crispus		A	I
	Western dock	Rumex occidentalis	R. obtusifolius	N	I
	Narrowleaf dock	Rumex salicifolius		N	I
Portulacaceae (Miner's lettuce)	Branching montia	Montia diffusa	Claytonia d.	N	I
	Spring-beauty	Montia parvifolia	Claytonia p.	N	I
	Candy flower	Montia sibirica	Claytonia s.	N	I
Primulaceae (Primroses)	Scarlet pimpernel	Anagallis arvensis		A	M
	Shootingstar	Dodecatheon jeffreyi		N	M
	Sea milk-wort	Glaux maritima		N	I
	Loosestrife	Lysimachus thyrsifolia		A	M
	Northern starflower	Trientalis arctica	T. europaea	N	I

APPENDIX A (continued)

FAMILY	COMMON NAME	OLD/CURRENT BINOMIAL	SYNONYM	NATIVE/ALIEN	MAINLAND/ISLAND
Ranunculaceae (Buttercups)	Red columbine	*Aquilegia formosa*		N	I
	Marsh marigold	*Caltha biflora*		N	M
	Fern-leaf goldthread	*Coptis asplenifolia*		N	M
	Three-leaf goldthread	*Coptis trifoliata*		N	M
	Western buttercup	*Ranunculus occidentalis*		N	I
	European buttercup	*Ranunculus repens*		A	I
	Little buttercup	*Ranunculus unciniatus*		N	I
Rhamnaceae	Cascara	*Rhamnus purshiana*		N	I
Rosaceae (Roses, Cherries, Raspberries, etc.)	Chokecherry	*Amelanchier alnifolia*		N	I
	Goat's-beard	*Aruncus sylvester*	*A. dioicus*	N	I
	Cotoneaster	*Cotoneaster sp.*		A	I
	Hawthorn	*Crategus douglasii*		N	I
	Beach strawberry	*Fragaria chiloensis*		N	I
	Yellow avens	*Geum macrophyllum*		N	I
	Ocean spray	*Holodiscus discolor*		N	I
	Nine-bark	*Physocarpus capitatus*		N	I
	Silverweed	*Potentilla pacifica*	*Argentina egedii*	N	I
	Marsh silverweed	*Potentilla palustris*	*Comarum palustre*	N	M
	Wooly cinquefoil	*Potentilla villosa*		N	I
	Bitter cherry	*Prunus emarginata*		N	I
	Domestic plum	*Prunus communis*		A	I
	Crab apple	*Pyrus fuscus*	*Malus fusca*	N	I
	Nootka rose	*Rosa nutkana*		N	I
	Clustered wild rose	*Rosa pisocarpa*		N	I
	Evergreen blackberry	*Rubus laciniatus*		A	I
	Thimbleberry	*Rubus parviflorus*		N	I
	Blackberry	*Rubus procerus*	*R. discolor*	A	I

	Salmonberry	Rubus spectabilis		N	I
	Trailing blackberry	Rubus ursinus		N	I
	Great burnet	Sanguisorba officinalis		N	M
	Rowan	Sorbus aucuparia		A	I
	Sitka mountain-ash	Sorbus sitchensis		N	I
	Douglas' spiraea	Spiraea douglasii	S. menziesii	N	M
Rubiaceae (Madders, Bedstraw)	Cleavers	Galium aparine		N	I
	Small bedstraw	Galium trifidum	G. cymosum	N	I
	Sweet-scented bedstraw	Galium triflorum		N	I
Salicaceae (Willows)	Hooker's willow	Salix hookeriana		N	I
	Scouler's willow	Salix scouleriana		N	I
	Sitka willow	Salix sitchensis		N	I
Saxifragaceae (Saxifrages)	Coast boykinia	Boykinia elata	B. occidentalis	N	I
	Alumroot	Heuchera micrantha		N	I
	Grass-of-Parnassus	Parnassia fimbriata[a]		N	M
	Alaska saxifrage	Saxifraga ferruginea		N	I
	Fringecup	Tellima grandiflora		N	I
	False mitrewort	Tiarella trifoliata	T. laciniata	N	I
	Piggy-back plant	Tolmiea menziesii		N	I
Scrophulariaceae (Foxgloves etc.)	Red paintbrush	Castilleja miniata		N	I
	Blue-eyed mary	Collinsia grandiflora	C. parviflora	N	I
	Common foxglove	Digitalis purpurea		A	I
	Yellow monkeyflower	Mimulus guttatus		N	I
	Musk-flower	Mimulus moschatus		N	I
	Owl-clover	Orthocarpus castillejoides		N	M
	California figwort	Scrophularia california	Castilleja ambigua	N	I
	American brooklime	Veronica americana	S. oregana	N	M
	Ivy speedwell	Veronica hederaefolia		N	I
	Purslane speedwell	Veronica peregrina		N	M
	Marsh speedwell	Veronica scutullata		N	M
Solanaceae	Black nightshade	Solanum americanum[a]		N	I

APPENDIX A *(continued)*

FAMILY	COMMON NAME	OLD/CURRENT BINOMIAL	SYNONYM	NATIVE/ALIEN	MAINLAND/ISLAND
Urticaceae	Stinging nettle	*Urtica dioica*[a]		N	I
Valerianaceae	Sea-blush	*Plectritis congesta*		N	I
	Scouler's valerian	*Valeriana scouleri*	?*V. sitchensis*	N	I
Violaceae	Marsh violet	*Viola palustris*		N	I
Araceae (Arums)	Skunk-cabbage	*Lysichiton americanum*		N	I
Cyperaceae (Sedges)	Star sedge	*Carex angustior*	*C. echinata*	N	M
	Club sedge	*Carex buxbaumii*		N	M
	Gray sedge	*Carex canescens*		N	I
	Dewey's sedge	*Carex deweyana s.l.*	*C. leptopoda*[a]	N	I
	Gmelin's sedge	*Carex gmelinii*		N	I
	Kellogg's sedge	*Carex kelloggii*	*C. lenticularis s.l.*	N	I
	Bristle-stalked sedge	*Carex leptalea*		N	I
	Pale sedge	*Carex livida*		N	I
	Lyngby's sedge	*Carex lyngbyei*		N	I
	Large-headed sedge	*Carex macrocephala*		N	M
	Slough sedge	*Carex obnupta*		N	I
	Sand-dune sedge	*Carex pansa*		N	I
	Few-flowered sedge	*Carex pauciflora*		N	M
	Stellate sedge	*Carex phyllomanica*		N	I
	Russet sedge	*Carex physocarpa*	*C. saxatalis*	N	I
	Several-flowered sedge	*Carex pluriflora*		N	M
	Beaked sedge	*Carex rostrata*	*C. utricularia*	N	I
	Sitka sedge	*Carex sitchensis*	*C. aquatilis*	N	M
	Sawbeak sedge	*Carex stipata*[a]		N	I
	Long-styled sedge	*Carex stylosa*		N	M
	Green sedge	*Carex viridula*		N	I
	Dulichium	*Dulichium arundinaceum*		N	I

	Creeping spike-rush	Eleocharis palustris		N	I
	Cotton-grass	Eriophorum angustifolium	E. polystachion	N	M
	White beak-rush	Rhynchospora alba		N	I
	Tufted club-rush	Scirpus cespitosus	Trichophorum c.	N	M
	Low club-rush	Scirpus cernuus	Isolepis cernuua	N	M
	Small-flowered bulrush	Scirpus sylvaticus	S. microcarpus	N	M
	Water club-rush	Scirpus subterminalis	Schoenoplectus s.	N	M
	Hard-stemmed bulrus	Scirpus vallidus	S. lacustris	N	M
Juncaceae (Rushes)					
	Tapered rush	Juncus acuminatus		N	M
	Arctic rush	Juncus arcticus	J. balticus	N	I
	Toad rush	Juncus bufonius		N	I
	Common rush	Juncus effusus		N	I
	Dagger-leaf rush	Juncus ensifolius		N	I
	Sickle-leaf rush	Juncus falcatus		N	I
	Spreading rush	Juncus oregonus	J. supiniformis	N	I
	Many-flowered wood-rush	Luzula multiflora	L. campestris	N	I
	Small-flowered wood-rush	Luzula parviflora	L. fastigiata	N	I
Juncaginaceae (Arrow-grasses)					
	Graceful arrow-grass	Triglochin concinnum		N	I
	Seaside arrow-grass	Triglochin maritimum		N	I
Iridaceae (Irises)					
	Yellow-flag	Iris ?pseudacorus		A	I
	Blue-eyed gras	Sisyrinchium angustifolium		N	I
	Golden-eyed grass	Sisyrinchium californicum	S. littorale	N	I
Liliaceae (Lillies)					
	Nodding onion	Allium cernuum		N	I
	Common camas	Camassia quamash		N	M
	Pink fawn lily	Erythronium revolutum		N	I
	Northern rice-root	Fritillaria camschatcensis		N	I
	Tiger lily	Lilium columbianum		N	I
	False lily-of-the-valley	Maianthemum dilatatum		N	I
	Twisted-stalk	Streptopus amplexifolius		N	I
	False asphodel	Tofieldia glutinosa		N	M
Orchidaceae (Orchids)					
	Rattlesnake orchid	Goodyeara oblongifolia		N	I
	White bog-orchid	Habenaria dilatata	Plantathera d.	N	I

APPENDIX A (continued)

FAMILY	COMMON NAME	OLD/CURRENT BINOMIAL	SYNONYM	NATIVE/ALIEN	MAINLAND/ISLAND
	Green bog-orchid	*Habenaria hyperborea*	*Plantathera h.*	N	M
	Northwest twayblade	*Listera caurina*		N	I
	Ladies' tresses	*Spiranthes romanzoffiana*		N	I
Poaceae (Grasses)	Redtop	*Agrostis alba*	*A. stolonifera*	A	I
	Spike bentgrass	*Agrostis exarata*		N	I
	Ticklegrass	*Agrostis geminata*	*A. scabra*	N	I
	Small-leaf bentgrass	*Agrostis microphylla*	?*A. exarata*	N	I
	Oregon redtop	*Agrostis oregonensis*	*A. aequivalvis*	N	I
	Dune bentgrass	*Agrostis pallens*		N	I
	Creeping bentgrass	*Agrostis stolonifera*		A	I
	Silvery hairgras	*Aira caryophyllea*		A	I
	Little hairgrass	*Aira praecox*		A	I
	Foxtail	*Alopecurus geniculatus*		N	M
	Sweet vernal grass	*Anthoxanthum odoratum*	*Hierochloe odorata*	A	I
	Oats	*Avena sativa*		A	I
	Soft brome	*Bromus mollis*		A	I
	Pacific brome	*Bromus pacificus*		N	I
	Alaska brome	*Bromus sitchensis*		N	I
	Barren brome	*Bromus sterilis*		A	I
	Downy brome	*Bromus tectorum*		A	I
	Canada reedgrass	*Camalgrostis canadensis*		N	I
	Thurber's reedgrass	*Calamagrostis crassiglumis*		N	I
	Pacific reedgrass	*Calamagrostis nootkaensis*		N	I
	Woodreed	*Cinna latifolia*		N	I
	Orchard grass	*Dactylis glomerata*		A	I
	California oatgrass	*Danthonia californica*		N	I
	Poverty oatgrass	*Danthonia spicata*		N	I
	Tufted hairgrass	*Deschampsia cespitosa*		N	I
	Slender hairgrass	*Deschampsia elongata*		N	I

Common name	Scientific name		Origin	Status
Blue wildrye	*Elymus glaucus*		N	I
Hairy wildrye	*Elymus hirsutus*		N	I
Beach rye	*Elymus mollis*	*Leymus m.*	N	I
Hybrid rye	*Elymus "glaucus x mollis"*		A	I
Reed fescue	*Festuca arundinacea*		A	I
Barren fescue	*Festuca bromoides*	*Vulpia b.*	A	I
Rat-tail fescue	*Festuca myuros*	*Vulpia m.*	A	I
Western fescue	*Festuca occidentalis*		N	I
English fescue	*Festuca pratensis*		A	I
Red fescue	*Festuca rubra*		N	I
Bearded fescue	*Festuca subulata*		N	I
Range fescue	*Festuca subulifera*		N	I
Yorkshire-fog	*Holcus lanatus*		A	I
Meadow barley	*Hordeum brachyantherum*		N	I
Barley	*Hordeum vulgare*		A	I
Hybrid rye	*"Hordelymus"*	*Hordeum X Elymus*		
Perennial ryegrass	*Lolium perenne*		A	I
Western panicum	*Panicum occidentale*		N	I
Canarygrass	*Phalaris arundinacea*		N	M
Timothygrass	*Phleum pratensis*		A	I
Annual bluegrass	*Poa annua*		A	I
Canadian bluegrass	*Poa compressa*		N	I
Coastline bluegrass	*Poa confinis*		N	I
Howell's bluegrass	*Poa howelli*		N	I
Withered bluegrass	*Poa marcida*		N	I
Fowl bluegrass	*Poa palustris*		A	I
Kentucky bluegrass	*Poa pratensis*		A	I
Rough bluegrass	*Poa trivialis*		A	I
Alaska alkaligrass	*Puccinellia nutkaensis*		N	I
Nuttall's alkaligrass	*Puccinellia nuttalliana*		N	I
Dwarf alkaligrass	*Puccinellia pumila*		N	I
Nodding trisetum	*Trisetum cernuum*		N	I
Wheat	*Triticum aestivum*		A	I
Corn	*Zea mays*		A	I

APPENDIX A (continued)

FAMILY	COMMON NAME	OLD/CURRENT BINOMIAL	SYNONYM	NATIVE/ALIEN	MAINLAND/ISLAND
Potamogetoneaceae	Grass-leaf pondweed	*Potamogeton grammineus*		N	I
	Floating-leaf pondweed	*Potamogeton natans*		N	I
	Small pondweed	*Potamogeton pusillus*		N	M
Sparganiaceae (Bur-reeds)	Narrow-leaf bur-reed	*Sparganium emersum*	*S. angustifolium*	N	I
	Small bur-reed	*Sparganium minimum*		N	M
Zosteraceae (Eelgrass, Surfgrass)	Surfgrass	*Phyllospadix torreyi*		N	(I)
	Japanese eelgrass	*Zostera japonica*		A	?
	Eelgrass	*Zostera marina*		N	(I)

[a] Bell and Harcombe 1973 record only.

Note: Number of species with island record is 302; number of additional mainland species is 62; proportion of alien species is 68/364 = 0.187.

APPENDIX B

Gazetteer of Barkley Sound Islands

ISLAND NUMBER	NAME	LATITUDE (°N)	LONGITUDE (°W)	AREA$_{FL}$ (m²)	ELEV (m)	D_{MLD} (km)	CYRS (y)	AVERAGE #SPP	CUMULATIVE #SPP
A1	N Hosie	48.912	125.033	7920	17.5	1.370	9	66.3	86
A2	Mid Hosie (see Fig. B)	48.910	125.034	2200	8.5	1.580	10	65.2	82
A3	Hosie Rks a	48.911	125.034	668	6	1.540	10	35.2	50
A4	Hosie Rks b	48.911	125.034	115	1	1.535	10	6.8	13
A5	Hosie Rks c	48.911	125.034	137	1.5	1.530	10	3.7	7
A6	Hosie Rks d	48.911	125.034	120	1.5	1.528	10	1.7	6
A7	S Hosie	48.909	125.036	42900	67	1.720	7	64.0	87
A8	San Jose NW	48.901	125.056	11800	3.5	1.920	10	42.2	58
A9	San Jose NE	48.901	125.053	3470	7.5	1.880	10	43.0	56
A10	San Jose SE	48.900	125.053	1140	5.5	1.800	10	27.8	39
A11	Nanat (see Fig. B)	48.885	125.075	9190	21	0.090	9	33.7	49
A12	Danvers (see Fig. B)	48.878	125.089	2940	9	0.010	9	33.8	44
A13	Ellis	48.862	125.106	3270	12	0.120	10	30.3	39
A14	Dixon	48.853	125.119	108000	40	0.080	8	77.6	97
A15	Brady Bch 1b	48.828	125.151	378	19.5	0.020	13	32.1	45
A16	Brady Bch 2c	48.828	125.151	1120	26	0.016	13	35.9	52
A17	Brady Bch 3d	48.827	125.154	1160	14	0.088	12	40.8	56
A18	Brady Bch 4e	48.827	125.154	378	12	0.035	13	12.0	21
A19	Brady Bch 5h	48.827	125.153	265	4	0.042	12	6.1	14
A20	Brady Bch 6g	48.827	125.153	45.8	4	0.050	12	5.2	11
A21	Brady Bch 7a	48.830	125.150	393	15.8	0.001	12	30.2	43
A22	Brady Bch 8f	48.827	125.153	100	1.5	0.034	11	3.6	8
A23	Nanette	48.885	125.071	211	1.5	0.050	7	9.6	11
A24	Cape Beale	48.788	125.214	111529	28	0.050	2	110.0	110
A25	Dixon S Stk	48.852	125.121	36.5	3.5	0.140	5	5.2	6
A26	Dixon W Rk	48.854	125.120	399	2	0.420	5	2.4	4
A27	Dixon SW Rk	48.852	125.122	908	1.5	0.150	4	2.0	2
B1	Wizard	48.858	125.159	6290	3.4	2.480	9	19.7	27
B2	Helby Self Pt	48.849	125.162	2820	19	1.810	7	35.3	48
B3	Ohiat (see Fig. B)	48.855	125.183	5170	24	3.460	10	44.8	63

APPENDIX B (continued)

ISLAND NUMBER	NAME	LATITUDE (°N)	LONGITUDE (°W)	AREA$_{FL}$ (m^2)	ELEV (m)	D_{MLD} (km)	CYRS (y)	AVERAGE #SPP	CUMULATIVE #SPP
B4	Ohiat N	48.855	125.183	3800	12.5	3.535	10	33.4	40
B5	Dodger NE	48.846	125.203	8370	8.5	3.920	6	40.2	52
B6	Seppings	48.840	125.206	34400	25	3.200	8	45.7	60
B7	Taylor	48.827	125.196	7500	21	1.840	8	42.0	59
B8	Helby Blackfish	48.847	125.165	787	8.5	1.920	7	25.0	32
B9	Helby NE (see Fig. B)	48.852	125.162	561	18	2.000	6	23.8	31
B10	Helby NEE	48.852	125.162	172	18	2.040	6	14.2	17
B11	Helby NNE	48.852	125.163	155	9.5	2.120	6	19.5	23
B12	Haines SE	48.827	125.196	18000	3	2.240	9	63.0	98
B13	Haines SE Rks b	48.831	125.195	877	2.7	2.220	9	22.3	41
B14	Haines SE Rks c	48.830	125.195	197	3.3	2.200	8	4.8	9
B15	Haines SE Rks dLo	48.830	125.196	858	3.3	2.200	9	16.1	27
B16	Haines SE Rks dHi	48.830	125.196	94	4.5	2.200	8	14.5	19
B17	Haines SE Rks e	48.830	125.196	51	2.1	2.180	9	1.0	1
B18	Haines SE Rks f	48.830	125.196	587	3.3	2.180	9	15.7	28
B19	Haines SE Rks g (see Fig. B)	48.830	125.197	1030	7.2	2.180	9	39.0	55
B20	Bordelais	48.818	125.226	9050	24	2.820	5	29.0	32
B21	Seppings SS	48.837	125.202	3800	3.6	3.080	5	24.2	34
B22	Seppings S Rk	48.838	125.203	503	3	3.160	5	6.4	11
B23	Seppings S Pt	48.839	125.203	253	3.9	3.360	7	11.4	18
B24	Seppings SE BchRk	48.839	125.203	49.7	3	3.360	6	3.1	5
B25	Bordelais S Rks	48.818	125.230	14500	11	2.760	5	21.2	25
B26	Diana Ddgr SE aN	48.831	125.192	993	4.2	2.120	6	32.3	39
B27	Diana Ddgr SE bS	48.831	125.192	1010	6.5	2.120	6	27.7	35
B28	Leach	48.831	125.240	28400	25	4.300	4	42.8	49
B29	Leach N	48.831	125.241	3830	12	4.460	4	27.8	31
B30	Folger			180	XX	3.480	1	Incomplete	Incomplete
B31	Seppings SSE Rk	48.839	125.203		XX			Incomplete	Incomplete
B32	Seppings SEE Rk	48.839	125.203	101	1.8	3.480	4	1.0	1
B33	Seppings Int Rk	48.839	125.203	291	2.8	3.480	2	0.5	1
C1	Sandford SE	48.866	125.164	1730	4.5	3.280	9	11.0	14
C2	Ross SSE Rk	48.868	125.159	1720	4.5	3.200	10	19.6	26
C3	Ross SSE	48.870	125.160	10300	4.5	3.360	12	54.1	73
C4	Middle Ross (see Fig. B)	48.872	125.161	2820	3	3.560	11	37.8	46
C5	Ross NNE	48.875	125.163	1140	9.5	3.920	9	26.3	34
C6	Ross/Sandford Adj (see Fig. B)	48.873	125.164	13300	18	3.680	8	63.9	84
C7	Ross E Cl a	48.874	125.158	1390	5.5	3.640	11	26.2	32
C8	Ross E Cl b	48.874	125.158	2540	4.5	3.600	11	34.1	46
C9	Ross E Cl c	48.873	125.158	1700	6	3.560	11	33.9	43
C10	Ross E Cl d	48.873	125.158	1910	4.5	3.600	11	39.9	53
C11	Ross E Cl e	48.873	125.159	3560	12	3.600	11	35.2	48
C12	Ross E Cl e'	48.873	125.157	1310	5	3.600	11	29.6	40
C13	Ross E Cl e"	48.873	125.157	483	3.5	3.680	10	16.1	25
C14	Ross E Cl f	48.874	125.157	8740	7	3.700	11	45.2	62
C15	Ross E Cl f'	48.875	125.160	963	5	3.760	11	23.3	28
C16	Ross E Cl cRk	48.873	125.157	115	2.5	3.520	10	3.2	5
C17	Ross E Adj (see Fig. B)	48.875	125.159	2920	17	3.720	8	23.1	31
C18	S Rock (see Fig. B)	48.875	125.157	131	3.5	3.700	10	18.4	21
C19	Ross Low Rk	48.874	125.156	251	1.5	3.680	9	4.6	5
C20	Ross Tall Rk	48.874	125.155	225	3	3.600	9	6.7	11

ISLAND NUMBER	NAME	LATITUDE (°N)	LONGITUDE (°W)	AREA$_{FL}$ (m^2)	ELEV (m)	D_{MLD} (km)	CYRS (y)	AVERAGE #SPP	CUMULATIVE #SPP
C21	Ross Mid Rks E	48.874	125.157	118	3.5	3.600	9	6.1	8
C22	Ross Mid Rks W	48.874	125.157	238	3	3.600	9	2.4	3
C23	Fleming E	48.884	125.123	5990	23	2.520	7	23.7	30
C24	Fleming E Rk	48.883	125.122	1050	6.7	2.420	8	26.1	30
C25	Fleming E NW Rk	48.884	125.123	161	2.5	2.600	6	7.0	7
C26	Robbers N End	48.899	125.118	6250	8.5	3.240	7	37.3	48
C27	Robbers NW Rks a	48.897	125.119	2250	5	3.200	8	26.0	37
C28	Robbers Nend WRk	48.897	125.118	2280	3.5	3.320	6	16.7	21
C29	Robbers NW Rks b	48.897	125.118	161	2	3.080	8	5.6	10
C30	Robbers NW Rks c	48.897	125.117	421	3.5	3.040	8	21.6	28
C31	Robbers S End Rk	48.893	125.106	888	2.7	2.440	6	16.8	23
C32	Tzartus SE Rk	48.894	125.105	673	2.1	2.120	6	9.8	10
C33	Robbers NE	48.898	125.116	1080	2.4	3.080	6	16.3	25
C34	Fleming W	48.895	125.134	3360	7.5	3.880	7	37.1	47
C35	Ross E Cl e"'	48.873	125.158	815	3.5	3.720	7	15.6	19
C36	Sandford	48.868	125.166	290000	93	3.320	6	113.2	139
D6	Mussel Rk	48.923	125.110	695	4.5	5.040	8	2.3	3
D7	Poison Ivy (see Fig. B)	48.924	125.110	1960	12	5.080	8	20.9	28
D8	Long Low Rf	48.926	125.108	959	2.5	5.200	8	4.0	6
D9	Geer N	48.930	125.111	15500	16.5	4.400	7	39.9	50
D10	Geer S	48.928	125.110	9300	13.5	4.480	6	33.8	43
D11	Friend	48.936	125.096	23700	24	4.040	6	73.5	97
D12	Meade N (see Fig. B)	48.926	125.119	18600	6	4.800	6	48.7	66
D13	Meade Rf aS	48.925	125.122	1090	3	5.480	6	19.3	26
D14	Meade Rf bM	48.925	125.121	1640	3.5	5.440	6	16.5	26
D15	Meade S	48.925	125.123	7800	5	5.480	6	38.2	49
D16	Meade NN Rk	48.929	125.120	1130	3	5.180	7	3.3	6
D17	Meade N Rk	48.928	125.119	339	3	5.140	6	4.0	5
D18	S Stud	48.941	125.094	6930	13	3.480	8	43.4	54
D19	S Stud E Rk	48.941	125.094	332	3.5	3.560	6	13.8	18
D20	M Studs aE	48.944	125.095	377	5.5	3.400	7	9.3	14
D21	M Studs W Rk	48.943	125.095	700	5.5	3.400	7	17.3	24
D22	M Studs W Rk E Adj	48.943	125.095	404	2.5	3.400	6	3.0	4
D23	M Studs Big (see Fig. B)	48.944	125.096	3280	6.5	3.360	8	37.9	50
D24	Big Stud	48.944	125.091	14800	12	3.200	8	59.8	83
D25	Studs NC Rk	48.945	125.093	310	3.5	3.120	8	8.8	13
D26	W Studs aE	48.945	125.095	411	4.5	3.160	8	10.3	16
D27	W Studs bS	48.945	125.096	640	5.5	3.160	8	23.6	29
D28	W Studs cNW	48.945	125.097	1340	9	3.120	8	28.6	36
D29	Big Stud SE Rk	48.942	125.092	598	6.2	3.440	7	20.6	25
D30	Big Stud W Rk	48.944	125.092	1040	3	3.400	8	26.5	37
D31	Meade Rf cN	48.925	125.121	300	2.4	5.440	3	2.7	3
D32	Big Stud NE Rk	48.943	125.091	182	4.5	3.320	7	21.7	25
D33	Big Stud SW Rk	48.943	125.093	194	3	3.360	7	10.7	13
D34	Big Stud E Rk	48.943	125.091	228	4.5	3.320	7	22.9	29
D35	Diplock N	48.936	125.103	3900	4.2	4.080	4	28.5	35
D36	Weld	48.950	125.089	220559	50	2.280	1	75.0	75
E1	Wiebe SW Rk	48.894	125.279	2010	3	6.280	6	1.0	2
E2	Wiebe SE Rk	48.896	125.277	2020	7	6.000	6	19.3	25
E3	Groundsel	48.907	125.285	1020	3.5	5.160	6	18.3	24
E4	Coon	48.905	125.285	15800	4	5.360	5	37.6	49
E5	Wee Pinnacle	48.906	125.288	256	4	5.400	5	9.4	10

APPENDIX B *(continued)*

ISLAND NUMBER	NAME	LATITUDE (°N)	LONGITUDE (°W)	AREA$_{FL}$ (m^2)	ELEV (m)	D_{MLD} (km)	CYRS (y)	AVERAGE #SPP	CUMULATIVE #SPP
E6	Mullins E	48.908	125.287	3300	7	5.160	5	30.2	42
E7	Mullins SE (see Fig. B)	48.906	125.284	3920	20	5.280	5	31.0	43
E8	Coon E Stk	48.905	125.286	277	4	5.440	4	23.3	27
E9	Elbow	48.902	125.272	3130	16	5.360	7	27.9	31
E10	Elbow Low	48.904	125.277	1980	5	5.400	6	12.7	16
E11	SW Twin	48.904	125.279	1980	6	5.240	7	18.4	23
E12	NE Twin	48.905	125.279	1980	6.5	5.200	7	20.6	22
E13	Faber NE Rk	48.894	125.300	423	4	7.000	7	3.9	7
E14	Faber NW Rks a	48.892	125.302	220	3.5	7.120	6	2.3	3
E15	Faber NW Rks bS	48.892	125.301	557	9	7.160	7	10.1	11
E16	Faber SW Rks	48.891	125.299	2730	10	7.240	7	16.1	17
E17	Faber E Rk	48.894	125.297	393	4	7.080	7	2.0	2
E18	Onion	48.904	125.293	46200	22	5.760	5	45.6	60
E19	Tiny Mid E	48.919	125.304	8170	10	5.160	6	20.5	29
E20	Tiny SW	48.917	125.307	345	3	5.480	6	16.8	19
E21	Tiny SW NW Rk	48.917	125.307	134	1.5	5.480	6	5.2	8
E22	Tiny SE	48.917	125.305	4410	2.5	5.400	6	29.0	40
E23	Tinier Tim	48.915	125.308	296	3.6	5.640	5	12.8	16
E24	Tiny W	48.920	125.306	2570	6.5	5.240	6	32.5	39
E25	Tiny Rf aE	48.916	125.306	1840	3.3	5.480	6	12.2	15
E26	Tiny Rf b	48.916	125.307	1840	2.5	5.480	6	17.8	25
E27	Tiny Rf c	48.916	125.307	1840	4	5.480	6	18.0	23
E28	Tiny Rf dW	48.917	125.308	674	4.5	5.480	6	19.3	23
E29	Tiny SW NN Rk	48.918	125.307	120	1.5	5.480	5	3.8	7
E30	S Groundsel	48.907	125.285	1020	2.2	5.200	4	9.5	12
E31	Little Rk	48.907	125.288	151	2.5	5.240	4	4.8	5
E32	Faber NE Rks S	48.893	125.302	168	2.5	7.120	4	1.8	2
E33	Faber NE Rks midW	48.894	125.301	227	2.7	7.080	4	1.3	2
E34	Tiny Big N	48.921	125.304	18600	8	5.000	5	43.0	51
E35	Tiny SW NNW Rk	48.920	125.307	92.5	1.8	5.480	4	2.6	7
E36	Tiny Tim	48.916	125.307	627	5.6	5.600	4	23.5	25
E37	Exotica	48.913	125.307	2190	2.8	5.840	5	21.4	28
E38	Onion Rks d	48.902	125.297	1300	6.5	6.040	3	19.3	21
E39	Onion Rks a	48.903	125.296	3190	6.5	6.200	2	28.0	31
E40	Onion Rks c	48.902	125.297	1430	6.5	6.000	3	18.3	21
E41	Onion Rks b	48.903	125.295	693	3.5	6.120	1	12.0	12
E42	Marchant	48.916	125.299	15600	7	5.080	4	50.5	60
E43	Marchant SE Sat	48.916	125.299	1560	3.9	5.200	4	18.0	22
E44	Walsh	48.918	125.320	85000	62	5.480	5	75.4	96
E45	Walsh WSW	48.915	125.321	9220	9	5.600	4	36.8	43
E46	Walsh SSW (see Fig. B)	48.917	125.322	2350	4.5	5.600	4	34.8	43
E47	Chalk SE	48.917	125.313	5900	5.5	5.440	4	32.8	42
E48	Chalk SW	48.917	125.313	5900	6	5.440	4	34.3	43
E49	Clarke	48.890	125.376	236000	90	7.480	3	109	136
E50	Clarke N	48.893	125.378	2530	7.58	7.440	3	26.3	30
E51	Clarke N'	48.893	125.377	513	5.5	7.400	3	18.0	19
E52	Clarke N"	48.893	125.379	1090	5.45	7.400	3	26.0	30
E53	Clarke E	48.889	125.368	1530	5.5	8.120	4	20.0	23
E54	Owen	48.894	125.373	67100	14.5	7.320	2	62.0	88
E55	Owen W Rk	48.896	125.376	1260	6.4	7.160	4	17.8	22
E56	Liz Rk	48.897	125.374	754	4.24	7.520	3	4.3	6
E57	Puffin (see Fig. B)	48.903	125.382	2340	12.5	6.880	4	20.5	22

ISLAND NUMBER	NAME	LATITUDE (°N)	LONGITUDE (°W)	AREA$_{FL}$ (m^2)	ELEV (m)	D_{MLD} (km)	CYRS (y)	AVERAGE #SPP	CUMULATIVE #SPP
Æ58	Big Pig	48.892	125.383	21700	23	7.080	3	54.0	69
Æ59	Little Pig E	48.892	125.380	9140	40	7.360	3	46.3	58
Æ60	Little Pig W	48.893	125.381	4280	15	7.320	3	33.7	42
Æ61	Little Pig EE Rk	48.892	125.379	1523	2.5	7.400	3	23.3	33
Æ62	Piglet	48.891	125.382	483	6	7.160	3	20.0	23
Æ63	Pig Gully	48.891	125.383	2460	8	7.360	3	24.3	30
Æ64	Big Pig E Rk	48.891	125.381	506	2.2	7.400	3	16.7	21
Æ65	Big Pig SE Sat	48.891	125.383	859	6.5	7.360	3	25.7	30
Æ66	Big Pig W Rks	48.892	125.385	1540	5.45	7.120	3	15.3	19
Æ67	Otter	48.905	125.285	2660	22	5.440	3	13.0	16
Æ68	Keith W	48.912	125.291	5380	2.7	5.080	4	52.5	65
Æ69	E Turtle Rk	48.907	125.305	170	2.7	6.200	3	0.0	0
Æ70	Village Rf	48.891	125.288	2120	2.2	7.160	2	0.0	0
Æ71	Faber NE Rks farSW	48.893	125.302	60.5	2.1	7.240	2	0.0	0
Æ72	Chalk N	48.922	125.317	1700	4.5	4.800	3	21.0	24
Æ73	Chalk N/S	48.923	125.317	997	3	4.880	3	13.0	15
Æ74	Clam Rk	48.899	125.351	507	6	8.040	4	1.8	2
Æ75	Ginnit	48.890	125.377	1670	9	7.400	2	18.0	20
Æ76	Turtle	48.910	125.322	1146105	73	5.720	1	101	101
Æ77	Willis	48.915	125.340	474559	78	5.840	1	92.0	92
Æ78	Willis W End	48.918	125.346	10892	12	5.840	2	33.5	40
Æ79	Willis NW	48.920	125.346	2945	8	5.960	2	50.5	59
Æ80	Dodd SW Rks aE	48.921	125.342	104	3.2	5.480	2	5.5	7
Æ81	Dodd SW Rks bM	48.922	125.343	207	3.3	5.440	2	11.0	12
Æ82	Dodd SW Rks cW	48.923	125.344	532	7.5	5.400	2	8.5	9
Æ83	Starlight Rf	48.881	125.484	6295	20	4.120	1	4.0	4
Æ84	Baeria Rks	48.955	125.151	5105	20	3.480	2	5.0	6
Æ85	Faber NE Rks midE	48.893	125.300	188	3	7.160	2	2.0	2
Æ86	Faber SW Rks bN	48.892	125.300	173	2.1	7.200	2	5.0	5
Æ87	Bryant	48.953	125.362	19623	50	3.360	1	Incomplete	Incomplete
Æ88	Lovett	48.906	125.372	XXX	25	4.160	1	Incomplete	Incomplete
Æ89	Curwen	48.947	125.359	5606	76	4.200	1	Incomplete	Incomplete
Æ90	Faber NE Rks midNE	48.893	125.300	63	2.1	7.160	1	0	0
Æ91	Faber NNW Rk	48.893	125.302	96	2.1	7.200	1	0	0

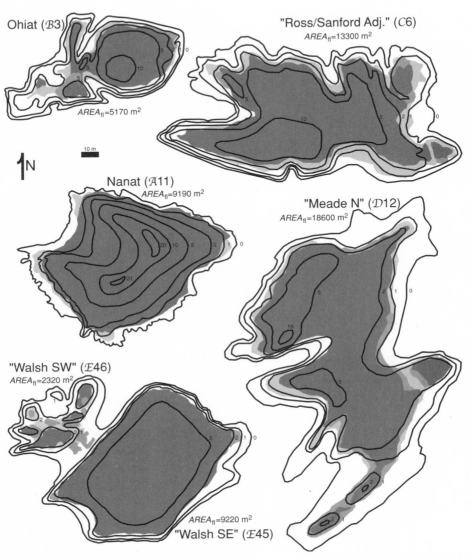

FIGURE B Representative islands with elevation contours at 1, 2, 5, 10, and 20 m intervals. Three levels of shading indicate, lightest to darkest, extent of vascular plants, area with continuous vegetation, and area covered by forest or shrub vegetation <1 m high.

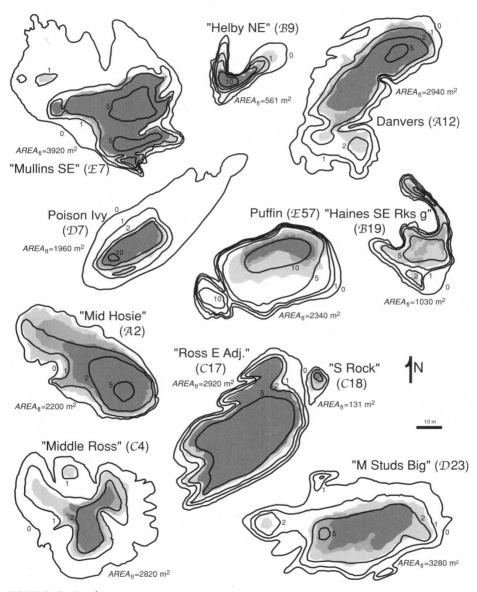

FIGURE B (*Continued*)

APPENDIX C
Island Historical Phylogenies

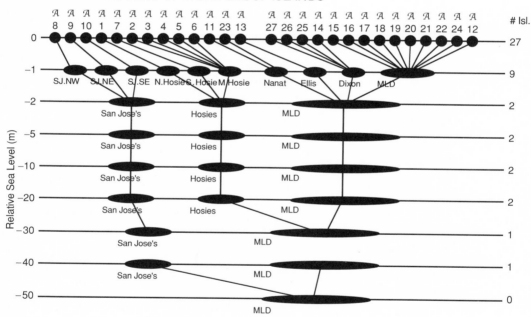

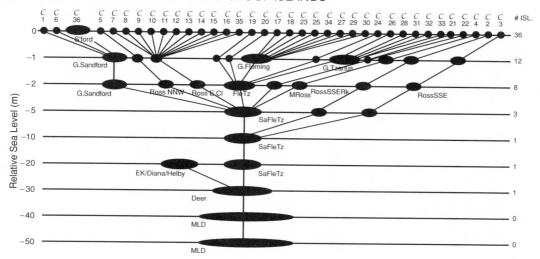

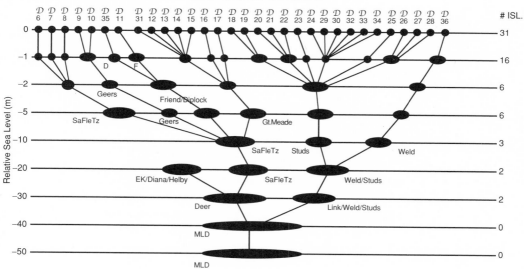

APPENDIXES 239

APPENDIX D

Bog Plants Found at 12 Mainland and 3 Islands Sites in the Bamfield Area and Barkley Sound

SPECIES	BAMFIELD MEADOWS	BAMFIELD #1 BOG	TEPALTOS	BAMFIELD LAKE	GRAPPLER	KICHHA LAKE	NETTLE ISLAND	HUU-AY-AHT BOG	EFFINGHAM ISLAND	LIGHTHOUSE	CALAMITY LAKE	NINEMILE LAKE	EIGHT-MILE BOG	CAPE BEALE	TURTLE ISLAND
	82500	12	1.93	13	1.6	3	2	0.95	1.5	0.94	2	5	0.11	1.17	0.2
Primitives															
Mosses (Sphagnaceae) *Sphagnum* (ca. 20 spp.)	1	1	1	1	1	1	1	1	1	1	1	1	1	1	1
Club-mosses (Lycopodiaceae) *Lycopodium clavatum*	1	1		1		1	1	1	1	1	1		1	1	
Lycopodium selago/inundatum	1	1		1	1	1	1	1	1						
Grape-ferns (Ophioglossaceae) *Botrychium multifidum/ selaginiformis*					1	1	1								
Total	3	3	1	3	3	3	3	2	3	2	2	1	1	2	1
Sedges (Cyperaceae)															
Carex angustior/echinata		1		1	1									2	
Carex buxbaumii		1								1			1		
Carex canescens												1			1
Carex leptalea	1		1		1		1			1	1		1		1
Carex livida	1	1	1	1	1	1		1	1	1	1		1	1	
Carex lyngbyei									1						1
Carex obnupta	1	1	1	1	1	1	1	1	1	1	1	1			1
Carex pauciflora	1														
Carex phyllomanica	1	1	1	1	1	1	1	1	1	1	1		1	1	1

Sedges (Cyperaceae)

Species	1	2	3	4	5	6	7	8	9	10	11	12	13	14	15
Carex physocarpa/saxatalis	1	1													
Carex pluriflora	1	1	1												
Carex rostrata/utricularia	1	1	1	1	1		1				1		1		
Carex saxatalis	1		1		1										
Carex sitchensis	1	1		1	1		1								
Carex stylosa					1										
Carex viridula	1	1	1	1	1	1	1	1	1	1	1		1		
Dulichium arundinaceum	1	1	1	1	1	1			1		1		1		
Eleocharis palustris				1	1		1				1		1		1
Eriophorum angustifolium/polystachion					1			1							
Rhynchospora alba	1	1	1	1	1	1	1		1		1		1		
Scirpus caespitosus	1	1	1	1	1	1	1								
Scirpus microcarpus/sylvaticus			1	1	1										
Scirpus subterminalis	1	1	1	1	1		1						1		
Scirpus vallidus	1	1				1								1	
Total	**15**	**13**	**12**	**11**	**14**	**7**	**9**	**7**	**9**	**7**	**9**	**6**	**9**	**6**	**6**

Rushes (Juncaceae)

Species	1	2	3	4	5	6	7	8	9	10	11	12	13	14	15
Juncus acuminatus		1	1				1		1		1		1	1	1
Juncus arcticus			1	1	1				1				1		
Juncus brevicaudatus		1	1	1				1	1						
Juncus effusus	1				1	1						1	1		
Juncus ensifolius	1	1	1	1	1		1		1		1	1	1		1
Juncus falcatus	1	1	1	1	1				1				1		
Juncus oregonus/supiniformis	1	1	1	1		1		1			1				
Total	**4**	**5**	**5**	**5**	**4**	**3**	**2**	**2**	**5**	**0**	**3**	**2**	**5**	**1**	**2**

Grasses (Poaceae)

Species	1	2	3	4	5	6	7	8	9	10	11	12	13	14	15
Agrostis aequivalvis	1	1	1	1	1	1		1	1		1		1	1	
Calamagrostis crassiglumis	1	1	1		1	1	1	1	1				1		
Deschampsia caespitosa	1	1		1	1	1		1					1		1
Panicum occidentale	1		1	1	1	1	1				1		1		
Phalaris arundinacea										1					
Total	**4**	**3**	**3**	**3**	**4**	**4**	**2**	**3**	**2**	**1**	**2**	**0**	**3**	**2**	**1**

APPENDIX D (continued)

SPECIES	BAMFIELD MEADOWS	BAMFIELD #1 BOG	TEPALTOS	BAMFIELD LAKE	GRAPPLER	KICHHA LAKE	NETTLE ISLAND	HUU-AY-AHT BOG	EFFINGHAM ISLAND	LIGHTHOUSE	CALAMITY LAKE	NINEMILE LAKE	EIGHT-MILE BOG	CAPE BEALE	TURTLE ISLAND
	82500	12	1.93	13	1.6	3	2	0.95	1.5	0.94	2	5	0.11	1.17	0.2
Aquatics															
Bur-reeds (Sparganiaceae)															
Sparganium emersum	1	1										1			1
Sparganium minimum	1											1			
Arums (Araceae)															
Lysichiton americanum		1	1	1	1	1	1	1	1	1	1		1	1	1
Haloragaceae															
Myriophyllum sibiricum	1														
Bladderworts (Lentibulariaceae)															
Utricularia intermedia			1												
Utricularia minor	1		1	1		1	1					1			
Utricularia vulgaris	1		1	1		1						1			
Buck-beans (Menyanthaceae)															
Menyanthes trifoliata	1	1	1			1	1	1	1		1	1		1	1
Nephrophyllidium crista-galli	1									1					
Waterlilies (Nymphaceae)															
Brassenia schreberi	1	1	1	1		1					1	1			
Nuphar polycephalum	1	1	1		1		1		1		1	1		1	1
Pondweeds (Potamogetonaceae)															
Potamogeton gramineus	1	1	1	1	1	1		1				1		1	1
Potamogeton natans	1	1	1	1	1	1	1		1		1	1			
Potamogeton pusillus												1			
Total	9	6	8	6	4	8	5	2	4	2	4	10	1	4	5
Dicot and monocot forbs															
Parsleys (Apiaceae)															
Cicuta douglasii	1	1		1	1		1	1	1	1					1
Oenanthe sarmentosa			1		1			1	1				1		
Comps (Asteraceae)															
Apargidium boreale	1	1	1	1	1	1	1	1	1	1					
Dogwoods (Cornaceae)															
Cornus unalaskensis	1	1	1	1	1	1	1	1	1	1		1			
Sundews (Droseraceae)															
Drosera anglica	1	1	1	1	1	1	1	1	1	1	1		1	1	
Drosera rotundifolia	1	1	1	1	1	1	1	1	1	1	1		1	1	1

Family	Species	S1	S2	S3	S4	S5	S6	S7	S8	S9	S10	S11	S12	S13	S14	S15
Gentians (Gentianaceae)	*Gentiana douglasiana*	1	1		1		1	1	1		1			1		
	Gentiana sceptrum	1	1	1	1	1	1	1	1		1					1
Hypericaceae	*Hypericum anagalloides*	1	1	1	1	1	1	1	1	1	1	1	1		1	1
Irises (Iridaceae)	*Sisyrinchium californicum*	1	1	1	1	1	1	1	1	1	1		1	1	1	
Mints (Lamiaceae)	*Lycopus uniflorus*	1	1		1	1	1							1		
Lilies (Liliaceae)	*Camassia quamash*	1	1	1	1	1	1		1							
Orchids (Orchidaceae)	*Toefieldia glutinosa*	1	1		1	1	1		1		1	1	1	1	1	
	Plantathera dilatata	1	1													
	Plantathera hyperborea	1							1							
Primroses (Primulaceae)	*Dodecatheon jeffreyi*	1							1							
	Trientalis arctica	1	1		1		1		1	1	1	1	1			1
Buttercups (Ranunculaceae)	*Caltha biflora*	1	1													
	Coptis asplenifolia	1							1							
	Coptis trifoliata								1							
Roses et al. (Rosaceae)	*Potentilla palustris*	1					1		1					1		
	Sanguisorba officinalis	1	1	1	1	1			1		1	1	1	1	1	
	Spiraea douglasii/menziesii	1	1	1	1		1	1	1		1	1	1			
Saxafragaceae	*Parnassia fimbriata*		1													
Scrophulariaceae	*Veronica americana*	1	1			1		1								
	Veronica scutullata				1											
Violets (Violaceae)	*Viola palustris*	1	1	1		1	1	1	1					1	1	
	Total	**22**	**18**	**12**	**12**	**14**	**9**	**11**	**13**	**6**	**11**	**5**	**6**	**8**	**5**	**6**

Shrubs

Family	Species	S1	S2	S3	S4	S5	S6	S7	S8	S9	S10	S11	S12	S13	S14	S15
(Empetraceae)	*Empetrum nigrum*	1	1	1	1		1	1	1		1		1		1	1
Cranberries etc. (Ericaceae)	*Kalmia occidentalis*	1	1	1	1	1	1	1	1	1	1	1	1		1	1
	Ledum groenlandicum	1	1	1	1		1	1	1	1	1	1	1	1		1
	Vaccinium alaskense	1	1	1			1		1		1					
	Vaccinium oxycoccus	1	1	1	1	1	1	1	1		1	1	1	1	1	
	Vaccinium uliginosum	1	1	1	1	1	1	1	1	1	1				1	
Gales (Myricaceae)	*Myrica gale*	1	1	1	1		1	1	1		1				1	1
Willows (Salicaceae)	*Salix sitchensis*					1	1							1		
	Total	**7**	**7**	**7**	**5**	**3**	**7**	**6**	**6**	**3**	**7**	**3**	**3**	**2**	**5**	**4**

APPENDIX D (continued)

	SPECIES	BAMFIELD MEADOWS	BAMFIELD #1 BOG	TEPALTOS	BAMFIELD LAKE	GRAPPLER	KICHHA LAKE	NETTLE ISLAND	HUU-AY-AHT BOG	EFFINGHAM ISLAND	LIGHTHOUSE	CALAMITY LAKE	NINEMILE LAKE	EIGHT-MILE BOG	CAPE BEALE	TURTLE ISLAND
		82500	12	1.93	13	1.6	3	2	0.95	1.5	0.94	2	5	0.11	1.17	0.2
Trees																
Pines (Pinaceae)	*Pinus contorta*	1	1	1	1	1	1	1	1	1					1	
	Pinus monticola	1	1		1		1		1							
Cedars (Cupressaceae)	*Chamaecyparis nootkatensis*	1	1		1		1		1	1	1	1				
	Total	3	3	1	3	1	3	1	3	1	2	1	0	0	1	0
Summary	Primitives	3	3	1	3	3	3	3	2	3	2	2	1	1	2	1
	Carices	15	13	12	11	14	7	9	7	9	7	9	6	9	6	6
	Rushes	4	5	5	5	4	3	2	2	5	0	3	2	5	1	2
	Grasses	4	3	3	3	3	4	4	2	3	1	3	2	3	2	1
	Aquatics	9	6	8	6	4	8	5	2	4	2	4	10	1	4	5
	Docit/monot forbs	22	18	12	12	14	9	11	13	6	11	5	6	8	5	6
	Shrubs	7	7	7	5	3	7	6	6	3	7	3	3	2	5	4
	Trees	3	3	1	3	1	3	1	3	1	2	1	0	0	1	0
	Grand Total	67	58	49	48	46	44	41	37	34	32	30	30	29	26	25
	Area (ha):	82500	12	1.93	13	1.6	3	2	0.95	1.5	0.94	2	5	0.11	1.17	0.2

APPENDIX E

Common Summer Birds of the Barkley Sound Area

First are listed the birds dependent on marine environments, which are commonly seen while boating among the islands. Second are listed species of terrestrial environments, including seven species (those followed by the designation "*" in the right-hand column) that we have recorded on the mainland but not on the islands. The other twenty-seven species of terrestrial birds have been recorded (by us) on the islands as well as on the mainland. These species are conspicuously "nested" in island distributions (see Ch. 5), with species recorded on smaller islands being to a large degree subsets of those on larger islands, where additionally there are species absent from the smaller islands. The numbers in the right-hand column reflect the rank-order of species as they would be encountered in moving from small to large islands (i.e. beginning with Song Sparrow, Rufous Hummingbird, Northwest Crow, Fox Sparrow, etc.).

Although the land birds can be ordered in this way, there are three pairs of related species that appear to affect each other's island distributions. These are Song Sparrow/Fox Sparrow, Wilson's Warbler/Orange-crowned Warbler, and Swainson's Thrush/Hermit Thrush. These species occur generally in low scrub, mid-height woodland, and tall forest, respectively; the first species of the pair is generally found in lower and more open habitat than the second and is replaced by the second in taller and denser habitat. The first species of the pair is more generally distributed on smaller islands, the second more restricted to larger islands, but either species may occur on a wide range of islands of intermediate size. On the islands of intermediate size, one can find one or the other member of the replacement pair, but rarely both, and often the sequence, with increasing island size, is sp.A → spA or spB → sp.B.

FAMILY	SPECIES	
Birds of marine environments		
Laridae (Gulls)	Glaucous-winged Gull *Larus glaucescens*	
Alcidae (Auks)	Pigeon Guillemot *Cepphus columba*	
	Common Murre *Uria aalge*	
	Marbled Murrelet *Brachyramphus marmoratus*	
	Tufted Puffin *Fratercula cirrhata*	
	Rhinoceros Auklet *Cerorhinca monocerata*	
Alcenidae (Kingfishers)	Belted Kingfisher *Ceryle alcyon*	
Accipitridae (Eagles, Hawks)	Bald Eagle *Haliaeetus leucocephalus*	
Birds of terrestrial habitats		
Accipitridae (Accipiters)	Goshawk *Accipiter gentilis*	*
Columbidae (Pigeons)	Band-tailed Pigeon *Columba fasciatus*	*
Trochilidae (Hummingbirds)	Rufous Hummingbird *Selasphorus rufus*	2
Picidae (Woodpeckers)	Pileated Woodpecker *Dryocopus pileatus*	*
	Hairy Woodpecker *Picoides villosus*	19
	Northern Flicker *Colaptes auratus*	16
Tyrannidae (Flycatchers)	Pacific Coast Flycatcher *Empidonax difficilis*	12
	Olive-sided Flycatcher *Contopus borealis*	26
Hirundinidae (Swallows)	Violet-green Swallow *Tachycineta thalassina*	*
	Barn Swallow *Hirundo rustica*	24
Corvidae (Crows)	Common Raven *Corvus corax*	*
	Northwest Crow *Corvus caurinus*	3
	Steller's Jay *Cyanositta stelleri*	10
Bombycillidae (Waxwings)	Cedar Waxwing *Bombycilla cedrorum*	27
Paridae (Titmice)	Chestnut-backed Chickadee *Parus rufescens*	7
Aegithalidae (Bushtits)	Common Bushtit *Psaltriparus minimus*	22
Certhidae (Creepers)	Brown Creeper *Certhia americana*	15
Sittidae (Nuthatches)	Red-breasted Nuthatch *Sitta canadensis*	17
Sylviidae (Old World Warblers)	Golden-crowned Kinglet *Regulus satrapa*	8
Troglodytidae (Wrens)	Winter Wren *Troglodytes troglodytes*	6
Turdidae (Thrushes)	American Robin *Turdus americanus*	9
	Varied Thrush *Ixoreus naevius*	21
	Hermit Thrush *Catharus guttatus*	23
	Swainson's Thrush *Catharus ustulatus*	5
Parulidae (New World Warblers)	Townsend's Warbler *Dendroica townsendi*	11
	Wilson's Warbler *Wilsonia pusilla*	20
	Orange-crowned Warbler *Vermivora celata*	13
Emberizidae (Sparrows)	Song Sparrow *Melospiza melodia*	1
	Fox Sparrow *Passerella iliaca*	4
	White-crowned Sparrow *Zonotrichia Leucophrys*	*
Fringillidae (Finches)	Red Crossbill *Loxia curvirostra*	18
	Pine Siskin *Spinus pinus*	25
	Cassin's Finch *Carpodacus cassinii*	14
	Purple Finch *Carpodacus purpureus*	*

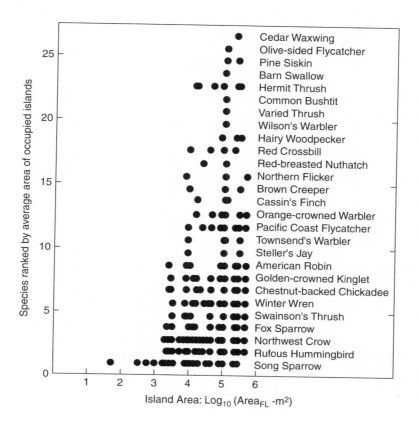

REFERENCES

Abbott, I. 1977. Species richness, turnover and equilibrium in insular floras near Perth, Western Australia. *Austr. J. Bot.* 25: 193–208.

Abbott, I., and R. Black. 1980. Changes in composition of floras on islets near Perth, Western Australia. *J. Biogeogr.* 7: 399–410.

Angerbjörn, A. 1985. The evolution of body size of mammals on islands: Some comments. *Am. Natur.* 125: 304–9.

Arima, E.Y. 1983. *The west coast (Nootka) people.* Spec. publ. 6. Victoria: British Columbia Provincial Museum.

Arrhenius, O. 1921. Species and area. *J. Ecol.* 9: 95–99.

Arrhenius, O. 1923. On the relation between species and area: A reply. *Ecology* 4: 90–91.

Baldwin, B., D.W. Kyhos, J. Dvorak, and G.D. Carr. 1991. Chloroplast DNA: Evidence for the North American origin of the Hawaiian silversword alliance (Asteraceae). *Proc. Nat. Acad. Sci. USA* 88: 1840–43.

Beilman, D.W. 2001. Plant community and diversity change due to localized permafrost dynamics. *Can. J. Bot.* 79: 983–93.

Bell, M.A.M., and A.P. Harcombe. 1973. *Flora and vegetation of Pacific Rim National Park: Phase II, Broken Group Islands.* Victoria, BC: BioCon Research.

Bentham, G. 1878. *Flora Australiensis.* Vol. 7, *Roxburghlaceae to Filices.* Ashford, Kent, UK: L. Reeve and Co..

Bigley, R.E. and J.L. Barreca. 1982. Evidence for synonymizing *Zostera americana* Den Hartog with *Zostera japonica* Aschers and Graebn. *Aquat. Bot.* 14: 349–56.

Billings, W., and H.A. Mooney. 1968. The ecology of arctic and alpine plants. *Biol. Rev.* 43: 481–529.

Boecklen, W.J., and N.J. Gotell. 1984. Island biogeographic theory and conservation practice: Species area or specious area relationships. *Biol. Conserv.* 29: 63–80.

Boyd, R.S., and M.G. Barbour. 1993. Replacement of *Cakile edentula* by *C. maritima* in the strand habitat of California. *Am. Midl. Nat.* 130: 209–28.

Brockman, C.F. 1947. *Flora of Mount Ranier National Park.* U.S. Govt. Printing Office, Washington, DC: National Park Service.

Brown, J.H., and A. Kodric-Brown. 1977. Turnover rates in insular biogeography: Effect of immigration on extinction. *Ecology* 58: 445–49.

Brown, J.H., and M.V. Lomolino. 1998. *Biogeography.* 2nd ed. Sunderland, MA: Sinauer Assoc.

Brown, K.J., and R.J. Hebda. 2002. Origin, development, and dynamics of coastal temperate conifer rainforests of southern Vancouver Island, Canada. *Can. J. Forest. Res.* 32: 353–72.

Buckingham, N.M., E.C. Schreiner, T.N. Kaye, J.E. Burger, and E.L. Tisch. 1995. *Flora of the Olympic Peninsula.* Seattle, WA: Northwest Interpretive Assoc.

Buckley, R.C., and S.B. Knedlhans. 1986. Beachcomer biogeography: Interception of dispersing propagules by islands. *J. Biogeogr.* 13: 68–70.

Bullock, J.M., R.E. Kenward, and R.S. Hails, eds. 2002. *Dispersal ecology.* Oxford: British Ecological Society, Blackwell.

Burns, K.C. 2002. Seed dispersal facilitation and geographic consistency in bird-fruit abundance pattern. *Global Ecol. Biogeogr.* 11: 253–59.

Burns, K.C. 2003. Broad-scale reciprocity in an avian seed dispersal mutualism. *Global Ecol. Biogeogr.* 12: 421–26.

Burns, K.C. 2004a. Scale and macroecological patterns in seed dispersal mutualisms. *Global Ecol. Biogeogr.* 13: 289–93.

Burns, K.C. 2004b. Patterns in specific leaf area and the structure of a temperate heath community. *Diversity Distrib.* 10: 105–12.

Burns, K.C., and J.L. Dalen. 2002. Foliage color contrasts and adaptive fruit color variation in a bird-dispersed plant community. *Oikos* 96: 463–69.

Cain, A.J., and P.M. Sheppard. 1950. Selection in the polymorphic land snail *Cepaea nemoralis*. *Heredity* 4: 275–94.

Carlquist, S. 1965. *Island life: A natural history of the islands of the world.* New York: Natural History Press.

Carlquist, S. 1970. *Hawaii: A natural history.* Garden City, NY: Natural History Press.

Carlquist, S. 1974. *Island biology.* New York: Columbia Univ. Press.

Carlquist, S., B.G. Baldwin, and G.D. Carr. 2003. *Tarweeds and silverswords: Evolution of the Madiinae (Asteraceae).* St. Louis, MO: Missouri Botanical Garden Press.

Carreño, A.L., and J. Helenes. 2002. Geology and ages of the islands. In *A new island biogeography in the Sea of Cortés,* ed. T. Case, M.L. Cody, and E. Ezcurra, 14–40. Oxford: Oxford Univ. Press.

Case, T.J., and M.L. Cody. 1983. *Island biogeography in the Sea of Cortez.* Berkeley and Los Angeles: Univ. of California Press.

Case, T.J., and M.L. Cody. 1987. Testing theories of island biogeography. *Am. Sci.* 75: 402–11.

Case, T., M.L. Cody, and E. Ezcurra, eds. 2002. *A new island biogeography in the Sea of Cortés.* Oxford: Oxford Univ. Press.

Cates, R.G., and G.H. Orians. 1975. Successional status and the palatability of plants to generalized herbivores. *Ecology* 56: 410–18.

Chang, A.S., R.T. Patterson, and R. McNeeley. 2003. Seasonal sediment and diatom record from late Holocene laminated sediments. *Palaios* 18: 477–94.

Chang, A.S., A. Prokoph, and R.T. Patterson, and H.M. Roe. 2005. Solar influences on climate and diatoms in the northeast Pacific. In *Geologic problem solving with microfossils,* ed. R.F. Waszczak and T.D. Demchuk, 44. Houston, TX: North American Micro-paleological Section, Society for Sedimentary Geology Symposium Abstracts.

Clark, C.W., and M.L. Rosenzweig. 1994. Extinction and colonization processes: Parameter estimates from sporadic surveys. *Am. Natur.* 143: 583–96.

Clobert, J., E. Danchin, A.A. Dhondt, and J.D. Nichols, eds. 2001. *Dispersal.* Oxford: Oxford Univ. Press.

Cody, M.L. 1973. Towards a theory of continental species diversities. In *Ecology and evolution of communities,* ed. M.L. Cody and J.M. Diamond, 214–57. Cambridge, MA: Belknap Press of Harvard Univ. Press.

Cody, M.L. 1983a. Continental diversity patterns and convergent evolution in bird communities. In *Ecological studies 43,* ed. F. Kruger, D.T. Mitchell, and J.U.M. Jarvis, 347–402. Vienna: Springer-Verlag.

Cody, M.L. 1983b. Bird species diversity and density in Afromontane woodlands. *Oecologia* 59: 210–15.

Cody, M.L. 1986a. Diversity and rarity in Mediterranean ecosystems. In *Conservation biology: The science of scarcity and diversity,* ed. M. Soulé, 122–52. Sunderland, MA: Sinauer Assoc.

Cody, M.L. 1986b. Structural niches in plant communities. In *Community ecology,* ed. J. Diamond and T.J. Case, 381–405. New York: Harper and Row.

Cody, M.L. 1999. Assembly rules in plant and bird communities. In *Ecological assembly rules: Advances and retreats,* ed. P. Keddy, and E. Wieher, 165–205. Cambridge: Cambridge Univ. Press.

Cody, M.L., and T.W.D. Cody. 2003. Morphology and spatial distribution of alien Sea-rockets (*Cakile* spp.) on South Australian and western Canadian beaches. *Austr. J. Bot.* 52: 175–83.

Cody, M.L., R. Moran, J. Rebman, and H.J. Thompson. 2002. Plants. In *A new island biogeography in the Sea of Cortés,* ed. T. Case, M.L. Cody, and E. Ezcurra, 63–111. Oxford: Oxford Univ. Press.

Cody, M.L., and J.M. Overton. 1996. Short-term evolution of reduced dispersal in island plant populations. *J. Ecol.* 84: 53–61.

Cody, M.L., and M.E. Velarde. 2002. The land birds. In *A new island biogeography in the Sea of Cortés*, ed. T. Case, M.L. Cody, and E. Ezcurra, ch. 8. Oxford: Oxford Univ. Press.

Connor, E.F., and E.D. McCoy. 1979. Statistics and biology of the species-area relationship. *Am. Natur.* 113: 791–833.

Cook, J. 1785. *A Voyage to the Pacific Ocean, undertaken by the command of His Majesty, for making discoveries in the Northern Hemisphere; Performed under the direction of Captains Cook, Clerk, and Gore . . . in the years 1776, 1777, 1778, 1779, 1780.* 3 vols. Vols. 1 and 2 written by Captain James Cook. Vol. 3, by Captain James King. London.

Cook, R.R. and J.F. Quinn. 1998. An evaluation of randomization models for nested species subsets analysis. *Oecologia* 113: 584–92.

Cox, G.W. 1999. *Alien species in North America and Hawaii: Impacts on natural ecosystems.* Washington, DC: Island Press.

Crawley, M.J. 1986. Life history and environment. In *Plant ecology*, ed. M.J. Crawley, ch. 8. Oxford: Blackwell.

Curtis, J.T. 1959. *The vegetation of Wisconsin.* Madison, WI: Univ. of Wisconsin Press.

Damman, A.W.H. 1978. Distribution and movement of elements in ombrotrophic peat bogs. *Oikos* 30: 480–95.

Darlington, P.J., Jr. 1957. *Zoogeography: The geographical distribution of animals.* New York: J. Wiley.

Davidson, A., and B. Clarke. 2000. History or current selection? A molecular analysis of "area effects" in the land snail *Cepaea nemoralis*. *Proc. Roy. Soc. Lon. Ser. B Biol. Sci.* 267: 1399–1405.

De Pamphilis, C.W., N.D. Young, and A.D. Wolfe. 1997. Evolution of plastid gene *rps2* in a lineage of hemiparasitic and holoparasitic plants: Many losses of photosynthesis and complex patterns of rate variation. *Proc. Nat. Acad. Sci. USA* 94: 7367–72.

Diamond, J.M. 1972. Biogeographic kinetics: Estimation of relaxation times for avifaunas of southwest Pacific islands. *Proc. Nat. Acad. Sci. USA* 69: 3199–3203.

Diamond, J.M. 1975a. Assembly of species communities. In *Ecology and evolution of communities*, ed. M.L. Cody and J.M. Diamond, 342–444. Cambridge, MA: Belknap Press of Harvard Univ. Press.

Diamond, J.M. 1975b. The island dilemma: Lessons of modern biogeographic studies for the design of nature reserves. *Biol. Conserv.* 7: 129–46.

Diamond, J.M., and M.E. Gilpin. 1982. Examination of the null model of Connor and Simberloff for species co-occurrence on islands. *Oecologia* 52: 64–74.

Diamond, J.M., and R.M. May. 1977. Species turnover on islands: Dependence on census interval. *Science* 197: 266–70.

Diamond, J.M., and R.M. May. 1981. Island biogeography and the design of nature reserves. In *Theoretical ecology*, ed. R.M. May, 228–52. Oxford: Blackwell.

Dony, J.G. 1970. *Species-area relationships.* Unpublished report to the Natural Environmental Research Council, London.

Douglas, G.W., G.B. Straley, D.E. Meidinger, and J. Pojar. 1998–2002. *Illustrated flora of British Columbia*, 8 vols. Ministry of Environment, Lands, and Parks, Victoria, BC: Ministry of Forests, British Columbia Provincial Govt.

Duff, W. 1997. *The Indian history of British Columbia: The impact of the white man.* Victoria: Royal British Columbia Museum.

Eldridge, L.G. 1999. Numbers of Hawaiian species. *Bishop Mus. Occ. Papers* 58 (Suppl. 4): 72–78.

Environment Canada. 1995. *The state of Canada's climate: Monitoring variability and change.* State of the environment report no. 95-1. Ottawa, Environment Canada.

Esselman, E.J., D.J. Crawford, S. Brauner, T. Stuessy, C.J. Anderson, and M. Silva-O. 2000. RAPD marker diversity within and divergence among species of *Dendroseris* (Asteraceae: Lactuceae). *Am. J. Bot.* 87: 591–96.

Estrada, A., and T.H. Fleming, eds. 1986. *Frugivores and seed dispersal.* Tasks in vegetation science 15. Dordrecht: Junk.

Flint, S.D., and I.G. Palmblad. 1978. Germination dimorphism and developmental flexibility in the ruderal weed *Heterotheca grandiflora*. *Oecologia* 36: 33–44.

Foster, J.B. 1964. Evolution of mammals on islands. *Nature (London)* 202: 234–35.

Friele, P.A., and I. Hutchinson. 1992. Holocene sea-level change on the central west coast of Vancouver Island, British Columbia. *Can. J. Earth Sci.* 30: 832–40.

Gauch, H.G. 1982. *Multivariate studies in community ecology*. Cambridge studies in ecology 1. New York: Cambridge Univ. Press.

Gignac, L.D., L.A. Halsey, and D.H. Vitt. 2000. A bioclimatic model for the distribution of *Sphagnum*-dominated peatlands in North America under present climatic conditions. *J. Biogeogr.* 27: 1139–51.

Gilpin, M.E., and J.M. Diamond. 1976. Calculations of extinction and immigration curves from the species-area-distance relation. *Proc. Nat. Acad. Sci. USA* 73: 4130–34.

Gilpin, M.E., and J.M. Diamond. 1981. Immigration and extinction probabilities for individual species: Relation to incidence functions and species colonization curves. *Proc. Nat. Acad. Sci. USA* 78: 392–96.

Gilpin, M.E., and J.M. Diamond. 1982. Factors contributing to non-randomness in species co-occurrences on islands. *Oecologia* 52: 75–84.

Gleason, H. 1922. On the relation between species and area. *Ecology* 3: 158–62.

Gordon, D.G. 1994. *Field guide to the slug*. Seattle, WA: Western Society of Malacologists and Sasquatch Books.

Grime, J.P. 1979. *Plant strategies and vegetation processes*. New York: John Wiley and Sons.

Grubb, P.J. 1977. The maintenance of species-richness in plant communities: The importance of the regeneration niche. *Biol. Rev.* 52: 107–45.

Haila, Y. 1990. Towards an ecological definition of an island: A northwest European perspective. *J. Biogeogr.* 17: 561–68.

Hall, E.R. 1981. *The mammals of North America*. Vols. 1 and 2. 2nd ed. New York: John Wiley and Sons.

Hanski, I. 1998. Metapopulation dynamics. *Nature (London)* 396: 41–49.

Harrison, P.G., and R.E. Bigley. 1982. The recent introduction of the seagrass *Zostera japonica* Aschers and Graebn to the Pacific coast of North America. *Can. J. Fish. Aquat. Sci.* 39: 1642–48.

Herrera, C.M. 1998. Long-term dynamics of Mediterranean frugivorous birds and fleshy fruits: A 12-year study. *Ecol. Monogr.* 68: 511–38.

Herrera, C.M., and O. Pellmyr. 2002. *Plant-animal interactions*. Oxford: Blackwell.

Heubner, C. 1996. What are the effects of island area on terrestrial snails? Student paper, Island Biogeography, Bamfield Marine Science Center 1996.

Heusser, C.J. 1989. North Pacific coastal refugia: The Queen Charlotte Islands in perspective. In *The outer shores*, ed. G.G.E. Scudder and N. Gessler, 91–106. Skidegate, QC: Queen Charlotte Islands Museum Press, Skidegate.

Heusser, C.J., L.E. Heusser, and S.S. Streeter. 1980. Quaternary temperatures and precipitation for the north-west coast of North America. *Nature* 286: 702–4.

Hickman, J.C., ed. 1993. *The Jepson manual: Higher plants of California*. Berkeley and Los Angeles: Univ. of California Press.

Hitchcock, C.L. and A. Cronquist. 1973. *Flora of the Pacific Northwest: An illustrated manual*. Seattle: Univ. of Washington Press.

Hitchcock, C.L., A. Cronquist, and M. Ownbey. 1955–1969. *Vascular flora of the Pacific Northwest*. 5 vols. Seattle: Univ. of Washington Publications in Biology.

Hooker, J.D. 1849. On the vegetation of the Galapagos Archipelago. *Trans. Linn. Soc. London* 20: 235–62.

Hooker, J.D. *Flora Tasmaniae* (Botany of the Anarctic Voyage, pt. 3), London: Lovell Reeve.

Hooker, W.J. 1829, 1838, 1840. *Flora Boreali-Americana*. London: H.C. Bohn.

Hooker, W.J. 1854. *A century of ferns*. London: W. Pamplin.

Howe, H.F., and J. Smallwood. 1982. Ecology of seed dispersal. *ARES* 13: 201–28.

Johnson, N.K., J.A. Marten, and C.J. Ralph. 1989. Genetic evidence for the origin and relationships of the Hawaiian honeycreepers (Aves: Fringillidae). *Condor* 91: 379–96.

Keddy, P., and E. Weiher, eds. 2001. *Ecological assembly rules: Advances and retreats*. Cambridge: Cambridge Univ. Press.

Kelly, B.J., J.B. Wilson, and A.F. Mark. 1989. Cause of the species-area relation: A study of islands in Lake Manapouri, New Zealand. *J. Ecol.* 77: 1021–28.

Kim, H., S.C. Keeley, P.S. Vroom, and R.K. Jansen. 1998. Molecular evidence for an African origin of the Hawaiian endemic genus *Hesperomannia* (Asteraceae). *Proc. Nat. Acad. Sci. USA* 95: 15440–45.

Klinger, L.F. 1996. The myth of the classic hydrosere model of bog succession. *Arct. Alp. Res.* 28: 1–9.

Klinka, K., V.J. Krajina, A. Ceska, and A.M. Scagel. 1989. *Indicator plants of British Columbia*. Vancouver: Univ. of British Columbia Press.

Kodric-Brown, A., and J. H. Brown. 1993. Highly structured fish communities in Australian desert springs. *Ecology* 74: 1847–55.

Kopp, H., E. R. Flueh, P. Thierer, D. Klaeschen, F. Tilmann, and C. Gaedike. 2002. *Geophysical investigations of the central Chilean continental margin.* GEOMAR cruise reports. hkopp@geomar.de.

Krajina, V. J. 1973. *Biogeoclimatic zones of British Columbia.* Victoria: British Columbia Ecological Reserves Committee.

Lack, D. L. 1976. *Island biology illustrated by the land birds of Jamaica.* Studies in ecology 3. Berkeley and Los Angeles: Univ. of California Press.

Larsen, J. A. 1982. *Ecology of the northern lowland bogs and coniferous forests.* New York: Academic Press.

Lawlor, T. E. 1982. The evolution of body size in mammals: Evidence from insular populations in Mexico. *Am. Natur.* 119: 54–72.

Lawlor, T. E., D. J. Hafner, P. Stapp, B. R. Riddle, and S. T. Alvarez-Castaneda. 2002. The mammals. In *A new island biogeography in the Sea of Cortés,* ed. T. Case, M. L. Cody, and E. Ezcurra, 326–361. Oxford: Oxford Univ. Press.

Lawton, J. H. 1984. Non-competitive populations, non-convergent communities, and vacant niches: The herbivores of bracken. In *Ecological communities: Conceptual issues and the evidence,* ed. D. R. Strong Jr. et al. Princeton, NJ: Princeton Univ. Press.

Leonard, J. L. 2001. Sex in banana slugs: Different clades do different things. *Am. Zool.* 41: 1504.

Leonard, J. L., J. S. Pearse, and A. B. Harper. 2002. Comparative reproductive biology of *Ariolimax californicus* and *Ariolimax dolichophallus*. *Invert. Reprod. Devel.* 41: 83–93.

Levy, D. J., W. R. Silva, and M. Galetti, eds. 2002. *Seed dispersal and frugivores.* New York: CABL Publ.

Liu, J. P., and J. D. Milliman. 2003. Timing and rate of post-glacial sea-level rise during melt-water pulses 1A and 1B. http://tethys.vims.edu/papers/mwp-1a/, March 21, 2003.

Lloyd, R. M., and E. J. Klekowski. 1970. Spore germination and viability in Pteridophyta: Evolutionary significance of chlorophyllous spores. *Biotropica* 2: 129–37.

Lomolino, M. V. 1985. Body size of mammals on islands: The island rule re-examined. *Am. Natur.* 125: 310–16.

Lomolino, M. V., and R. Davis 1997. Biogeographic scale and biodiversity of montane forest mammals of western North America. *Global Ecol. Biogeogr. Lett.* 6: 57–76.

Losos, J. B., and D. Schluter. 2000. Analysis of an evolutionary species-area relationship. *Nature (London)* 408: 847–50.

Ludvigsen, R., and G. Beard. 1994. *West coast fossils: A guide to the ancient life of Vancouver Island.* Vancouver, BC: Whitecap Books.

Lynch, J. F., and N. K. Johnson. 1974. Turnover and equilibria in insular faunas, with special reference to the Channel Islands. *Condor* 76: 370–84.

MacArthur, R. H. 1958. Population ecology of some warblers of northeastern coniferous forests. *Ecology* 39: 599–619.

MacArthur, R. H., and E. O. Wilson. 1963. An equilibrium theory of insular zoogeography. *Evolution* 17: 373–87.

MacArthur, R. H., and E. O. Wilson. 1967. *The theory of island biogeography.* Monographs in population biology 1. Princeton, NJ: Princeton Univ. Press.

MacHutchon, A. G. 1989. Spring and summer food habits of black bears in the Pelly River Valley, Yukon, Canada. *Northwest Sci.* 63: 116–18.

Maisels, F. G., M. Check, and C. Wild. 2000. Rare plants on the summit of Mt. Oko, Cameroon. *Oryx* 34: 136–40.

Mandelbrot, B. B. 1977. *The fractal geometry of nature.* New York: W. H. Freeman and Co.

Marchand, P. J. 1975. Apparent ecotypic differences in the water relations of some northern bog Ericaceae. *Rhodora* 77: 53–63.

McPhee, J. 1993. *Assembling California.* New York: Noonday Press, Farrar, Straus, and Giroux.

McQueen, C. B. 1992. The genus *Sphagnum* in Costa Rica. *Brenesia* 38: 77–81.

Mooney, H. A. 1989. Chaparral physiological ecology: Paradigms revisited. In *The California chaparral: Paradigms re-examined,* ed. J. A. Keeley, 85–90. Los Angeles County Natural History Museum science series 34. Los Angeles: Natural History Museum of Los Angeles County.

Mooney, H. A., and E. E. Cleland. 2001. The evolutionary impact of invasive species. *Proc. Nat. Acad. Sci. USA* 98: 5446–51.

Mooney, H. A., and S. L. Gulmon. 1982. Constraints on leaf structure and function in reference to herbivory. *Bioscience* 32: 198–206.

Mooney, H.A., and R.J. Hobbs, eds. 2000. *Invasive species in a changing world*. Washington, DC: Island Press.

Moore, P. 1980. The advantages of being evergreen. *Nature* 285: 535.

Muller, J.E. 1974. *Geology of Pacific Rim National Park*. Unpublished report to Parks Canada by Geological Survey of Canada.

Nathan, R., G.G. Katul, H.S. Horn, S.M. Thomas, R. Oren, R. Avissar, S.W. Pacala, S.A. Levin. 2002. Mechanisms of long-distance dispersal of seeds. *Nature (London)* 418: 409–13.

Naveh, Z., and J. Dan. 1973. The human degradation of Mediterranean landscapes. In *Mediterranean-type ecosystems: Origin and structure*. Ecological studies 7, ed. F. di Castri and H.A. Mooney, 373–90. New York: Springer-Verlag.

Nekolâ, J.C., and P.S. White. 1999. The distance decay in biogeography and ecology. *J. Biogeogr.* 26: 867–78.

Nguyen-Phuc, T. 1996. Intra-island and inter-island variation in leaf morphology of *Gaultheria shallon* Pursh. Student papers, Island Biogeography, Bamfield Marine Science Center 1996.

Nilsson, I.N., and S.G. Nilsson. 1982. Turnover of vascular plant species on small islands in Lake Möckeln, South Sweden, 1976–1980. *Oecologia* 53: 128–33.

Nilsson, I.N., and S.G. Nilsson. 1985. Experimental estimates of census efficiency and pseudoturover on islands: Error trend and between observer variation when recording vascular plants. *J. Ecol.* 73: 65–70.

Ogilvie, R.T., and A. Ceska. 1984. Alpine plants of phytogeographic interest on northwestern Vancouver Island, Canada. *Can. J. Bot.* 62: 2356–62.

Patterson, B.D., and W. Atmar 1986. Nested subsets and the structure of insular mammal faunes and archipelagoes. *Biol. J. Linn. Soc.* 28: 65–82.

Pojar, J. 1980. *Brooks Peninsula: Possible Pleistocene glacial refugium on northwestern Vancouver Island*. Botanical Society of America misc. series publ. 158: 89. Vancouver: British Columbia Ministry of Forests and Lone Pine Publishing.

Polis, G.A., M.D. Rose, F. Sanchez-Pinero, P.T. Stapp, and W.B. Anderson. 2002. Island food webs. In *A new island biogeography in the Sea of Cortés*, ed. T. Case, M.L. Cody, and E. Ezcurra, 362–80. Oxford: Oxford Univ. Press.

Praeger, R.L. 1969. *The way that I went*. Dublin: Rivverrun, A. Figgis.

P'Yavchenko, N.I. 1964. *Peat bogs of the Russian forest steppe*. Jerusalem: Israel Program for Scientific Translations.

Reader, R.J. 1978. Contribution of overwintering leaves to the growth of three broad-leaved evergreen shrubs belonging to the Ericaceae family. *Can. J. Bot.* 56: 1248–61.

Ridley, H.N. 1930. *The dispersal of plants throughout the world*. Ashford, Kent: L. Reeve and Co..

Roden, C.M. 1998. Persistence, extinction and different species pools within the flora of lake islands in western Ireland. *J. Biogeogr.* 25: 301–10.

Rosenzweig, M.L. 1995. *Species diversity in space and time*. New York: Cambridge Univ. Press.

Rosenzweig, M.L., and C.W. Clark. 1994. Island extinction rates from regular censuses. *Conserv. Biol.* 8: 491–94.

Rousset, F. 2001. Genetic approaches to the estimation of dispersal rates. In *Dispersal*, ed. J.E. Clobert, E. Danchin, A.A. Dhondt, and J.D. Nichols, 18–28. Oxford: Oxford Univ. Press.

Sachs, O. 1997. *The island of the colorblind*. New York: A.A. Knopf.

Sanderson, J.G., M.P. Moulton, and R.G. Selfridge. 1998. Null matrices and the analysis of species co-occurrences. *Oecologia* 116: 275–83.

Sang, T., D.J. Crawford, T. Stuessy, and M. Silva-O. 1995. ITS sequences and phylogeny of the genus *Robinsonia* (Asteraceae). *Syst. Bot.* 20: 55–64.

Sanger, M., M.V. Lomolino, and D.R. Perault. 1997. Niche dynamics of deer mice in a fragmented, old-growth forest landscape. *J. Mammal.* 78: 1027–39.

Sauer, J.D. 1969. Oceanic islands and biogeographic theory: A review. *Geogr. Rev.* 59: 582–93.

Saulnier, T.P., and E.G. Reekie. 1995. Effect of reproduction on nitrogen allocation and carbon gain in *Oenothera biennis*. *J. Ecol.* 83: 23–29.

Saunders, D.A., R.J. Hobbs, and C.R. Margules. 1991. Biological consequences of ecosystem fragmentation: A review. *Conserv. Biol.* 5: 18–32.

Schilling, E.E., J.L. Panero, and U.H. Eliasson. 1994. Evidence from chloroplast DNA restriction site analysis on the relationships of *Scalesia* (Asteraceae: Heliantheae). *Am. J. Bot.* 81: 248–54.

Schoener, T.W. 1976. The species-area relation within archipelagoes: Models and evidence from island land birds. In *Proceedings of the Sixteenth International Ornithological Congress*, ed. H.J. Firth and J.H. Calaby, 629–42. Canberra: Australian Academy of Science.

Schoener, T.W., and D.A. Spiller. 1995. Effects of predators and area on invasion: An experimental study with island spiders. *Science* 267: 1811–13.

Scott, R.B. 1970. A short history of Barkley Sound. *Bamfield Surv. Rept.* 1: 14–67.

Seligman, N.G., and A. Perevolotsky. 1994. Has intensive grazing by domestic livestock degraded Mediterranean Basin rangelands? In *Plant-animal interactions in Mediterranean-type ecosystems,* ed. M. Arianoutsou and R.H. Groves, 93–103. Tasks for vegetation science 31. Dordrecht: Kluwer Academic.

Shmida, A. 1981. Mediterranean vegetation in California and Israel: Similarities and differences. *Isr. J. Bot.* 30: 105–23.

Siegel, S. 1956. *Nonparametric statistics for the behavioral sciences.* New York: McGraw-Hill.

Simberloff, D., and J.-L. Martin. 1991. Nestedness of insular avifaunas: Simple summary statistics masking complex patterns. *Ornis Fenn.* 68: 178–92.

Simberloff, D.S., and E.O. Wilson. 1969. Experimental zoogeography of islands: The colonization of empty islands. *Ecology* 50: 278–96.

Simberloff, D.S., and E.O. Wilson. 1970. Experimental zoogeography of islands: A 2 year record of colonization. *Ecology* 51: 934–37.

Skottsberg, C. 1953. The vegetation of the Juan Fernandez Islands. *Natur. Hist. Juan Fernandez Easter Isl.* 2: 793–960.

Small, E. 1972. Photosynthetic rates in relation to nitrogen recycling as an adaptation to nutrient deficiency in peat bog plants. *Can. J. Bot.* 50: 2227–33.

Smith, J., M.J. Taitt, C.M. Rogers, P. Arcese, L.F. Keller, A.L.E.V. Cassidy, and W.M. Hochachka. 1996. A metapopulation approach to the population biology of the Song Sparrow *Melospiza melodia*. *Ibis* 138: 120–28.

Smouse, P.E., and V.L. Sork. 2004. Measuring pollen flow in forest trees: An exposition of alternative approaches. *Forest Ecol. Manage.* 197: 21–38.

Speck, G. 1954. *Northwest explorations*. Portland, OR: Binfords and Mort.

Spiller, D.A., and T.W. Schoener. 2001. An experimental test for predator-mediated interactions among spider species. *Ecology* 82: 1560–70.

Sprent, J.I., R. Scott, and K.M. Perry. 1978. The nitrogen ecology of *Myrica gale* in the field. *J. Ecol.* 66: 657–68.

Spring, O., N. Heil, and U. Eliasson. 1999. Chemosystematic studies on the genus *Scalesia* (Asteraceae). *Biochem. Syst. Ecol.* 27: 277–88.

Sterk, A.A. 1969. Biosystematic studies on *Spergularia media* and *S. marina* in the Netherlands. 4. Reproduction, dissemination, karyogenetics and taxonomy. *Acta Bot. Neerl.* 18: 639–50.

Sterk, A.A., and L. Dijkhuisen. 1972. The relation between the genetic determination and the ecological significance of the seed wing in *Spergularia media* and *S. marina*. *Acta Bot. Neerl.* 21: 481–90.

Strong, W.L. 2002. Lodgepole pine/Labrador tea type communities of western Canada. *Can. J. Bot.* 80: 151–65.

Stuessy, T., C. Morticana, R. Rodriguez-R., D.J. Crawford, and M. Silva-O. 1992. Endemism in the vascular flora of the Juan Fernandez Islands. *Aliso* 13: 297–308.

Stuessy, T., U. Swenson, D.J. Crawford, G. Anderson, and M. Silva-O. 1998. Plant conservation in the Juan Fernandez archipelago, Chile. *Aliso* 16: 89–102.

Thornton, I.W.B. 1996. *Krakatau: The destruction and reassembly of an island ecosystem*. Cambridge, MA: Harvard Univ. Press.

Toft, C.A., and T.W. Schoener. 1983. Abundance and diversity of orb spiders on 106 Bahamian islands: Biogeography at an intermediate trophic level. *Oikos* 41: 411–26.

Van der Pijl, L. 1982. *Principles of dispersal in higher plants*. Berlin: Springer-Verlag.

Van Valen, L. 1965. Morphological variation and the width of the ecological niche. *Am. Natur.* 99: 377–89.

Wagner, W.L., and W. Gagne. 1988. A summary of the derivation, speciation, and endemism of the Hawaiian biota. *Hawaiian Bot. Soc. Newslett.*

Wagner, W.L., D.R. Herbst, and S.H. Sohmer. 1999. *Manual of the flowering plants of Hawaii,* vols. 1 and 2. Honolulu: Univ. of Hawai'i Press, Bishop Museum.

Wallace, A.R. 1876. *The geographical distribution of animals*. London: MacMillan.

Wallace, A.R. 1880. *Island life*. London: MacMillan.

Wallace, A.R. 1881. *Island life or, the phenomena and causes of insular faunas and floras including a revision and attempted solution to the problem of geological climates*. New York: Harper and Brothers.

Whinan, J. 1995. Successional sequences in the Tasmanian valley *Sphagnum* peatlands. *J. Veg. Sci.* 6: 675–82.

Whitehead, D.R., and C.E. Jones. 1969. Small islands and the equilibrium theory of insular biogeography. *Evolution* 23: 171–79.

Whittaker, R.J. 1998. *Island biogeography: Ecology, evolution, and conservation.* Oxford: Oxford Univ. Press.

Williamson, M. 1981. *Island populations.* Oxford: Oxford Univ. Press.

Wilson, E.O., and D.S. Simberloff. 1969. Experimental zoogeography of islands: Defaunation and monitoring techniques. *Ecology* 50: 267–78.

Wood, W.F., T.J. Clark, D.E. Bradshaw, B.D. Foy, D.L. Largent, and B.L. Thompson. 2004. Clitolactone, a banana slug antifeedant from *Clitocybe flaccida*. *Mycologia* 96: 23–25.

Wood, D.M., and R. del Moral. 2000. Seed rain during early primary succession on Mount St. Helens, Washington. *Madroño* 47: 1–9.

Wright, D.H., B.D. Patterson, and G.G. Mikkelson. 1998. A comparative analysis of nested subset patterns of species composition. *Oecologia* 113: 1–20.

Wright, I.J., P.B. Reich, M. Westoby, D.D. Ackerly, Z. Baruch, F. Bongers, J. Cavender-Bares, T. Chapin, J.H.C. Cornellssen, M. Diemer, J. Flexas, E. Garnier, P.K. Groom, J. Gullas, K. Hikosaka, B.B. Lamont, T. Lee, W. Lee, C. Lusk, J.J. Midgley, M.-L. Lavas, U.N. Ilnemets, J. Oleksyn, N. Osada, H. Poorter, P. Poot, L. Prior, V.I. Pyankov, C. Roumet, S.C. Thomas, M.G. Tjoelker, E.J. Veneldaas and R. Villar. 2004. The worldwide leaf economics spectrum. *Nature (London)* 428: 821–27.

Yeaton, R.I., and M.L. Cody. 1974. Competitive release in island song sparrow populations. *Theor. Pop. Biol.* 5: 42–58.

Yorath, C.J. 1990. *Where terranes collide.* Victoria, BC: Orca Books.

INDEX

Numbers in boldface are the page numbers of figures and tables; others are simply page references. A selection of the more important plant species is included in this index, although many more plant species are discussed in the text and appear in graphs and tables.

Achillea millefolium (Yarrow), 47, **48**, **51**, **102**, 108, **109**, 110, **111**, **112**, **113**, **169**, 170, **172**, **184**, 214, 219
Ageratum conyzoides, 1, 166
Alien plants on islands, 25ff, 70, 86, 102, 124, 139, 151, 156, 161, **162**, 163, 168, 181ff, **183**, **184**, 206ff, 209
Alien slugs, 195
Anemochory, 1, 139, 163ff, 166ff, 169, 214
Annual bluegrass (*Poa annua*), 26, 86, **87**, **88ff**, 97, 125, **184**, 208, 230
Anthropogenic influence, 24, 25, 102, 181ff, 206ff
Arctostaphylos columbianum (Manzanita), 24, 29, 100, 135, **135**, **151**, **153**, 222
Ariolimax columbianum (Banana slug), 194, 195ff, **196**, **197**, **199**, **200**, **201**, **202**, **203**
Assembly rules, **4**, 55, 74ff
Asteraceae (Daisy family), 1, 47, **48**, 108, **109**, 166ff, **169**, **172**, **173**, **175**, **176**, 180, 183, **184**, 185, 214, 219, 220, 243

Banana slug (*Ariolimax columbianum*), 194, 195ff, **196**, **197**, **199**, **200**, **201**, **202**, **203**
  spot patterns vary with island temperatures, 202
Barkley Sound Islands
  Ages, 18ff
  areas, 13ff, **15**, **17**, **18**, **21**, **232ff**
  climate, 5, 10ff, **11**, 27ff, 65, 72ff, 115, 141, 206
  elevation, 14ff, **17**, 24, 68ff, 79, 80ff, **100**, **105**, 133, 206ff

  geography, 6
  geology, 9ff
  history, 5, 6ff, 9, 17ff, **19**, **20**, **21**, 70, 73, 121, 133ff, 205
  isolation, 13, **22**, **23**
  topography, 9ff, 13ff, 18, 56, 68ff, 74, 132, **133**, 187, **237**, **238**
  vegetation, 5, 8, 10, 14ff, 23ff, **29**, 31, 65ff, 205ff, 237ff
Birds
  morphological changes on islands, 195
  on Barkley Sound islands, 3, 4, 246, **247ff**
  on other islands, 39, 53ff, **54**, 74, 187
  seabird colonies, 13, 17, 71, 72, 164, 168, 187
  seed transport by, 1, 102, 139, 150ff, **152–54**, 155, 161, 170, 188, 215
Bittercherry (*Prunus emarginata*), 25, 100, 113ff, **114**, **154**, 187, 212ff, 225
Bogs
  on Hawaii, 187ff
  on the islands, 30, 189
  on the mainland, 189
  species composition, 142, 151, **152ff**, 168, 187ff, **189**, **190ff**, 192, 208, 241
  vegetation, 24, 25, 27, 30

Cakile spp. (Sea-rockets), 26, 86, **87**, **88ff**, 161, **162**, 182, 208, **220**
Carlquist, C., 2, 5, 139, 141, 161, 166, 167, 182
Census techniques, 30ff, **32**

256 INDEX

Chamaecyparis nootkatensis, 24, 218, 245
Coast strawberry (*Fragaria chiloensis*), 25, 102, 151, **154**, 225
Colonization
  asynchrony with extinction, 117
  balance with extinction, 57, **58**, **93**, 116, 158
  episodic, 97
  followed by extinction, with time lag, 117
  rates affected by sunspot cycles, 117
  rates are taxon-specific, 5, Ch.4, Ch.5, 117
  rates as functions of island size, 95, 124
  rates as functions of species numbers, isolation, 4, 36ff, 41ff, **42**, **44**, **45**, **48**, 116ff
  rates on continental versus oceanic islands, 3, 30, 53
  rates vary among years, 57, 117
  redundant, 33, 95, 116, 158
  synchrony in rates among neighboring islands, 95ff
  trade-off with persistence, 115, 157ff, 160ff, 168, 212, 214ff
Competitive release
  in birds, 186
  in bog plants, 193
  in mammals, 195
Cook, James, 6

Daisy family (Asteraceae), 1, 47, **48**, **108**, **109**, 166ff, **169**, 172, 173, 175, 176, 180, 183, **184**, 185, 214, 219, 220, 243
Darwin, C. R., 1, 2
Deer mice, 194
  gigantism in, 194
Density compensation, 181, 186ff, 213
Dispersal
  in general, 1, 2, 3, 12, 25, 34, 38, 47, 53, 55, 85, 94, 102, 108, 113, 117, 128, 134, 138ff, 180, 181, 182, 188, 194, 196, 205ff
  limitation by distance, 94, 139ff, 141, 144, 147, 162, 164
  reduced potential in islands populations, 1, 166ff, 174, **178**, 179, 180, 181
Douglas fir (*Pseudotsuga menziesii*), 10, 24, 29, 46, **47**, **51**, 76ff, **77**, **81**, 83ff, **85**, 218

Endemics, endemism, 1, 2, 3, 10, 36, 53, **54**, 141, 149, 166, 167, 177, 180, 194, 206, 215
Epilobium angustifolium (Fireweed), 102, 167, **170**, **172**, 212, 215, 224
Epilobium watsonii (Purple-leaved willowherb), 102, 167, 168, **169**, **170**, **172**, 212, 215, 224
Equilibrium concept
  and historical legacies, 50ff, **52**
  island numbers supporting specific plants, 39ff
  species numbers on islands, 36ff, 56, 59ff, 122ff, 127, 206ff, **212**

Evergreen huckleberry (*Vaccinium ovatum*), 24, 68, 100, 103, 104, **105**, **106**, **107**, 153, **156**, **157**, **159**, 160, 193, 222
Extinction, see Colonization page references
  of forest plants on islands, 208, **212**
  of small slug populations, 198

False azalea (*Menziesia ferruginea*), 24, **46**, 47, **51**, 76, **77**, **81**, 139ff, 193, 222
Festuca rubra (Red fescue), 25, 87, **88ff**, 98, **100**, **184**, 230
Ferns, 1, 24, 35, 46, **47**, 76, 82, 141ff, **142**, 189, 209, 216, 217, 241
Fireweed (*Epilobium angustifolium*), 102, 167, **170**, **172**, 212, 215, 224
Fractal dimension (of shorelines), 14, **16**
Fragaria chiloensis (Coast strawberry), 25, 102, 151, **154**, 225
Fucus line, 14, 16, 66, 70, 72, 198

Galapagos Islands, 1, 2, 139, 166
Gaultheria shallon (Salal), 24, 68, 100, 102, **103**, 104, **105**, **106**, **107**, 151, **151**, **155**, **156**, **157**, 158, **159**, 193, 208, 222
Genetic turnover among islands, over time, 33, 138, 139, 166, 180ff, 186, 193ff, 199, 204, 215ff
Grass family (Poaceae), 49, 106, 108, 110, 139, 167, **183**, **184**, 185, 229, 243
Growth rates, trees, 84ff, **85**, **86**

Hairy cat's-ear (*Hypochaeris radicata*), 26, 35, 102, 108, **109**, 110, **111**, **112**, **113**, 126, **129**, 169, 172, **174**, **175**, **176**, **177**, **178**, 179, **184**, 216
Hairy cinquefoil (*Potentilla villosa*), 25, 87, 88, **93**, **94**, 210, 225
Hawaiian Islands, 1, 2, 36, 139, 141, 161, 166, 182, 187, 188, 191, 205
Historical legacies in island floras, 18, 21, 36, 50ff, 56, 63, 68ff, 121, 133ff, 135ff
Hordeum brachyanthemum (Meadow barley), 25, 35, 87, **88ff**, 98ff, **100**, 125, **184**, 230
Hooker, J. D., 1, 4, 5, 139, 166
Hooker, W. J., 141, 151
Hydrochory, 85, 139, 161, 163, 164
Hypochaeris radicata (Hairy cat's-ear), 26, 35, 102, 108, **109**, 110, **111**, **112**, **113**, 126, **129**, 169, 172, **174**, **175**, **176**, **177**, **178**, 179, **184**, 216
Hysteresis
  in equilibration of landbridge islands, **52**
  in species' incidence functions, 110ff, **111**, 214

Incidence functions
  and dispersal, 139ff

and fern morphology, 146, 148
and zoochorous shrubs, 159, 160
as a dynamic equilibrium, 40
of area, 39ff, 40
of isolation, 50, 51, 140ff, 140
over time, 118ff
Incidence gaps
  among island groups, 142, 143, 152ff, 172
  among transients versus permanent residents, 144, 145, 148
  in shoreline species, 162, 163, 164
Incidence shifts and interspecific influences, 104ff, 106
Interspecific influences on incidence, distribution, 3, 79, 84, 102, 104ff, 106, 108, 112, 113, 114, 213ff, 246

Juan Fernandez Islands, 1, 141, 167, 177, 180, 182

Lactuca (Mycelis) muralis (Wall lettuce), 29, 109, 110, 111, 112, 113, 126, 129, 167, 169, 173, 174, 175, 176, 177, 178, 184, 187, 214, 219
Leatherleaf fern (Polypodium scouleri), 68, 142, 143, 144, 145, 146, 147ff, 149, 150, 215, 218
Long-term resident species, 118, 119, 121, 122, 131, 132, 158, 179, 209

MacArthur-Wilson (M/W) theory, 5, 36ff, 37, 55, 124, 207
Malus (Pyrus) fuscus (Western crab-apple), 24, 100, 154, 113ff, 114, 154, 212, 215, 225
Manzanita (Arctostaphylos columbianum), 24, 29, 100, 135, 135, 151, 153, 222
Meadow barley (Hordeum brachyanthemum), 25, 35, 87, 88ff, 98ff, 100, 125, 184, 230
Menziesia ferruginea (False azalea), 24, 46, 47, 51, 76, 77, 81, 139ff, 193, 222

Neighbor effects
  measurement, 21, 22, 23
  on incidence, 62, 68, 70ff, 101, 103, 121, 126ff, 134ff, 164, 165, 206, 209ff
  on shoreline plants, 69, 80, 94ff, 101
Nestedness, 74ff, 191ff, 246ff
Niche partitioning, 186, 213ff
Niche shifts, 3, 181, 185ff, 191, 206, 213, 215

Origins of islands
  general, 2
  Sea of Cortés, 2

Pacific yew (Taxus brevifolius), 24, 76, 77, 80, 83, 152, 168, 207, 218
Pearlwort (Sagina crassicaulis), 25, 44, 45, 51, 86, 87, 88ff, 93ff, 95, 96, 97, 99, 100, 162, 211, 221
Picea sitchensis (Sitka spruce), 6, 10, 23, 24, 34, 46, 47, 68, 76ff, 77, 79, 81, 84, 85, 86, 187, 213, 218
Pinus contorta (Shore pine), 24, 29, 76, 80ff, 81, 82, 84, 100, 121, 142, 218, 246
Plantago maritima (Sea plantain), 35, 44, 45, 51, 86, 87, 88ff, 93, 125ff, 210, 224
Poa annua (Annual bluegrass), 26, 86, 87, 88ff, 97, 125, 184, 208, 230
Poaceae (Grass family), 49, 106, 108, 110, 139, 167, 183, 184, 185, 229, 243
Polypodium scouleri (Leatherleaf fern), 68, 142, 143, 144, 145, 146, 147ff, 149, 150, 215, 218
Potentilla villosa (Hairy cinquefoil), 25, 87, 88, 93, 94, 210, 225
Prunus emarginata (Bittercherry), 25, 100, 113ff, 114, 154, 187, 212ff, 225
Pseudotsuga menziesii (Douglas fir), 10, 24, 29, 46, 47, 51, 76ff, 77, 81, 83ff, 85, 218
Pseudoturnover, 33
Pseudoextinction, 34
Purple-leaved willowherb (Epilobium watsonii), 102, 167, 168, 169, 170, 172, 212, 215, 224
Pyrus (Malus) fuscus (Western crab-apple), 24, 100, 154, 113ff, 114, 154, 212, 215, 225

Red elderberry (Sambucus racemosus), 47, 48, 51, 68, 108, 152, 221
Red fescue (Festuca rubra), 25, 87, 88ff, 98, 100, 184, 230
Red huckleberry (Vaccinium parvifolium), 24, 46, 47, 51, 153, 155, 156, 157, 159, 193, 196, 222
Refuging species, 87, 94, 97, 211
Relictual distributions
  plants, 3, 4, 47, 54, 73, 124, 140, 141, 181, 189, 194, 207ff
  slugs, 196, 197
Rescue effects, 4, 33, 38ff, 38, 42, 54, 95, 116, 118, 134, 141, 147, 158, 159, 168
Ribes divaricata (Wild gooseberry), 102ff, 103, 104, 105, 106, 107, 153, 159, 160, 223
Rubus spectabilis (Salmonberry), 24, 151, 154, 157, 102, 103, 104, 105, 106, 151, 155, 156, 157, 158, 159, 225

Sagina crassicaulis (Pearlwort), 25, 44, 45, 51, 86, 87, 88ff, 93ff, 95, 96, 97, 99, 100, 162, 211, 221
Salal (Gaultheria shallon), 24, 68, 100, 102, 103, 104, 105, 106, 107, 151, 151, 155, 156, 157, 158, 159, 193, 208, 222

Salmonberry (*Rubus spectabilis*), 24, 151, **154**, **157**, 102, 103, 104, **105**, **106**, 151, **155**, **156**, **157**, 158, **159**, 225,
Sambucus racemosus (Red elderberry), 47, **48**, 51, 68, 108, **152**, 221
Sea of Cortés Islands, 2, 36, 53, 54, 65, **66**, 74, 149, 187, 194
Sea-level changes, 2, 17ff, **19**, **20**, 239ff
Sea-rockets (*Cakile* spp.), 26, 86, **87**, **88ff**, 161, **162**, 182, 208, **220**
Sea plantain (*Plantago maritima*), 35, 44, **45**, 51, 86, **87**, **88ff**, 93, 125ff, 210, 224
Senecio sylvaticus (Wood Groundsel), 102, **109**, **110**, **111**, **112**, **113**, 167, **169**, 172ff, **173**, **176**, 177, 182, 184, 214, **220**
Shore pine (*Pinus contorta*), 24, 29, 76, 80ff, **81**, **82**, 84, 100, 218, 246
Sitka spruce (*Picea sitchensis*), 6, 10, 23, 24, 34, **46**, 47, 68, 76ff, **77**, **79**, **81**, 84, **85**, **86**, 187, 213, 218
SLOSS, 68, 75
Snails, 195, 204
Species richness
  cumulative species over time, 62ff, **63**
  influence of elevation, **69**
  influence of isolation, **69**
  island plants, 27, 63ff, **67**
  regional, 26
  trends over time, 57ff, **58**, **59**, **60**, **61**
  versus island area, 63ff, **65**, **67**, **69**
Species turnover
  between island twins, 132ff, **133**, **134**
  in space, 3, 4, 128ff, **130**, **131**, **132**
  in time, 4, 33, **37**, 38ff, 47, 50, 55, 57, 93ff, 98ff, **103**, 116ff, **124**, 124ff, **125**, **127**, **128**, 206ff, **212**
Sunspot cycles, 12, 117, 206

Target effects, **4**, **38**, 42, 55, 62, 70, 75, 124, 133, 144, 165, 208, 210ff
  Taxon cycles, 3, **4**
Taxus brevifolius (Pacific yew), 24, 76, **77**, 80, **83**, 152, 168, 207, 218

Thuja plicata (Western red cedar), 23, 24, 34, 68, 76ff, **77**, **79**, **81**, 83, 84, **85**, **86**, 187, 213, 218
Transient populations, **48**, **49**, 144, **146**, 147ff, **148**, 158, 160, 162, 208ff
Tsuga heterophylla (Western hemlock), 10, 23, 24, 34, 68, 77ff, **79**, **81**, 83ff, **85**, **86**, 187, 213, 218
Tsunamis, 16–17, 207
Turnover noise, 37, 57

Vaccinium ovatum (Evergreen huckleberry), 24, 68, 100, 103, 104, **105**, **106**, **107**, 153, **156**, **157**, **159**, 160, 193, 222
Vaccinium parvifolium (Red huckleberry), 24, **46**, 47, 51, 153, **155**, **156**, **157**, **159**, 193, 196, 222

Wall lettuce (*Lactuca* (*Mycelis*) *muralis*), 29, **109**, 110, **111**, **112**, **113**, 126, **129**, 167, **169**, **173**, **174**, **175**, **176**, **177**, 178, **184**, 187, 214, 219
Weeds-to-trees, 1, 166, 185
Western crab-apple (*Pyrus* [*Malus*] *fuscus*), 24, 100, **154**, 113ff, **114**, 154, 212, 215, 225
Western hemlock (*Tsuga heterophylla*), 10, 23, 24, 34, 68, 77ff, **79**, **81**, 83ff, **85**, **86**, 187, 213, 218
Western red cedar (*Thuja plicata*), 23, 24, 34, 68, 76ff, **77**, **79**, **81**, 83, 84, **85**, **86**, 187, 213, 218
Wild gooseberry (*Ribes divaricata*), 102ff, **103**, 104, **105**, **106**, **107**, 153, 159, **160**, 223
Wood groundsel (*Senecio sylvaticus*), 102, **109**, **110**, **111**, **112**, **113**, 167, **169**, 172ff, **173**, **176**, 177, 182, 184, 214, 220

Yarrow (*Achillea millefolium*), 47, **48**, 51, 102, 108, **109**, **110**, **111**, **112**, **113**, **169**, 170, **172**, **184**, 214, 219

Zoochory, 151ff, 161ff, **169**